THE CITY & GUILDS TEXTBOOK

LEVEL 3 DIPLOMA IN

SITE CARPENTRY

AND BENCH JOINERY

THE CITY & GUILDS TEXTBOOK

LEVEL 3 DIPLOMA IN SITE CARPENTRY AND BENCH JOINERY

MARTIN BURDFIELD

STEPHEN REDFERN

COLIN FEARN

SERIES TECHNICAL EDITOR
MARTIN BURDFIELD

City&
Guilds

About City & Guilds

City & Guilds is the UK's leading provider of vocational qualifications, offering over 500 awards across a wide range of industries, and progressing from entry level to the highest levels of professional achievement. With over 8500 centres in 100 countries, City & Guilds is recognised by employers worldwide for providing qualifications that offer proof of the skills they need to get the job done.

Equal opportunities

City & Guilds fully supports the principle of equal opportunities and we are committed to satisfying this principle in all our activities and published material. A copy of our equal opportunities policy statement is available on the City & Guilds website.

First edition 2015
Reprinted 2016
ISBN 978-0-85193-304-7

Commissioning Manager: Charlie Evans
Content Project Manager: Jennie Pick
Production Editor: Lauren Cubbage

Cover design by Design Deluxe, Bath
Illustrations by Barking Dog Art and Palimpsest Book Production Ltd
Typeset by Palimpsest Book Production Ltd, Falkirk, Stirlingshire
Printed in Croatia by Zrinski

British Library Cataloguing in Publication Data

A catalogue record is available from the British Library.

Publications

For information about or to order City & Guilds support materials, contact 0844 543 0000 or centresupport@cityandguilds.com. Calls to our 0844 numbers cost 5 pence per minute plus your telephone company's access charge.

Every effort has been made to ensure that the information contained in this publication is true and correct at the time of going to press. However, City & Guilds' products and services are subject to continuous development and improvement and the right is reserved to change products and services from time to time. City & Guilds cannot accept liability for loss or damage arising from the use of information in this publication.

City & Guilds
1 Giltspur Street
London EC1A 9DD

www.cityandguilds.com

publishingfeedback@cityandguilds.com

CONTENTS

FOREWORD

Whether in good times or in a difficult job market, I think one of the most important things for young people is to learn a skill. There will always be a demand for talented and skilled individuals who have knowledge and experience. That's why I'm such an avid supporter of vocational training. Vocational courses provide a unique opportunity for young people to learn from people in the industry, who know their trade inside out.

Careers rarely turn out as you plan them. You never know what opportunity is going to come your way. However, my personal experience has shown that if you haven't rigorously learned skills and gained knowledge, you are unlikely to be best placed to capitalise on opportunities that do come your way.

When I left school, I went straight to work in a butcher's shop, which was a fantastic experience. It may not be the industry I ended up making my career in, but being in the butcher's shop, working my way up to management level and learning from the people around me was something that taught me a lot about business and about the working environment.

Later, once I trained in the construction industry and was embarking on my career as a builder, these commercial principles were vital in my success and helped me to go on to set up my own business. The skills I had learned gave me an advantage and I was therefore able to make the most of my opportunities.

Later still, I could never have imagined that my career would take another turn into television. Of course, I recognise that I have had lucky breaks in my career, but when people say you make your own luck, I think there is definitely more than a grain of truth in that. People often ask me what my most life-changing moment has been, expecting me to say winning the first series of *Big Brother*. However, I always answer that my most life-changing moment was deciding to make the effort to learn the construction skills that I still use every day. That's why I was passionate about helping to set up a construction academy in the North West, helping other people to acquire skills and experience that will stay with them for their whole lives.

After all, an appearance on a reality TV show might have given me a degree of celebrity, but it is the skills that I learned as a builder that have kept me in demand as a presenter of DIY and building shows, and I have always continued to run my construction business. The truth is, you can never predict the way your life will turn out, but if you have learned a skill from experts in the field, you'll always be able to take advantage of the opportunities that come your way.

Craig Phillips

City & Guilds qualified bricklayer, owner of a successful construction business and television presenter of numerous construction and DIY shows

ABOUT THE AUTHORS

MARTIN BURDFIELD

CHAPTERS 2, 6 AND 7
SERIES TECHNICAL EDITOR

I come from a long line of builders and strongly believe that you will find a career in the construction industry a very rewarding one. Be proud of the work you produce; it will be there for others to admire for many years.

As an apprentice I enjoyed acquiring new knowledge and learning new skills. I achieved the C&G Silver Medal for the highest marks in the Advanced Craft Certificate and won the UK's first Gold Medal in Joinery at the World Skills Competition. My career took me on from foreman, to estimator and then works manager with a number of large joinery companies, where I had the privilege of working on some prestigious projects.

Concurrent with this I began working in education. I have now worked in further education for over 35 years enjoying watching learners' skills improve during their training. For 10 years I ran the Skillbuild Joinery competitions and was the UK Training Manager and Chief Expert Elect at the World Skills Competition, training the UK's second Gold Medallist in Joinery.

Working with City & Guilds in various roles over the past 25 years has been very rewarding.

I believe that if you work and study hard anything is possible.

STEPHEN REDFERN
CHAPTERS 3–5

I was born and grew up in the Midlands, where I continue to live. I am married with two children and four grandchildren.

On leaving school at 16, I managed to get an indentured apprenticeship with a joinery manufacturer. I have spent the better part of 40 years working in joinery and the construction industry, 26 of which were at a further education college from which I have now retired as a course leader for Joinery. During my time as course leader, I delivered Wood Machining, Carpentry and Joinery courses from Level 1 through to Level 3 to apprentices, full-time students and adult learners.

In my spare time, other than undertaking construction projects, I like fishing, working my spaniels and clay shooting.

COLIN FEARN
CHAPTER 1

I was born, grew up and continue to live in Cornwall with my wife, three children and Staffordshire bull terrier.

As a qualified carpenter and joiner, I have worked for many years on sites and in several joinery shops.

I won the National Wood Award for joinery work and am also a Fellow of the Institute of Carpenters, holder of the Master Craft certificate and have a BA in Education and Training.

I was until recently a full-time lecturer at Cornwall College, teaching both full-time students and apprentices.

I now work full-time as a writer for construction qualifications, practical assessments, questions and teaching materials for UK and Caribbean qualifications.

In my spare time I enjoy walks, small antiques and 'keeping my hand in' with various building projects.

HOW TO USE THIS TEXTBOOK

Welcome to your City & Guilds Level 3 Diploma in Site Carpentry and Bench Joinery textbook. It is designed to guide you through your Level 3 qualification and be a useful reference for you throughout your career. Each chapter covers a unit from the Level 3 qualifications, and covers everything you will need to understand in order to complete your written or online tests and prepare for your practical assessments.

Please note that not all of the chapters will cover the learning outcomes in order. They have been put into a logical sequence as used within the industry and to cover all skills and techniques required.

Throughout this textbook you will see the following features:

Dihedral angle

The angle formed on the top edge of the hip rafter running along its length from the eaves to the ridge

Useful words – Words in bold in the text are explained in the margin to help your understanding.

INDUSTRY TIP

When cutting timber, cut the longest lengths required first, using the straightest timber. Any bent, sprung or twisted timber can be used for shorter lengths where these defects will be less noticeable.

Industry tips – Useful hints and tips related to working in the carpentry and joinery industries.

ACTIVITY

What geometry is required to draw a curtail-ended step? Draw the plan shape of the riser with a scroll projection of 150mm.

Activities – These are suggested activities for you to complete.

FUNCTIONAL SKILLS

A customer requires 48 balusters for a staircase. If each baluster costs £9.75, what would be the total cost including 20% VAT?

Work on this activity can support FM2 (C2.5).

Answer: £561.60

Functional Skills – These are activities that are tied to learning outcomes for the Functional Skills Maths, English and ICT qualifications.

Look at the project drawings for 'Our House'. The owners have decided to convert the loft and want a bullseye window inserting into the gable wall. Draw up a full-size rod for a fixed glazed window 675mm in diameter.

'Our House' – These are activities that tie in directly with 'Our House' on SmartScreen to help you put the techniques in the book in context. Ask your tutor for your log-in details.

STEP 1 Using a fine triangular file to suit the auger, file the end face of the auger maintaining the original grinding angle.

STEP 2 File the wing spurs just lightly enough to ensure they are sharp.

Step-by-steps – These steps illustrate techniques and procedures that you will need to learn in order to carry out carpentry and bench joinery tasks.

Case Study: Daniel

Daniel has been asked by his supervisor to carry out two tasks on a local building project. The first task is to construct a cut roof. Daniel has been asked how the edge of the roof is to be finished and whether anything can be done to help stop the rain missing the guttering during heavy rain storms.

- What type of eaves finish would you recommend and why?

Case Studies – Each chapter ends with a case study of an operative who has faced a common problem in the industry. Some of these will reveal the solution and others provide you with the opportunity to solve the problem.

Half-turn stairs

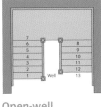

Open-well

Half-turn stairs turn through 180°, normally by way of a half-space landing. They can take two forms: dog-leg or open-well.

Trade dictionary – This feature lists key terms and tools that you will pick up from reading this book.

At the end of every chapter are some 'Test your knowledge' questions. These questions are designed to test your understanding of what you have learnt in that chapter. This can help with identifying further training or revision needed. You will find the answers at the end of the book.

INTRODUCTION

This book has been written to support students studying site carpentry and bench joinery at Level 3. By studying this book, you should gain the more advanced skills and knowledge required to complete your course and either progress to further study, enter the workforce or, where you are already working, to progress to a supervisory level. You will learn in more depth how carpentry and joinery work is acquired, the contract documents used, and how the requirements of planning for time, labour and materials help allow the build or project to be completed efficiently.

Site carpenters will learn about advanced first fixing of roof structures and stairs, the second fixing of double doors, and installing curved and raking mouldings. Bench joiners will study how to set out and manufacture shaped frames and stairs incorporating turns. Both trades will study how to use a range of machines required to carry out your work.

In addition to the features listed on the previous pages, which are there to help you retain the information you will need to become a proficient carpenter or joiner, this textbook includes a large trade dictionary. Use this for reference in class and in the workshop. Become familiar with the terms and techniques, and pay attention to the skills you need to master.

Carpenters who study at this level are often promoted to site supervisors and, when further experienced, site managers. Joiners are likely to be promoted to supervisors, setters out, estimators and, again when further experienced, joinery works managers. You will be proud of the many buildings that you have helped to build. They will stand as a testament to your skills and knowledge, and for others to admire for a long time.

ACKNOWLEDGEMENTS

To my gorgeous wife Clare, without whose constant support, understanding and patience I would not have been able to continue. To Matthew and Eleanor, for not being there on too many occasions; normal service will be resumed. Finally, my parents, to whom I will always be grateful.

Martin Burdfield

I would like to thank my wife Sharon and my two children, Katie and Shaun, for all their support and patience, without whose help these chapters would not have been completed. I would also like to thank Brian, a former tutor of mine and now a dear friend of around 35 years, who has continually encouraged and pushed me to achieve the success I have.

Stephen Redfern

I would like to thank my dear wife Helen for her support in writing for this book. I dedicate my work to Matt, Tasha and Daisy, and not forgetting Floyd and Mrs Dusty.

Colin Fearn

City & Guilds would like to sincerely thank the following:

For invaluable carpentry and joinery expertise

Andrew Barnes, Anthony Brindley and Julian Walden.

For freelance editorial support

Cambridge Editorial Partnership Ltd (www.camedit.com), Ben Gardiner, Anna Carroll and Richard Churchman.

For their help with photoshoots

Kane Bramhall and Ian Vanes-Jones, Burton and South Derbyshire College. Models: Taylor Colebrook, Daniel Scallo, Stephen Knight, James Tayler, Matthew Uppington and Tom Warrington.

Building Crafts College. Models: Luke Wood and Scott Kerr.

For taking college photos

Andrew Buckle Photography and Martin Burdfield.

TRADE DICTIONARY

Industry term	Definition and regional variations
Access equipment	Equipment that will enable you to gain access to work at a higher level than you can reach from the floor.
Annealed glass	The standard type of glass readily available for doors that have glazing.
Approved Code of Practice (ACoP)	The ACoP gives practical advice for those involved in the construction industry in relation to using machinery safely. The ACoP has a special legal status and employers and employees are expected to work within its guidelines.
Architect	A trained professional who designs structures and represents the client who is building the structure. They are responsible for the production of the working drawings and supervise the construction of buildings and structures.
Architectural technician	A draftsperson who works in an architectural practice. They usually prepare the drawings for a building.

Industry term	Definition and regional variations
Architrave 	A moulded section that is fixed around a door lining that covers the gap between the lining and the wall.
Auger bit 	Rotating cutting tool used in cordless drills and hand braces to bore holes, such as for fitting cylinder and mortice locks.
Back saw 	A saw with either a strip of steel or brass along the top edge that keeps the blade straight. Back saws are used for the accurate cutting of joints and can be used to cut with and across the grain. *Regional variation: backed saw*
Backer off	When using fixed and transportable machinery, this is someone who removes material from the machine after processing. *Regional variation: taker off*
Barefaced tenon 	A tenon that is shouldered on only one face of the joint.
Belt sander 	A portable sanding tool with a power-driven abrasive-coated continuous belt.
Bench hook 	Used in conjunction with a tenon saw. This simple piece of equipment hooks over the edge of the bench, or preferably fits into a vice.

Industry term	Definition and regional variations
Bevel-edged chisel 	A chisel with the edges bevelled. Used for light work, eg when chopping out the sockets when dovetailing.
Bill of quantities 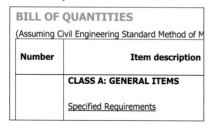	Produced by the quantity surveyor and describes everything that is required for the job based on the drawings, specification and schedules. It is sent out to contractors and ensures that all the contractors are pricing for the job using the same information.
Birdsmouth cut 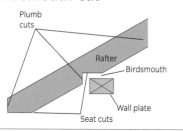	A plumb cut (vertical cut) and seat cut (horizontal cut) together at the bottom of a rafter.
Bisecting line 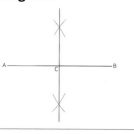	A line that divides another line into two equal parts.
Black japanned 	Finished with a black enamel lacquer, originally associated with products from Japan.
Block plane 	A small plane that can easily be used with one hand. It has a much lower frog angle, improving the finish of the cut when used on end grain or laminate. Not suitable for cleaning up faces as it will likely tear the grain.

Industry term	Definition and regional variations
Blue stain 	A blue fungal discolouration in sap wood.
Boxed heart 	Boxing (a term used in timber conversion) the heart refers to eliminating the heartwood from the boards that would otherwise produce shakes or may even be rotten. This can be achieved by either tangential or radial cutting.
Box square 	Used to square around timber that has already been profiled because it is very difficult to mark around profiled timber with a standard try or combination square. There are simple box squares that are little more than a piece of angled metal, and more complex ones that will mark mitres.
Bridle joint Corner bridle joint	Similar to a mortice and tenon, in that a tenon is cut on the end of one member and a mortice is cut into the other to accept it. The distinguishing feature is that the tenon and the mortice are cut to the full width of the tenon member.
British Standards Institute (BSI)	The organisation that develops and publishes standards for the UK.
Building Regulations	A series of documents that set out legal requirements for the standard of building work.
Butt joint 	Simple joint in which two pieces of wood are placed against each other but held with nails, screws, dowels, glue or other fastenings.

Industry term	Definition and regional variations
CAD	Short for computer-aided design, used to create drawings when setting out. Items drawn up with CAD can be processed by a CNC (Computer Numerical Control) router, saving much time and labour.
Cavity wall	Walls built in two separate skins/leaves (usually of different materials) with a void held together by wall ties.
Centre	A temporary structure made by a carpenter or joiner to allow a bricklayer to turn a brickwork arch around them. When the mortar has set, the centre may be 'struck' (removed).
Centres	The distance from the centre of one point to the centre of the next, ie used with rafters, floor joists and studwork.
Centrifugal force	A force that is generated by the rotating cutter block in surface planers and thicknesser. It forces objects away from the centre of rotation, thereby wedging the cutter tight in the block.
Chop saw	Tool used for making accurate square, angled and compound cuts (a cut incorporating two angles). Chop saws are used on site for a wide variety of tasks including cuts to architraves and skirting, stair spindles, studwork, rafter cuts and any other job where a straight clean cut is required. *Regional variation: mitre saw*
Circular table saw	A fixed machine that is generally used to cut timber and sheet materials to size. Some are designed just to rip planks down to smaller useable sizes. Other saws have sliding tables that support material when cutting large boards or lengths of timber.

Industry term	Definition and regional variations
Claw hammer	A claw hammer is used for driving nails, pins, wedges and dowels. Claw hammers can have a straight or curved claw, used to remove nails from timber.
Combination gauge	Very similar to a mortice gauge but with an additional pin on the opposite side of the stem, allowing it to be used as a marking gauge.
Combination plane	Similar to a rebate plane, but has additional cutters enabling it to cut grooves and create moulding profiles. Although a router is commonly used to perform the jobs that a combination plane does, the moulds produced by the plane are considered better by some, as they are free of the 'pitch marks' which routers create.
Combination square	A square that marks both 90° and 45° angles. Also used to produce parallel lines on a workshop rod.
Compass	Used to set out smaller curves. Better-quality ones have a wheel in the centre that increases the accuracy (known as springbow compasses).
Compound cut	A cut that consists of two angles.
Contact adhesive	Used for sticking plastic laminates to timber. A thin layer is applied to each surface, left to dry, and then brought together for a permanent bond.
Conversion Quarter sawn	The process of cutting logs into pieces of timber. Four different methods used are boxed heart, tangential, through-and-through and quarter sawn.

Industry term	Definition and regional variations
Coping saw	A type of 'frame saw' with a thin, narrow blade used for removing waste material (eg when cutting dovetail or bridle joints) and cutting scribed moulding profiles.
Crosscut saw	A saw used to cut across the board at right angles to the grain, ie cutting across the grain.
Cutting gauge	Similar to a marking gauge except that instead of a pin it has a cutting knife and a wedge to hold it in place. The knife severs the fibres of the grain, leaving a clean finish.
Cut roof	A construction technique that builds up the roof without the use of trusses.
Cut string	Where the shape of the string is cut to the step shape.
Cutter block	The machine block that holds the cutting knives or cutters.
Cutting list	A cutting list contains all the material required for a job: the finished sizes of timber (width and thickness) and also the ex sizes (the size of the sawn section required to produce the planed size, typically 5–6mm larger than the planed size). Also listed is the length – depending on the job and the machinery being used to produce the joinery, this may include waste.

Industry term	Definition and regional variations
Damp proof course (DPC) 	A layer or strip of watertight material placed in a joint of a wall to prevent the passage of water. Fixed at a minimum of 150mm above finished ground level.
Damp proof membrane (DPM) 	A layer or sheet of watertight material, incorporated into a solid floor to prevent the rise of moisture.
Datum point 	A fixed point or height from which to take reference levels. They may be permanent Ordnance Bench Marks (OBMs) or Temporary Bench Marks (TBMs). The datum point is used to transfer levels across a building site. It can represent the finished floor level (FFL) on a building.
Decibel (dB(A))	Decibels are used to measure sound levels, while the A weighting is applied to instrument-measured sound levels to try to account for the relative loudness perceived by the human ear.
Defect 	Issues with timber, that can either be natural or from seasoning. Shakes and knots are types of defects.
Demountable power feed unit	Automatic feed systems that can be moved out of the way. Mainly used for continuous feeding of material during cutting or profiling operations.
Detector 	An electronic detector can be used to scan over wall surfaces; when pipes or wires are detected it will emit a loud, high-pitched noise.

Industry term	Definition and regional variations
Dog-leg stairs	A type of half-turn stairs. The outer strings of the upper and lower flights are joined into the same newel immediately above each other. These stairs take their name from the appearance of their sectional elevation.
Door leaf	One half of a pair of double doors.
Doors	One of the many joinery items you will come across. There are countless different door designs that a joiner may be required to produce. Doors may be: fully glazed, half glazed, panelled, solid or matchboarded (made with tongue and groove).
Dormer	An opening in the roof surface housing a window where a secondary roof projects away from the plane of the existing roof. Typically used to give greater head room within the primary roof and add light to the roof space/loft.
Double roof	Double roof designs allow greater spans to be covered than single roofs. They require longer rafters to be used but without having to use excessively large sectional sizes of timbers. This is achieved by providing intermediate support along the rafters, called 'purlins', which are usually positioned midway along the rafter.
Dovetail joint Lapped dovetail joint	An attractive joint with interlocking teeth used for drawers and in fine furniture. The slope or pitch of the dovetails resists the two parts of the joint being pulled apart under load. Dovetail joints are held together using adhesive alone – no additional fixings should be required.

Industry term	Definition and regional variations
Dovetail saw	A small fine-toothed saw used for cutting dovetails. It is a type of backed saw.
Dovetail square	A special square used to mark dovetail lines when marking out.
Dowelled joint	An alternative to using mortice and tenon joints to joint a door. Dowels are not commonly used in joinery shops, but are used as the jointing method for mass-produced doors.
Drawer lock chisel	Drawer lock chisels are used to cut out housings or mortices (for instance, mortices for locks) into the framing of cabinets where there is not enough room to use a standard chisel.
Edge joint Edge joint with biscuits	Edge joints are used to make timber boards wider. They can be 'rubbed' (where glue is applied and the two pieces rubbed together to ensure good coverage of the joint), tongued, biscuited or doweled to increase strength.
Façade	The exterior face of a building, generally referred to as the 'front face' of the building.

Industry term	Definition and regional variations
Fair ending	Trimming the end of a board to remove defects such as splits and to square it up.
F cramp	This cramp has an adjustable arm which can be set to accommodate any width with the distance of the main bar. So called because of its shape.
Finished floor level (FFL)	This is the height of the finished floor level in a property. Represented with a datum point, a horizontal DPC is often installed at this height.
Fink truss	Commonly used truss which is symmetrical in shape, used to support roofs with a span of up to 16m. *Regional variation: W truss*
First-fix nail gun	A powered nailer used to fix studwork and for carcassing.
Folding wedges	A pair of wedges used 'back to back' to create a pair of parallel faces. By sliding one wedge against the other, the total parallel thickness of the folding wedges can be adjusted to pack out the required gap.
Forend face plate	The part of the lock that is seen when the lock is housed in the door.

Industry term	Definition and regional variations
Forest Stewardship Council (FSC®) 	An international not-for-profit organisation established in 1993 to promote responsible management of the world's forests.
G cramp 	A steel cramp, so called because the open cramp including the screw shaft looks like a capital letter G. Used to hold together materials under pressure.
Gap-filling cartridge adhesive 	Supplied in a tube, this adhesive is used for site work such as bonding timber to concrete, eg fixing panels to walls.
Gent saw 	A variety of back saw with a turned handle. It is sometimes even finer than a dovetail saw and is used for very fine work. *Regional variation: gent's saw*
Geometrical staircase 	A staircase that turns and rises without the use of landings – the stair strings are formed to rise and turn and are also known as 'wreathed strings'.
Georgian wired glass 	The name given to glass that has a thin wire mesh embedded into it, used in fire doors and where added safety is required.
Glue line	A term used to describe the amount of glue contact area of a joint. This is particularly important when designing widening joints. The greater the glue line, the stronger the joint.

Industry term	Definition and regional variations
Gouges Blade Blade ground on inside	Gouges are chisels with a hollow curved blade. There are two types: 1 Those which are ground on the inside face (in canalled). These are commonly used by carpenters and joiners to scribe joints (cutting out to fit over moulding profiles). 2 Those ground on the outside (externally canalled) used only by wood carvers.
Growth ring Annual rings	A ring that appears on the inside of a tree, one for every year it has grown. Each ring is split into spring and summer growth. *Regional variation: annual ring*
Half-turn stairs Open well	Half-turn stairs turn through 180°, normally by way of a half-space landing. They can take two forms: dog-leg or open-well.
Halving joint Corner halving joint	A joint where half of each of the two boards being joined is removed, so that the two boards join together flush with one another.
Hand-held circular saw	A versatile, portable power tool used to cut timber and sheet materials. Most models of hand-held circular saws have a riving knife and a fence attachment. They are available as corded or battery-powered versions.
Hardwood Ash	Timber that has been cut from deciduous trees (broad-leaved trees which lose their leaves in autumn). They are used for heavy structural work and fine joinery. Types of hardwood include beech, oak, ash, sapele, birch and mahogany.

Industry term	Definition and regional variations
Hatching	Marking waste timber with diagonal lines. This prevents mistakes, as only the hatched area is removed.
Haunch Double haunched mortice and tenon	A part of a tenon that is cut shorter and does not run right through the stile in doors. It allows more material to be left in the stile, avoiding a very long, weak mortice while preventing twist in the rail.
Heading joint 	A joint between two pieces of timber which are jointed end to end. It is a type of non-structural lengthening joint.
Health and Safety Executive (HSE)	Government body which provides advice on safety and publishes booklets and information sheets including the ACoP.
Housing joint Through housing joint	Joint consisting of a groove usually cut across the grain into which the end of another member is housed or fitted to form the joint. This type of joint is commonly used for making door linings and shelving units, as well as for stair construction. A housing joint is a strong joint that requires fixing with nails, screws, wedges or adhesive. *Regional variation: trenching joint*
Hypotenuse 	The side opposite the right angle and the longest side of a right-angled triangle
Improvement notice	Issued by an HSE or Local Authority inspector to formally notify a company that improvements are needed to the way it is working.
Insulation Fibreglass	Materials used to retain heat and improve the thermal value of a building. Can also be used in managing sound transfer.

Industry term	Definition and regional variations
Interlocking grain	Where grain spirals around the axis of a tree but reverses its direction regularly, causing a poor finish whichever way it is planed. *Regional variation: refractory grain*
Intumescent seal Smoke seal Intumescent strip Door leaf Combined intumescent strip and smoke seal (brush type) fitted in door leaf	A seal that swells as a result of exposure to heat, closing the gap between the frame and the door.
Ironmongery Butt hinge	Metal items on doors, windows, etc. Includes hinges, fasteners, deals, handles, escutcheon plates, locks, latches, etc. Most come in a variety of finishes including chrome, brass, stainless steel, bronze, black iron and painted steel.
Jack plane	A plane used for general straightening work and preparing sawn timber. A jack plane is not to be used for cleaning up, as its length will hit the high spots and prevent effective smoothing of joints. It is 375–400mm long, and its most common use on site is to shoot doors into a lining or frame.
Jig	A means of safely holding components while carrying out machining operations. Used to provide repeatability and accuracy in manufacturing components.
Jigsaw	A versatile power tool used to cut shaped work (curved lines). The type of work undertaken might include cutting out hob and sink apertures in worktops or curved shapes from plywood.

Industry term	Definition and regional variations
Joiner's hammer 	A joiner's hammer is very useful for driving wedges and smaller nails and pins. *Regional variation: Warrington hammer*
Joint line	The fit of a joint where two pieces are bonded to each other.
Joists	A series of spaced, parallel beams used to span the distance between walls to support a floor.
Kerfing 	The act of cutting a series of kerfs (the total width of saw cuts) into a piece of wood in close proximity to each other so that the wood can be bent.
Kinetic lifting 	A method of lifting that ensures the risk of injury is reduced.
Knives 	The name given to the part of the block that does the cutting. *Regional variations: cutters, blades*
Laminate 	A plastic sheet glued on top of a base layer like chipboard, as used in kitchen worktops.
Laminated glass	Two pieces of annealed glass with a thin plastic interlayer that acts as an adhesive holding the glass together in the event of breakage, typically used in car windscreens.

Industry term	Definition and regional variations
Laser level 	A tool used for setting out datum lines and transferring levels.
Lengthening joint  Heading joint	Used to join two pieces of timber together to gain a longer length. Lengthening joints can be categorised as either structural or non-structural. A heading joint is an example of a non-structural lengthening joint.
Local exhaust ventilation (LEV) 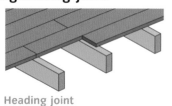	An engineering control system to reduce exposure to airborne contaminants such as dust in the workplace. It vacuums away dust from the work area, and is usually attached to the tool being used. *Regional variation: extraction*
Mallet 	A driving tool with a large wooden head, used to knock pieces together or to drive dowels or chisels. Commonly made from beech.
Manufactured boards 	Manufactured boards are man-made boards and can include plywood, MDF (medium-density fibreboard), blockboard, chipboard, hard board and OSB (oriented strand board).
Marking gauge 	A simple gauge with a single pin used in marking out. This tool is ideal for marking single lines, for example with housings or halving joints.
Marking knife 	A marking knife is often used for better-quality joinery; it is ground and sharpened on one side only, so it will sit tight up against the square when marking out, giving a very accurate line.

Industry term	Definition and regional variations
Mechanical joints	Joints that, due to their design, hold or pull themselves together.
Method statement	A description of the intended method of carrying out a task, often linked to a risk assessment.
Mitre box	A piece of equipment used with a saw to help cut accurate mitres.
Mitre joint	An internal or external junction of a moulding, similar to a butt joint, but where both pieces have been bevelled (usually at a 45° angle).
Mortice and tenon joint	A very strong joint which is formed by a tongue-like piece or tenon. The tenon then fits into a mortice or slot cut into a second piece.
Mortice chisel	Mortice chisels are a very strong type of chisel, specially designed for morticing. The blade section is square or rectangular, and deep. This not only increases the strength of the tool, it also helps ensure the mortice is clean as the wide sides act as a guide, keeping the chisel square to the mortice. Mortice chisels are generally available in sizes from 6mm up to 25mm.
Mortice gauge	Similar to a marking gauge but it has two pins rather than one with a mechanism to allow the adjustment of the gap between the pins. Primarily used in marking out mortice and tenon or bridle joints.

Industry term	Definition and regional variations
Morticer	A very useful piece of equipment in the joinery shop. It is a fixed machine used to cut mortices for mortice and tenon joints.
Moulded timber	Timber that has been moulded has a shape run along its length, such as an ovolo or chamfer. The shape can be referred to as a 'profile'.
Nail punch	A tool used to drive nail and pin heads below the surface.
Narrow band saw	A fixed machine with a blade of up to 50mm wide, used for a variety of jobs such as cutting wedges, ripping tenons and curved work.
Obscured glass	'Frosted' glass that provides a level of privacy by obscuring the view through the glass. Different patterns are available depending on the level of privacy required, graded from 1 to 5, where 1 is the least and 5 is the greatest. This type of glass is commonly used in bathroom windows and front entrances to residential properties.
Open-newel stairs	A type of half-turn stairs. Open-newel stairs, also known as open-well stairs, are where two newels are used at landing level. This separates the string of the upper and lower flights, thus creating a central space or well.

Industry term	Definition and regional variations
Orbital sander	An electric sander that moves the abrasive in an elliptical pattern. They are smaller machines than belt sanders and are used for finishing after the surface of the timber has been planed.
Ordnance Bench Mark (OBM)	An Ordnance Bench Mark has a given height on an Ordnance Survey map. This fixed height is described as a value, eg so many metres above sea level (as calculated from the average sea height at Newlyn, Cornwall).
Pad saw	A thin saw blade in a tool pad used for cutting holes. *Regional variation: plasterboard saw, keyhole saw*
Panel saw	A short hand saw with small teeth, typically 10–12 teeth per 25mm. This saw is intended for finer work such as cutting smaller sections – for example, floorboards, skirting and manufactured boards.
Partition	Partitions are non-load-bearing internal walls that are used to divide a space into a number of smaller rooms. They are commonly formed from either timber or metal. They are usually covered by plasterboard.
Pattern Chamfer Rebate	The first component marked out, from which all similar pieces are marked. This ensures uniformity of size in production.
Perimeter 2.2m 4.2m	The distance around an object or room.

Industry term	Definition and regional variations
Personal protective equipment (PPE) 	This is defined in the Personal Protective Equipment (PPE) at Work Regulations 1992 as 'all equipment (including clothing affording protection against the weather) which is intended to be worn or held by a person at work and which protects against one or more risks to a person's health or safety.' For example, safety helmets, gloves, eye protection, high-visibility clothing, safety foot wear and safety harnesses.
Phillips (PH)	Screw with a cross slot in a simple star shape driven by a matching screwdriver.
Pick-up	Where timber is fed into a cutter against the grain.
Pillar drill	A type of power drill. It is a drill incorporated into a stand holding the drill head, a table that is usually able to rise and fall and to which the components to be drilled can be attached, and a handle that lowers the drill head. These are particularly useful where repeated holes are required to be drilled accurately.
Pin hammer	A pin hammer is a small hammer used for light work such as driving pins.
Pitched roof	The most common type of roof, usually one with two slopes meeting at the central ridge.

Industry term	Definition and regional variations
Pitch line (roofing) 	A line measuring two-thirds of the way down from the top of the common rafter and which is used to set out the lengths of the rafter. *Regional variation: setting out line*
Pitch mark 	The name given to the small circular cuts produced by all rotary planing machines.
Polyurethane (PU) adhesive 	Commonly referred to as 'foam glue'. A yellow-brown resin that foams upon contact with air. It forms a strong water-resistant bond and has excellent gap-filling properties. This adhesive can be used to join damp timbers.
Polyvinyl acetate (PVA) adhesive 	Commonly referred to as 'white glue'. A resin dissolved in water. As the water evaporates the glue dries (goes off). Available in interior and exterior grades. It has good gap-filling properties and provides a strong permanent bond.
Portable appliance testing (PAT) 	PAT is a regular test carried out by a competent person (eg a qualified electrician) to ensure the tool is in a safe electrical condition. A sticker showing the result of the test is then placed on the tool.
Portable planer 	Planers are used to plane timber to produce a flat surface, reducing it in thickness or width. Planers are available in corded and battery-powered versions.

Industry term	Definition and regional variations
Pozidriv (P2)	Screw with a cross slot shaped in a double star which must be driven by a matching screwdriver.
Profiles Lambs tongue	A shape cut into timber along its length. Examples include rebate, groove, ovolo, chamfer, lambs tongue, scotia, ogee, bullnose.
Prohibition notice	Issued by an HSE or Local Authority inspector when there is an immediate risk of personal injury. Receiving one means you are breaking health and safety regulations.
Protractor	Used to measure angles when setting out.
Push stick	A push stick is used for safety reasons when guiding wood being cut through a circular saw. It is a piece of timber at least 450mm long.
Quarter sawn	Term used in timber conversion. Cutting at a 90° angle from the growth rings on a log to produce a vertical and uniform pattern grain. The grain on the face of a quarter-sawn board will be parallel lines that are straight, tight and run the length of the board.
Quick release cramp	A cramp for light, temporary jobs, such as holding down work. They are not very strong, so are not suitable when gluing.

Industry term	Definition and regional variations
Random orbital sander 	A power tool that leaves a better finish than an orbital sander, so should be used for pieces that will be varnished or stained.
Rebate	Rectangular recess in the edge of a board which holds a panel or glass in a door or picture-frame.
Rebate plane	A plane for cutting rebates in timber. There are two types: the rebate plane and the bench rebate plane, available in jack and smoothing sizes. The plane has a fence to control the width of the rebate and a depth stop to set the depth of the rebate.
Rip saw	A saw with between 2 and 5 teeth per 25mm, used for cutting lengthways with the grain.
Rise and going	In relation to stairs, rise refers to a vertical measurement and going to a horizontal measurement. These terms can be applied to the whole staircase, or to individual steps.
Risk assessment	An assessment of the hazards and risks associated with an activity and the reduction and monitoring of them.
Rod 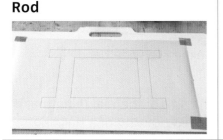	Full-size drawing of the item being made. Using a working drawing as a guide, overall measurements are transferred to the rod. This will include the length and width of the frame, along with sizes of the materials to be used.

Industry term	Definition and regional variations
Roofing square 	A setting out tool used with a batten or fence to accurately mark out stair treads and to determine the angles and cuts required for a traditional roof.
Router 	A versatile tool, it can be used to form grooves, rebates, housings, openings and profiles. Essentially, a router is a big motor attached directly to a cutter via a collet. Fences, guide bushes and stops control and adjust the depth and width of cut.
Rules Scale rule	An accurate means of measuring dimensions over shorter distances. Rules are available in a variety of lengths, from 100mm to 1m. Foldable 1m rules are also available; these conveniently fold away for easy storage. Scale rules are used to determine dimensions from scale drawings.
Saddle Workpiece Saddle	A means of holding and presenting materials for machining operations, allowing repetitive reproductions.
Sash cramp 	Cramp used for work, up to about 2m long. It is an adjustable steel bar with a bolt at one end and a fixed jaw at the other.
Saw doctor	A specialist who maintains machine tooling like wide band saw blades.
Scale Scale: 1:1250	The ratio of the size on a drawing to the size of the real thing that it represents. It is impossible to fit a full-sized drawing of a building onto a sheet of paper, so it is necessary to scale the size of the building to enable it to fit. Scale rules are used to draw scaled-down buildings on paper.

Industry term	Definition and regional variations
Scribed joint	A scribed joint refers to the shoulders of a frame component or end of a length of moulded timber that are shaped to neatly fit the contours of an abutting member.
Sealant	After frames are fixed they require sealing between the wall and the back of the frame with a sealant, to achieve adequate weather or fire protection, or to give an acoustic or thermal seal.
Second-fix nail gun	A powered nailer suitable for use on fixing moulding to stop beads to door linings.
Second seasoning	Where timber is sawn to the sizes required on a cutting list and left 'in stick' for as many days as time will allow. The freshly sawn faces will then acclimatise to the atmospheric conditions in the joiner's shop, minimising the amount the timber will distort after planing.
Secondary machining	Jointing and profiling of planed components.
Set and tee square	Set squares are used on a drawing board together with a tee square.
Set mitre square	A special square used to draw 45° lines when marking out.
Setter out	An experienced joiner who has proved their competency and is the person who produces the rods, cutting lists and orders for contracts.
Shakes Thunder shake	A type of timber defect. Shakes are cracks in the timber which appear due to excessive heat, frost or twisting due to wind during the growth of a tree. Depending upon their shape and position, shakes can be classified as thunder shake, star shake, cup shake, ring shakes or heart shakes.

Industry term	Definition and regional variations
Sheathing	The covering of the top of the rafters/roof surface with timber-based material.
Shoot in	The process of fitting a door or sash to an opening with a parallel gap that allows for fitting and finish clearance.
Short grain	Where the general direction of the wood fibres lies across the direction of cut, increasing the likelihood of the timber breaking up during cutting operations.
Single curvature	Where work is shaped in one view only, either elevation or plan.
Single roof	Single pitch roof designs rely on rafters that are only supported at either end. Single roof designs are not suitable for spans in excess of 5.5m because of the exceptional section size in length and width and the thickness of the timber required to support the weight of the roof.
Couple roof	Single roofs can be further divided into three types: couple roof, close-couple roof and collared roof.
Sliding bevel	A tool that can be set to different angles to aid marking out. It is constructed of a hardwood stock, a sliding blade and a locking screw. The blade can be loosened, adjusted and then tightened to the required angle.
Slotted drive system (SDS) hammer drill	Extremely powerful hammer drills used when extra power is required, eg for heavy-duty jobs, such as to drill stone and concrete down.
Smoothing plane	All-purpose plane used mainly for cleaning up and finishing work. Smoothing planes are not suitable for straightening timber due to their relative shortness (250mm).
Snatching	In a machining context, this is where the rotating cutter hooks into the split, causing it to break or the work to be thrown back at the operative. The operative could lose control or have an accident.

Industry term	Definition and regional variations
Softwood  Douglas fir	Softwood comes from coniferous trees, which are generally evergreen, have needle-like leaves and, in the case of pine trees, produce pinecones. Softwoods are generally used for carcassing (studwork, joists and rafters), high-class structural work and joinery. Examples include whitewood, redwood, Douglas fir, cedar and yellow pine.
Specification	A contract document that gives information about the quality of materials and standards of workmanship required.
Spelching	Part of the material breaking away in an uncontrolled manner, producing damaged components.
Spindle moulder	A fixed machine capable of producing rebates and mouldings. It consists of a bed, fence, motor and spindle upon which a cutter block can be mounted. The spindle can be raised or lowered by means of an adjusting wheel, and some models tilt. *Regional variation: spindle*
Spokeshaves	Used to shape curved surfaces, consists of a blade fastened between two handles. The sole can be flat or round bottomed for planing concave or convex shapes.
Squaring rod	A very useful tool for checking frames for square. It is simply made from a thin section of timber such as a bead or offcut. An angled cut is formed or a pin is driven in at the end, and this end is placed in one corner of the joinery to be tested for square.
Staircase 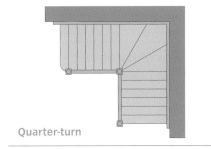 Quarter-turn	Staircases come in a wide variety of shapes and sizes. Three common ones are straight flight, quarter-turn and half-turn. Their purpose is to enable safe movement between floor levels. Joints used in a staircase include housing joints (used to join the treads to the string) and mortice and tenon, used for the strings into the newel posts as well as for the handrails.

Industry term	Definition and regional variations
Stair router jig 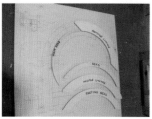	Used for marking out staircases. They have adjustable stops and can be set in a very similar way to a roofing square.
Stair string	An inclined board each side of the stair which carries the treads (the top or horizontal surface of a step) and risers (the board that forms the face of the step).
Storey rod (stair)	Traditionally, a lath of timber on to which the position of the landings was marked on site. The rod was taken back to the joinery shop and divided up to find the step rise for the stair. This is very seldom used in modern practice but the term is still used and shown on drawings.
Surface planer	Used to produce standard facing and edging work, but also to produce bevels, chamfers and, in older versions, rebates.
Take off	Taking the length, width and thickness of components and recording them on the cutting list.
Tangential conversion	Term used in timber conversion, where the timber is sawn at a tangent to the heart. Boards converted using this method are used for floor joists as they offer good resistance to deflection from vertical loads.
Templet	Also known as a 'template', this is a thin piece of hardboard, MDF or ply that is cut to the shape required and is then used to help reproduce the shape.

Industry term	Definition and regional variations
Temporary Bench Mark (TBM)	Unlike an OBM, this is only temporary and is set up on site. These can be timber pegs surrounded with concrete.
Tenon saw	Small saw used for the cutting of the shoulders of a tenon joint. A type of backed saw.
Thicknesser	The thicknesser planes the timber (that has been previously faced and edged on an overhead planer) to the required finished size.
Through-and-through	Term used in timber conversion. Produces mostly tangentially sawn timber and some quarter sawn boards. Through-and-through timber is the most economical form of timber conversion.
Tongue and groove joint V-jointed	A joint between two boards in which a raised area on the edge (tongue) of one board fits into a corresponding groove in the edge of the other to produce a flush surface. Commonly used in floorboards. Two different styles of this joint are V-jointed and bead-jointed. *Regional variation: T&G joint*
Tooling	The part of the machine that cuts the timber, eg a circular saw blade. This can be made from TCT (tungsten carbide tipped), HSS (high speed steel) or PCD (polycrystalline diamond).
Toughened glass	Standard annealed glass that is heated and rapidly cooled to give added strength. Used in doors that have glazing.

Industry term	Definition and regional variations
Trammel heads and beam 	Used for setting out larger curves that are too great for conventional compasses and dividers. Two trammel heads are attached to a batten (beam). The trammel heads can be fitted with a metal point or pencil.
Transportable circular saw bench 	Used on site to cut sheet material to width and to convert larger sections to the smaller sections required.
Triple roof  King post roof truss	This type of roof was traditionally used for very large spans. It is nowadays mainly found in renovation work, heritage buildings, barn conversions and oak-framed buildings. This roof design includes a truss, typically a 'king post truss', which is fixed at intervals along the roof to provide intermediate support for the purlins, which in turn provide support for the common rafters.
Try plane 	A try plane is a very long plane (550mm long) used to straighten joints, but is not in common use today. The length of the sole of the plane means it will always take off the high points before reaching any hollows in the edge of the timber, ensuring a straight length of timber is produced.
Try square 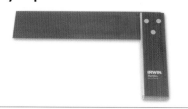	A square with a steel tongue in a wooden handle, used to mark or test a right angle. A try square is the most accurate type of square to use for marking out.
Turned finial 	A turned decorative finish to the top of a post.

Industry term	Definition and regional variations
Unit 	Used in kitchens, bathrooms, bedrooms and libraries. Although units are used widely, their construction is usually very similar: traditional framed or modern non-framed. Shelves and rails are usually included inside the unit. If on show, the shelves may be solid or veneered board; if hidden, they are usually made from melamine faced sheet material. Units are joined using biscuits, dowels and other proprietary means, eg stud and cam.
Urea-formaldehyde (UF) adhesive 	An adhesive commonly referred to as powdered resin glue. It must be mixed with water before use and forms a strong water-resistant bond when set.
Vice 	An essential tool used to hold timber as it is being cut. It is faced with plywood or timber to avoid damage to the timber being held.
Volatile organic compound (VOC) 	The measurement of volatile organic compounds shows how much pollution a product will emit into the air when in use.
Water level 	A tool used for setting out and transferring datum lines and levels.
Weathering	A means of preventing water from gaining entry, eg when double doors meet in the middle.

Industry term	Definition and regional variations
Winders 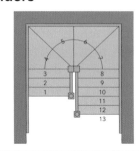	These are tapered steps used to save space by allowing extra risers to be incorporated. They generally turn through 90° or 180°.
Window 	The main functions of windows are to provide daylight, ventilation and a view of the outside. They can also be used as a means of escape in an emergency. The most common types of window design are: traditional casement windows, stormproof casement windows, sliding sashes, bay and bow windows, pivot-hung windows, tilt and turn windows and roof lights.
Wire brush 	If timber has been infected by dry rot and needs replacing, but is covered by plaster, a wire brush can be used to remove the plaster.

Chapter 1
Health, safety and welfare in construction

A career in the building industry can be a very rewarding one, both personally and financially. However, building sites and workshops are potentially very dangerous places; there are many potential hazards in the construction industry. Many construction operatives (workers) are injured each year, some fatally. Regulations have been brought in over the years to reduce accidents and improve working conditions.

By reading this chapter you will know about:

1 The health and safety regulations, roles and responsibilities.
2 Accident and emergency reporting procedures and documentation.
3 Identifying hazards in the workplace.
4 Health and welfare in the workplace.
5 Handling materials and equipment safely.
6 Access equipment and working at heights.
7 Working with electrical equipment in the workplace.
8 Using personal protective equipment (PPE).
9 The cause of fire and fire emergency procedures.

HEALTH AND SAFETY LEGISLATION

According to the Health and Safety Executive (HSE) figures, in 2012/13:

- Forty-six construction operatives were fatally injured. Twelve of these operatives were self-employed. This compares with an average of 50 fatalities over the previous five years, of which an average of 17 fatally injured construction operatives were self-employed.

- The rate of fatal injury per 100,000 construction operatives was 1.94, compared with a five-year average of 2.07.

- Construction industry operatives were involved in 27% of fatal injuries across all industry sectors and it accounts for the greatest number of fatal injuries in any industry sector.

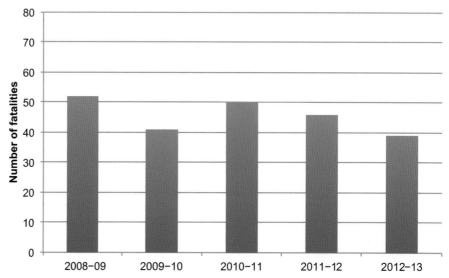

Number and rate of fatal injuries to workers in construction (RIDDOR)

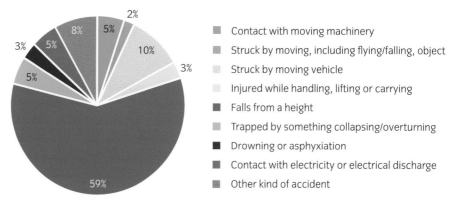

Proportion of fatalities in 2012/13 in construction

Health and safety legislation and great efforts made by the industry have made workplaces much safer in recent years. It is the responsibility of everyone involved in the building industry to continue to make it safer. Statistics are not just meaningless numbers – they represent injuries to real people. Many people believe that an accident will never happen to them, but it can. Accidents can:

- have a devastating effect on lives and families

- cost a lot financially in injury claims

- result in prosecution

- lead to job loss if an employee broke their company's safety policy.

Employers have an additional duty to ensure operatives have access to welfare facilities, eg drinking water, first aid and toilets, which will be discussed later in this chapter.

If everyone who works in the building industry pays close attention to health, safety and welfare, all operatives – including you – have every chance of enjoying a long, injury-free career.

FUNCTIONAL SKILLS

Using the HSE website, find the most recent health and safety statistics. Of the total number of accidents that resulted in three or more days off work, what proportion (as a percentage) are those who were killed during that year?

Work on this activity can support FM L2.3.1 and C2.4.

UK HEALTH AND SAFETY REGULATIONS, ROLES AND RESPONSIBILITIES

In the UK there are many laws (legislation) that have been put into place to make sure that those working on construction sites, and members of the public, are kept healthy and safe. If these laws and regulations are not obeyed then prosecutions can take place. Worse still, there is a greater risk of injury and damage to your health and the health of those around you.

Standard construction safety equipment

The principal legislation which relates to health, safety and welfare in construction is:

- Health and Safety at Work Act (HASAWA) 1974

- Control of Substances Hazardous to Health (COSHH) Regulations 2002

- Reporting of Injuries, Diseases and Dangerous Occurrences Regulations (RIDDOR) 2013

- Construction, Design and Management (CDM) Regulations 2015

- Provision and Use of Work Equipment Regulations (PUWER) 1998

- Manual Handling Operations Regulations 1992

- Personal Protective Equipment (PPE) at Work Regulations 1992

- Work at Height Regulations 2005 (as amended)
- Lifting Operations and Lifting Equipment Regulations (LOLER) 1998
- Control of Noise at Work Regulations 2005
- Control of Vibration at Work Regulations 2005.

HEALTH AND SAFETY AT WORK ACT (HASAWA) 1974

The Health and Safety at Work Act (HASAWA) 1974 applies to all workplaces. Everyone who works on a building site or in a workshop is covered by this legislation. This includes employed and self-employed operatives, subcontractors, the employer and those delivering goods to the site. It not only protects those working, it also ensures the safety of anyone else who might be nearby.

KEY EMPLOYER RESPONSIBILITIES

The key employer health and safety responsibilities under HASAWA are to:

- provide a safe working environment
- provide safe access (entrance) and egress (exit) to the work area
- provide adequate staff training
- have a written health and safety policy in place
- provide health and safety information and display the appropriate signs
- carry out risk assessments
- provide safe machinery and equipment and to ensure it is well-maintained and in a safe condition
- provide adequate supervision to ensure safe practices are carried out
- involve trade union safety representatives, where appointed, in matters relating to health and safety
- provide personal protective equipment (**PPE**) free of charge, ensure the appropriate PPE is used whenever needed, and that operatives are properly supervised
- ensure materials and substances are transported, used and stored safely.

PPE

This is defined in the Personal Protective Equipment at Work Regulations 1992 as 'all equipment (including clothing affording protection against the weather) which is intended to be worn or held by a person at work and which protects against one or more risks to a person's health or safety.'

Risk assessments and method statements

The HASAWA requires that employers must carry out regular **risk assessments** to make sure that there are minimal dangers to their employees in a workplace.

Risk assessment

An assessment of the hazards and risks associated with an activity and the reduction and monitoring of them

Risk Assessment

Activity / Workplace assessed: Return to work after accident
Persons consulted / involved in risk assessment
Date:
Reviewed on:

Location:
Risk assessment reference number:
Review date:
Review by:

Significant hazard	People at risk and what is the risk Describe the harm that is likely to result from the hazard (eg cut, broken leg, chemical burn etc) and who could be harmed (eg employees, contractors, visitors etc)	Existing control measure What is currently in place to control the risk?	Risk rating Use matrix identified in guidance note Likelihood (L) Severity (S) Multiply (L) * (S) to produce risk rating (RR)				Further action required What is required to bring the risk down to an acceptable level? Use hierarchy of control described in guidance note when considering the controls needed	Actioned to: Who will complete the action?	Due date: When will the action be completed by?	Completion date: Initial and date once the action has been completed
			L	S	RR	L/M/H				
Uneven floors	Operatives	Verbal warning and supervision	2	1	2	M	None applicable	Site supervisor	Active now	Ongoing
Steps	Operatives	Verbal warning	2	1	2	M	None applicable	Site supervisor	Active now	Ongoing
Staircases	Operatives	Verbal warning	2	2	4	M	None applicable	Site supervisor	Active now	Ongoing

		Likelihood		
		1 **Unlikely**	**2** **Possible**	**3** **Very likely**
Severity	**1** Slight/minor injuries/minor damage	1	2	3
	2 Medium injuries/significant damage	2	4	6
	3 Major injury/extensive damage	3	6	9

Likelihood
3 – Very likely
2 – Possible
1 – Unlikely

Severity
3 – Major injury/extensive damage
2 – Medium injury/significant damage
1 – Slight/minor damage

1 – Low risk, action should be taken to reduce the risk if reasonably practicable
2, 3, 4 – Medium risk, is a significant risk and would require an appropriate level of resource
6 & 9 – High risk, may require considerable resource to mitigate. Control should focus on elimination of risk, if not possible control should be obtained by following the hierarchy of control

123 type risk assessment

A risk assessment is a legally required tool used by employers to:

- identify work hazards

- assess the risk of harm arising from these hazards

- adequately control the risk.

Risk assessments are carried out as follows:

1 Identify the hazards. Consider the environment in which the job will be done. Which tools and materials will be used?

2 Identify who might be at risk. Think about operatives, visitors and members of the public.

3 Evaluate the risk. How severe is the potential injury? How likely is it to happen? A severe injury may be possible but may also be very improbable. On the other hand a minor injury might be very likely.

4 If there is an unacceptable risk, can the job be changed? Could different tools or materials be used instead?

5 If the risk is acceptable, what measures can be taken to reduce the risk? This could be training, special equipment and using PPE.

6 Keep good records. Explain the findings of the risk assessment to the operatives involved. Update the risk assessment as required – there may be new machinery, materials or staff. Even adverse weather can bring additional risks.

A **method statement** is required by law and is a useful way of recording the hazards involved in a specific task. It is used to communicate the risk and precautions required to all those involved in the work. It should be clear, uncomplicated and easy to understand as it is for the benefit of those carrying out the work (and their immediate supervisors).

Inductions and toolbox talks

Any new visitors to and operatives on a site will be given an induction. This will explain:

- the layout of the site
- any hazards of which they need to be aware
- the location of welfare facilities
- the assembly areas in case of emergency
- site rules.

Toolbox talks are short talks given at regular intervals. They give timely safety reminders and outline any new hazards that may have arisen because construction sites change as they develop. Weather conditions such as extreme heat, wind or rain may create new hazards.

KEY EMPLOYEE RESPONSIBILITIES

The HASAWA covers the responsibilities of employees and subcontractors:

- You must work in a safe manner and take care at all times.
- You must make sure you do not put yourself or others at risk by your actions or inactions.

Method statement
A description of the intended method of carrying out a task, often linked to a risk assessment

INDUSTRY TIP
The Construction Skills Certification Scheme (CSCS) was set up in the mid-'90s with the aim of improving site operatives' competence to reduce accidents and drive up on-site efficiency. Card holders must take a health and safety test. The colour of card depends on level of qualification held and job role. For more information see www.cscs.uk.com

ACTIVITY
Think back to your induction. Write down what was discussed. Did you understand everything? Do you need any further information? If you have not had an induction, write a list of the things you think you need to know.

INDUSTRY TIP
Remember, if you are unsure about any health and safety issue always seek help and advice.

- You must co-operate with your employer in regard to health and safety. If you do not you risk injury (to yourself or others), prosecution, a fine and loss of employment. Do not take part in practical jokes and horseplay.

- You must use any equipment and safeguards provided by your employer. For example, you must wear, look after and report any damage to the PPE that your employer provides.

- You must not interfere or tamper with any safety equipment.

- You must not misuse or interfere with anything that is provided for employees' safety.

FIRST AID AND FIRST-AID KITS

First aid should only be applied by someone trained in first aid. Even a minor injury could become infected and therefore should be cleaned and a dressing applied. If any cut or injury shows signs of infection, becomes inflamed or painful seek medical attention. An employer's first-aid needs should be assessed to indicate whether a first-aider (someone trained in first aid) is necessary. The minimum requirement is to appoint a person to take charge of first-aid arrangements. The role of this appointed person includes looking after the first-aid equipment and facilities and calling the emergency services when required.

First-aid kits vary according to the size of the workforce. First-aid boxes should not contain tablets or medicines.

INDUSTRY TIP

The key employee health and safety responsibilities are to:
- work safely
- work in partnership with your employer
- report hazards and accidents as per company policy.

INDUSTRY TIP

Employees must not be charged for anything given to them or done for them by the employer in relation to safety.

INDUSTRY TIP

In the event of an accident, first aid will be carried out by a qualified first aider. First aid is designed to stabilise a patient for later treatment if required. The casualty may be taken to hospital or an ambulance may be called. In the event of an emergency you should raise the alarm.

ACTIVITY

Your place of work or training will have an appointed first-aider who deals with first aid. Find out who they are and how to make contact with them.

ACTIVITY

Find the first-aid kit in your workplace or place of training. What is inside it? Is there anything missing?

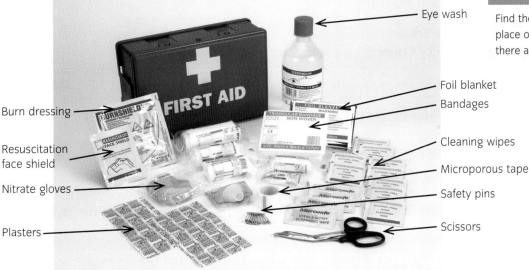

First-aid kit

SOURCES OF HEALTH AND SAFETY INFORMATION

Source	How they can help
Health and Safety Executive (HSE)	A government body that oversees health and safety in the workplace. It produces health and safety literature such as the **Approved Code of Practice** (ACoP).
Construction Skills	The construction industry training body produces literature and is directly involved with construction training.
The Royal Society for the Prevention of Accidents (ROSPA)	It produces literature and gives advice.
The Royal Society for Public Health	An independent, multi-disciplinary charity that is dedicated to the promotion and protection of collective human health and wellbeing.
Institution of Occupational Safety and Health (IOSH)	A chartered body for health and safety practitioners. The world's largest health and safety professional membership organisation.
The British Safety Council	It helps businesses with their health, safety and environmental management.

HEALTH AND SAFETY EXECUTIVE (HSE)

The HSE is a body set up by the government. The HSE ensures that the law is carried out correctly and has extensive powers to ensure that it can do its job. It can make spot checks in the workplace, bring the police, examine anything on the premises and take things away to be examined.

If the HSE finds a health and safety problem that breaks health and safety law it might issue an **improvement notice** giving the employer a set amount of time to correct the problem. For serious health and safety risks where there is a risk of immediate major injury, it can issue a **prohibition notice** which will stop all work on site until the health and safety issues are rectified. It may take an employer, employee, self-employed person (subcontractor) or anyone else

---

Approved Code of Practice

ACoP gives practical advice for those in the construction industry in relation to using machinery

INDUSTRY TIP

There are many other trade organisations, eg the Timber Research and Development Association (TRADA), which also offer advice on safe practices.

ACTIVITY

You have been asked to give a toolbox talk because of several minor injuries involving tripping on site. What topics would you include in this talk?

INDUSTRY TIP

To find out more information on the sources in the table, enter their names into a search engine on the internet.

Improvement notice

Issued by an HSE or Local Authority inspector to formally notify a company that improvements are needed to the way it is working

Prohibition notice

Issued by an HSE or Local Authority inspector when there is an immediate risk of personal injury. They are not issued lightly and if you are on the receiving end of one, you are clearly breaking a health and safety regulation

involved with the building process to court for breaking health and safety legislation.

The HSE provides a lot of advice on safety and publishes numerous booklets and information sheets. One example of this is the Approved Code of Practice (ACoP) which applies to wood working machinery. The ACoP has a special legal status and employers and employees are expected to work within its guidelines.

The duties of the HSE are to:

- give advice
- issue improvement and prohibition notices
- caution
- prosecute
- investigate.

The Approved Code of Practice booklet is available free online

CONTROL OF SUBSTANCES HAZARDOUS TO HEALTH (COSHH) REGULATIONS 2002

The Control of Substances Hazardous to Health (COSHH) Regulations 2002 control the use of dangerous substances, eg preservatives, fuels, solvents, adhesives, cement and oil-based paint. These have to be moved, stored and used safely without polluting the environment. It also covers hazardous substances produced while working, eg wood dust produced when sanding or drilling.

Hazardous substances may be discovered during the building process, eg lead-based paint or asbestos. These are covered by separate regulations.

When considering substances and materials that may be hazardous to health an employer should do the following to comply with COSHH:

- Read and check the COSHH safety data sheet that comes with the product. It will outline any hazards associated with the product and the safety measures to be taken.

- Check with the supplier if there are any known risks to health.

- Use the trade press to find out if there is any information about this substance or material.

- Use the HSE website, or other websites, to check any known issues with the substance or material.

When assessing the risk of a potentially dangerous substance or material it is important to consider how operatives could be exposed to it. For example:

Example of COSHH data sheet

- by breathing in gas or mist

- by swallowing it

- by getting it into their eyes

- through their skin, either by contact or through cuts.

Safety data sheets

Products you use may be 'dangerous for supply'. If so, they will have a label that has one or more hazard symbols. Some examples are given here.

These products include common substances in everyday use such as paint, bleach, solvent or fillers. When a product is 'dangerous for supply', by law, the supplier must provide you with a safety data sheet. Note: medicines, pesticides and cosmetic products have different legislation and don't have a safety data sheet. Ask the supplier how the product can be used safely.

Safety data sheets can be hard to understand, with little information on measures for control. However, to find out about health risks and emergency situations, concentrate on:

- Sections 2 and 16 of the sheet, which tell you what the dangers are;
- Sections 4-8, which tell you about emergencies, storage and handling.

Since 2009, new international symbols have been gradually replacing the European symbols. Some of them are similar to the European symbols, but there is no single word describing the hazard. Read the hazard statement on the packaging and the safety data sheet from the supplier.

European symbols

Toxic   Very toxic   Harmful   Irritant

Highly flammable   Extremely flammable   Explosive   Dangerous to the environment

Oxidising   Corrosive

New International symbols

Hazard checklist

☐ Does any product you use have a danger label?
☐ Does your process produce gas, fume, dust, mist or vapour?
☐ Is the substance harmful to breathe in?
☐ Can the substance harm your skin?
☐ Is it likely that harm could arise because of the way you use or produce it?
☐ What are you going to do about it?
  - Use something else?
  - Use it in another, safer way?
  - Control it to stop harm being caused?

CONTROL MEASURES

The control measures below are in order of importance.

1 Eliminate the use of the harmful substance and use a safer one. For instance, swap high **VOC** oil-based paint for a lower VOC water-based paint.

2 Use a safer form of the product. Is the product available ready-mixed? Is there a lower strength option that will still do the job?

INDUSTRY TIP

Product data sheets are free and have to be produced by the supplier of the product.

3 Change the work method to emit less of the substance. For instance, applying paint with a brush releases fewer VOCs into the air than spraying paint. Wet grinding produces less dust than dry grinding.

4 Enclose the work area so that the substance does not escape. This can mean setting up a tented area or closing doors.

5 Use extraction or filtration (eg a dust bag) in the work area.

6 Keep operatives in the area to a minimum.

7 Employers must provide appropriate PPE.

Paint with high VOC content

European symbols

| Toxic | Very toxic | Harmful | Irritant |

| Highly flammable | Extremely flammable | Explosive | Dangerous to the environment |

| Oxidising | Corrosive |

New International symbols

| Toxic | May explode when heated | Irritant |

| Causes fire | Explosive | Dangerous to the environment |

| Intensifies fire | Long term health hazard | Corrosive |

COSHH symbols. The international symbols will replace the European symbols in 2015.

INDUSTRY TIP

For more detailed information on RIDDOR visit the HSE webpage at www.hse.gov.uk/riddor.

REPORTING OF INJURIES, DISEASES AND DANGEROUS OCCURRENCES REGULATIONS (RIDDOR) 2013

Despite all the efforts put into health and safety, incidents still happen. The Reporting of Injuries, Diseases and Dangerous Occurrences Regulations (RIDDOR) 2013 state that employers must report to the HSE all serious accidents and injuries that result in an employee needing more than seven days off work (including non-working days). Diseases and dangerous occurrences must also be reported. A serious occurrence that has not caused an injury (a near miss) should still be reported because next time it happens things might not work out as well.

Below are some examples of injuries, diseases and dangerous occurrences that would need to be reported:

- A joiner cuts off a finger while using a circular saw.

- A plumber takes a week off after a splinter in her hand becomes infected.

- A ground operative contracts **leptospirosis**.

- A labourer contracts dermatitis (a serious skin problem) after contact with an irritant substance.

- A scaffold suffers a collapse following severe weather, unauthorised alteration or overloading but no one is injured.

Leptospirosis

Also known as Weil's disease, this is a serious disease spread by rats and cattle

The purpose of RIDDOR is to enable the HSE to investigate serious incidents and collate statistical data. This information is used to help reduce the number of similar accidents happening in future and to make the workplace safer.

INDUSTRY TIP

Accidents do not just affect the person who has the accident. Work colleagues or members of the public might be affected and so will the employer. The consequences may include:

- a poor company image (this may put potential customers off)
- loss of production
- insurance costs increasing
- closure of the site
- having to pay sick pay
- other additional costs.

New HSE guidelines require employers to pay an hourly rate for time taken by the HSE to investigate an accident. This is potentially very costly.

An F2508 injury report form

Although minor accidents and injuries are not reported to HSE, records must be kept. Accidents must be recorded in the accident book. This provides a record of what happened and is useful for future reference. Trends may become apparent and the employer may take action to try to prevent that particular type of accident occurring again.

CONSTRUCTION, DESIGN AND MANAGEMENT (CDM) REGULATIONS 2015

The Construction, Design and Management (CDM) Regulations 2015 focus attention on the effective planning and management of construction projects, from the design concept through to maintenance and repair. The aim is for health and safety considerations to be integrated into a project's development, rather than be an inconvenient afterthought. The CDM Regulations reduce the risk of harm to those that have to maintain or use the structure throughout its life, from construction through to **demolition**.

CDM Regulations protect workers from the construction to demolition of large and complex structures

The CDM Regulations apply to all projects, including business and domestic clients. Property developers also need to follow the CDM Regulations. These regulations will be in force from April 2015 and replace CDM 2007. At the time of publication, they are in draft form.

Under the CDM Regulations, the HSE must be notified where the construction work will take:

■ more than 30 working days

■ have more than 20 workers working 'at the same time' on the project

■ exceed 500 person days.

The CDM Regulations play a role in safety during demolition

Demolition

When something, often a building, is completely torn down and destroyed

DUTY HOLDERS

Under the CDM Regulations 2015 there are several duty holders, each with a specific role.

Duty holder	Role
Client	This is the person or organisation who wishes to have the work done. The client is fully accountable for health, safety and well-being for the whole project and must ensure that: ■ competent contractors are appointed ■ sufficient time and resources are arranged ■ other duty holders are provided with relevant information ■ principal designers and contractors fulfil their duties ■ suitable welfare facilities are available for the project. For domestic clients whose project falls under CDM 2015, their duties may transfer to the contractor, principal contractor or principal designer.
Principal designer	This role is responsible for the planning, managing, monitoring and coordination of health and safety, before construction starts. This involves identifying and managing risks, and ensuring that designers carry out their duties. A principal designer and principal contractor are required on all projects that have more than one contractor working on them. The principal designer's responsibilities include: ■ planning, managing, monitoring and coordinating health and safety in the pre-construction phase ■ managing and controlling all foreseeable risks ■ making sure that designers fulfil their duties.
Designer	Designers preparing or modifying designs or systems affecting the construction work must ensure that all foreseeable risks are eliminated, reduced or controlled. This includes the construction stages and also for the ongoing use and maintenance of the building. Designers must also ensure that: ■ they cooperate with the principal designer ■ the client is aware of the designers' duties ■ information about the risk resulting from their design is provided.
Principal contractor	The principal contractor will plan, manage and monitor the construction and coordinate any other contractors involved in the project. This involves developing a written plan and site rules before the construction begins. The principal contractor ensures that the site is made secure and suitable welfare facilities are provided from the start and maintained throughout construction. The principal contractor will also make sure that all operatives have site inductions and any further training that might be required to make sure the workforce is competent. The principal contractor must ensure that the client is aware of their CDM duties, liaise with all other duty holders during the project and cooperate and share information with the principal designer.

Duty holder	Role
Contractor	Subcontractors and self-employed operatives will plan, manage and monitor their own work and employees, cooperating with any main contractor in relation to site rules. Contractors will make sure that all workers are competent and comply with principal contractors or designers on the project. A contractor also reports any incidents under RIDDOR 2013 to the principal contractor.
Workers	Workers are not specifically duty holders for the project but of course they have duties under Section 7 of the Health and Safety at Work Act 1974, whereby they must take care of their own safety and that of others who might be affected by their actions. Workers must also report any hazards and cooperate with others involved with the project. Workers will be consulted on with regards to health, safety and welfare arrangements, be provided with an induction and have the necessary training for health and safety.

Note:
- Individuals or organisations can undertake more than one duty holder role as long as they are 'fully competent' to do so.
- The CDM 2007 coordinator role is replaced by the principal designer role in CDM 2015.

WELFARE FACILITIES REQUIRED ON SITE UNDER THE CDM REGULATIONS

The table below shows the welfare facilities that must be available on site.

Facility	Site requirement
Sanitary conveniences (toilets)	▪ Suitable and sufficient toilets should be provided or made available. ▪ Toilets should be adequately ventilated and lit and should be clean. ▪ Separate toilet facilities should be provided for men and women.
Washing facilities	▪ Sufficient facilities must be available, and include showers if required by the nature of the work. ▪ They should be in the same place as the toilets and near any changing rooms. ▪ There must be a supply of clean hot (or warm) and cold running water, soap and towels. ▪ There must be separate washing facilities provided for men and women unless the area is for washing hands and the face only.

Facility	Site requirement
Clean drinking water 	<ul><li>This must be provided or made available.</li><li>It should be clearly marked by an appropriate sign.</li><li>Cups should be provided unless the supply of drinking water is from a water fountain.</li></ul>
Changing rooms and lockers 	<ul><li>Changing rooms must be provided or made available if operatives have to wear special clothing and if they cannot be expected to change elsewhere.</li><li>There must be separate rooms for, or separate use of rooms by, men and women where necessary.</li><li>The rooms must have seating and include, where necessary, facilities to enable operatives to dry their special clothing and their own clothing and personal effects.</li><li>Lockers should also be provided.</li></ul>
Rest rooms or rest areas 	<ul><li>They should have enough tables and seating with backs for the number of operatives likely to use them at any one time.</li><li>Where necessary, rest rooms should include suitable facilities for pregnant women or nursing mothers to rest lying down.</li><li>Arrangements must be made to ensure that meals can be prepared, heated and eaten. It must also be possible to boil water.</li></ul>

ACTIVITY

What facilities are provided at your workplace or place of training?

PROVISION AND USE OF WORK EQUIPMENT REGULATIONS (PUWER) 1998

The Provision and Use of Work Equipment Regulations (PUWER) 1998 place duties on:

- people and companies who own, operate or have control over work equipment

- employers whose employees use work equipment.

Work equipment can be defined as any machinery, appliance, apparatus, tool or installation for use at work (whether exclusively or not). This includes equipment employees provide for their own use at work. The scope of work equipment is therefore extremely wide. The use of work equipment is also very widely interpreted and, according to the HSE, means 'any activity involving work equipment and includes starting, stopping, programming, setting, transporting, repairing, modifying, maintaining, servicing and cleaning.' It includes equipment such as diggers, electric planers, stepladders, hammers or wheelbarrows.

Under PUWER, work equipment must be:

- suitable for the intended use

- safe to use

- well maintained

- inspected regularly.

Regular inspection is important as a tool that was safe when it was new may no longer be safe after considerable use.

Additionally, work equipment must only be used by people who have received adequate instruction and training. Information regarding the use of the equipment must be given to the operator and must only be used for what it was designed to do.

Protective devices, eg emergency stops, must be used. Brakes must be fitted where appropriate to slow down moving parts to bring the equipment to a safe condition when turned off or stopped. Equipment must have adequate means of isolation. Warnings, either by signs or other means such as sounds or lights, must be used as appropriate. Access to dangerous parts of the machinery must be controlled. Some work equipment is subject to additional health and safety legislation which must also be followed.

Employers who use work equipment must manage the risks. ACoPs (see page 9) have been developed in line with PUWER. The ACoPs have a special legal status, as outlined in the introduction to the PUWER ACoP:

Following the guidance is not compulsory and you are free to take other action. But if you do follow the guidance you will normally be doing enough to comply with the law. Health and safety inspectors seek to secure compliance with the law and may refer to this guidance as illustrating good practice.

INDUSTRY TIP

Abrasive wheels are used for grinding. Under PUWER these wheels can only be changed by someone who has received training to do this. Wrongly fitted wheels can explode!

ACTIVITY

All the tools you use for your work are covered by PUWER. They must be well maintained and suitable for the task. A damaged head on a bolster chisel must be reshaped. A split shaft on a joiner's wood chisel must be repaired. Why would these tools be dangerous in a damaged condition? List the reasons.

MANUAL HANDLING OPERATIONS REGULATIONS 1992

Employers must try to avoid manual handling within reason if there is a possibility of injury. If manual handling cannot be avoided then they must reduce the risk of injury by means of a risk assessment.

LIFTING AND HANDLING

Incorrect lifting and handling is a serious risk to your health. It is very easy to injure your back – just ask any experienced builder. An injured back can be very unpleasant, so it's best to look after it.

Here are a few things to consider when lifting:

- Assess the load. Is it too heavy? Do you need assistance or additional training? Is it an awkward shape?

- Can a lifting aid be used, such as any of the below?

An operative lifting heavy bricks

Wheelbarrow

Gin lift

Scissor lift

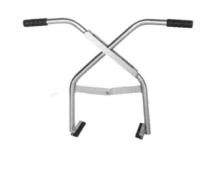

Kerb lifter

- Does the lift involve twisting or reaching?

- Where is the load going to end up? Is there a clear path? Is the place it's going to be taken to cleared and ready?

How to lift and place an item correctly

If you cannot use a machine, it is important that you keep the correct posture when lifting any load. The correct technique to do this is known as **kinetic lifting**. Always lift with your back straight, elbows in, knees bent and your feet slightly apart.

Kinetic lifting

A method of lifting that ensures that the risk of injury is reduced

Safe kinetic lifting technique

When placing the item, again be sure to use your knees and beware of trapping your fingers. If stacking materials, be sure that they are on a sound level base and on bearers if required.

Heavy objects that cannot easily be lifted by mechanical methods can be lifted by several people. It is important that one person in the team is in charge, and that lifting is done in a co-operative way. It has been known for one person to fall down and the others to then drop the item!

CONTROL OF NOISE AT WORK REGULATIONS 2005

Under the Control of Noise at Work Regulations 2005, duties are placed on employers and employees to reduce the risk of hearing damage to the lowest reasonable level practicable. Hearing loss caused by work is preventable. Hearing damage is permanent and cannot be restored once lost.

EMPLOYER'S DUTIES UNDER THE REGULATIONS

An employer's duties are:

- to carry out a risk assessment and identify who is at risk

- to eliminate or control its employees' exposure to noise at the workplace and to reduce the noise as far as practicable

- to provide suitable hearing protection

- to provide health surveillance to those identified as at risk by the risk assessment

■ to provide information and training about the risks to their employees as identified by the risk assessment.

EMPLOYEES' DUTIES UNDER THE REGULATIONS

Employees must:

■ make full and proper use of personal hearing protectors provided to them by their employer

■ report to their employer any defect in any personal hearing protectors or other control measures as soon as is practicable.

NOISE LEVELS

Under the Regulations, specific actions are triggered at specific noise levels. Noise is measured in decibels and shown as dB(A). The two main action levels are 80 dB(A) and 85 dB(A).

Requirements at 80 dB(A) to 85 dB(A):

■ Assess the risk to operatives' health and provide them with information and training.

■ Provide suitable ear protection free of charge to those who request ear protection.

Requirements above 85 dB(A):

■ Reduce noise exposure as far as practicable by means other than ear protection.

■ Set up an ear protection zone using suitable signage and segregation.

■ Provide suitable ear protection free of charge to those affected and ensure they are worn.

Ear defenders

Ear plugs

INDUSTRY TIP

The typical noise level for a hammer drill and a concrete mixer is 90 to 100 dB(A).

PERSONAL PROTECTIVE EQUIPMENT (PPE) AT WORK REGULATIONS 1992

Employees and subcontractors must work in a safe manner. Not only must they wear the PPE that their employers provide but they must also look after it and report any damage to it. Importantly, employees must not be charged for anything given to them or done for them by the employer in relation to safety.

ACTIVITY

Think about your place of work or training. What PPE do you think you should use when working with cement or using a powered planer?

The hearing and respiratory PPE provided for most work situations is not covered by these Regulations because other regulations apply to it. However, these items need to be compatible with any other PPE provided.

The main requirement of the Regulations is that PPE must be supplied and used at work wherever there are risks to health and safety that cannot be adequately controlled in other ways.

The Regulations also require that PPE is:

- included in the method statement

- properly assessed before use to ensure it is suitable

- maintained and stored properly

- provided to employees with instructions on how they can use it safely

- used correctly by employees.

An employer cannot ask for money from an employee for PPE, whether it is returnable or not. This includes agency workers if they are legally regarded as employees. If employment has been terminated and the employee keeps the PPE without the employer's permission, then, as long as it has been made clear in the contract of employment, the employer may be able to deduct the cost of the replacement from any wages owed.

Using PPE is a very important part of staying safe. For it to do its job properly it must be kept in good condition and used correctly. If any damage does occur to an article of PPE it is important that this is reported and it is replaced. It must also be remembered that PPE is a last line of defence and should not be used in place of a good safety policy!

A site safety sign showing the PPE required to work there

The following table shows the type of PPE used in the workplace and explains why it is important to store, maintain and use PPE correctly. It also shows why it is important to check and report damage to PPE.

PPE	Correct use
Hard hat/safety helmet 	Hard hats must be worn when there is danger of hitting your head or danger of falling objects. They often prevent a wide variety of head injuries. Most sites insist on hard hats being worn. They must be adjusted to fit your head correctly and must not be worn back to front! Check the date of manufacture as plastic can become brittle over time. Solvents, pens and paints can damage the plastic too.
Toe-cap boots or shoes  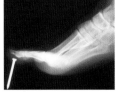 Safety boots A nail in a construction worker's foot	Toe-cap boots or shoes are worn on most sites as a matter of course and protect the feet from heavy falling objects. Some safety footwear has additional insole protection to help prevent nails going up through the foot. Toe caps can be made of steel or lighter plastic.
Ear defenders and plugs  Ear defenders Ear plugs	Your ears can be very easily damaged by loud noise. Ear protection will help prevent hearing loss while using loud tools or if there is a lot of noise going on around you. When using earplugs always ensure your hands are clean before handling the plugs as this reduces the risk of infection. If your ear defenders are damaged or fail to make a good seal around your ears have them replaced.
High-visibility (hi-viz) jacket 	This makes it much easier for other people to see you. This is especially important when there is plant or vehicles moving in the vicinity.
Goggles and safety glasses   Safety goggles Safety glasses	These protect the eyes from dust and flying debris while you are working. It has been known for casualties to be taken to hospital after dust has blown up from a dry mud road. You only get one pair of eyes: look after them!

PPE	Correct use
Dust masks and respirators   Dust mask Respirator	Dust is produced during most construction work and it can be hazardous to your lungs. It can cause all sorts of ailments from asthma through to cancer. Wear a dust mask to filter this dust out. You must ensure it is well fitted. Another hazard is dangerous gases such as solvents. A respirator will filter out hazardous gases but a dust mask will not! Respirators are rated P1, P2 and P3, with P3 giving the highest protection.
Gloves Latex glove Nitrile glove   Gauntlet gloves Leather gloves	Gloves protect your hands. Hazards include cuts, abrasions, dermatitis, chemical burns or splinters. Latex and nitrile gloves are good for fine work, although some people are allergic to latex. Gauntlets provide protection from strong chemicals. Other types of gloves provide good grip and protect the fingers. 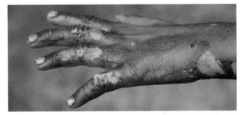 A chemical burn as a result of not wearing safety gloves
Sunscreen  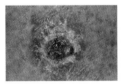 Suncream Melanoma	Another risk, especially in the summer months, is sunburn. Although a good tan is sometimes considered desirable, over-exposure to the sun can cause skin cancer such as melanoma. When out in the sun, cover up and use sunscreen (ie suncream) on exposed areas of your body to prevent burning.
Preventing HAVS 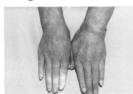	Hand–arm vibration syndrome (HAVS), also known as vibration white finger (VWF), is an industrial injury caused by using vibrating power tools (such as a hammer drill, vibrating poker and vibrating plate) for a long time. This injury is controlled by limiting the time such power tools are used. For more information see page 31.

ACTIVITY

You are working on a site and a brick falls on your head. Luckily, you are doing as you have been instructed and you are wearing a helmet. You notice that the helmet has a small crack in it. What do you do?

1 Carry on using it as your employer will charge you for a new one: after all it is only a small crack.
2 Take it to your supervisor as it will no longer offer you full protection and it will need replacing.
3 Buy a new helmet because the old one no longer looks very nice.

INDUSTRY TIP

The most important pieces of PPE when using a disc cutter are dust masks, glasses and ear protection.

WORK AT HEIGHT REGULATIONS 2005 (AS AMENDED)

The Work at Height Regulations 2005 (as amended by the Work at Height (Amendment) Regulations 2007) put several duties upon employers:

- Working at height should be avoided if possible.

- If working at height cannot be avoided, the work must be properly organised with risk assessments carried out.

- Risk assessments should be regularly updated.

- Those working at height must be trained and competent.

- A method statement must be provided.

Workers wearing safety harnesses on an aerial access platform

Several points should be considered when working at height:

- How long is the job expected to take?

- What type of work will it be? It could be anything from fitting a single light bulb, through to removing a chimney or installing a roof.
  - How is the access platform going to be reached? By how many people?
  - Will people be able to get on and off the structure safely? Could there be overcrowding?

- What are the risks to passers-by? Could debris or dust blow off and injure anyone on the road below?

- What are the conditions like? Extreme weather, unstable buildings and poor ground conditions need to be taken into account.

A cherry picker can assist you when working at height

ACCESS EQUIPMENT AND SAFE METHODS OF USE

The means of access should only be chosen after a risk assessment has been carried out. There are various types of access.

Ladders

Ladders are normally used for access onto an access platform. They are not designed for working from except for light, short-duration work. A ladder should lean at an angle of 75°, ie one unit out for every four units up.

Strong upper resting point

Adequate lap on extension ladders

Ground back slope not exceeding 6°

Ground side slope not exceeding 16°, clean and free of slippery algae and moss

Using a ladder correctly

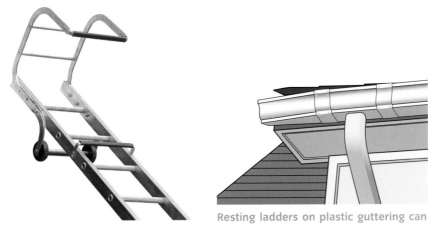

Roof ladder

Resting ladders on plastic guttering can cause it to bend and break

The following images show how to use a ladder or stepladder safely.

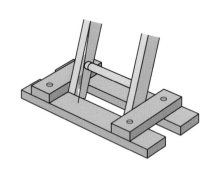

A ladder secured at the base.

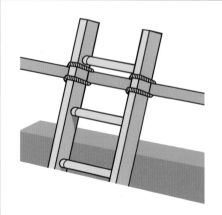

A ladder secured at the top of a platform for working from.

Access ladders should extend 1m above the landing point to provide a strong handhold.

Certain stepladders are unsafe to work from the top three rungs.

Don't overreach, and stay on the same rung.

Grip the ladder when climbing and remember to keep three points of contact.

INDUSTRY TIP

Always complete ladder pre-checks. Check the stiles (the two uprights) and rungs for damage such as splits or cracks. Do not use painted ladders because the paint could be hiding damage! Check all of the equipment including any stays and feet.

Stepladders

Stepladders are designed for light, short-term work.

Working from the side can make stepladders unstable. Do not overreach

Don't stand on the top three steps

Stepladder is fully open

Locked open firm and level on the ground

Using a stepladder correctly

FUNCTIONAL SKILLS

Using information on stepladders on the HSE website, write down two examples of what sort of job stepladders could be used for and two jobs they should not be used for.

Work on this activity can support FICT 2.A and FE 2.2.2.

Trestles

This is a working platform used for work of a slightly longer duration.

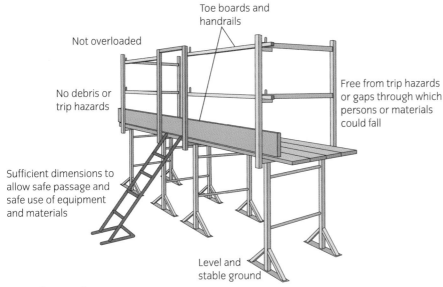

Toe boards and handrails

Not overloaded

No debris or trip hazards

Free from trip hazards or gaps through which persons or materials could fall

Sufficient dimensions to allow safe passage and safe use of equipment and materials

Level and stable ground

Parts of a trestle

Tower scaffold

These are usually proprietary (manufactured) and are made from galvanised steel or lightweight aluminium alloy. They must be erected by someone competent in the erection and dismantling of mobile scaffolds.

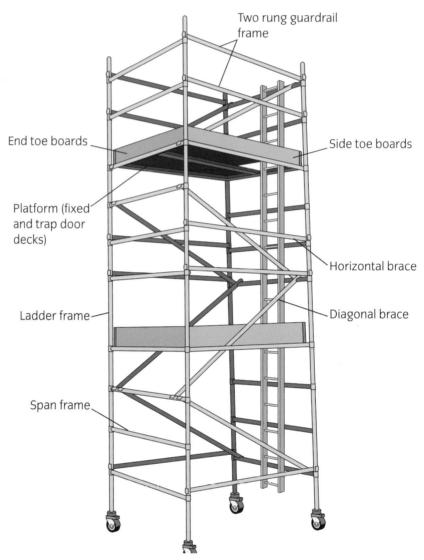

Two rung guardrail frame

End toe boards

Side toe boards

Platform (fixed and trap door decks)

Horizontal brace

Ladder frame

Diagonal brace

Span frame

Parts of a tower scaffold

To use a tower scaffold safely:

- Always read and follow the manufacturer's instruction manual.

- Only use the equipment for what it is designed for.

- The wheels or feet of the tower must be in contact with a firm surface.

- Outriggers should be used to increase stability. The maximum height given in the manufacturer's instructions must not be exceeded.

- The platform must not be overloaded.

- The platform should be unloaded (and reduced in height if required) before it is moved.

- Never move a platform, even a small distance, if it is occupied.

Tubular scaffold

This comes in two types:

- independent scaffold has two sets of standards or uprights

- putlog scaffold is built into the brickwork.

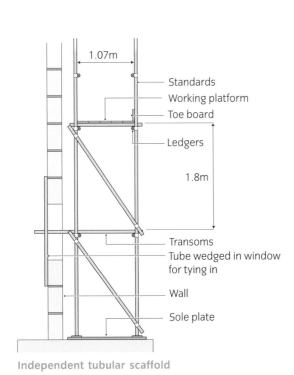

Independent tubular scaffold

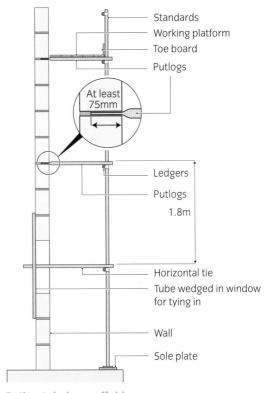

Putlog tubular scaffold

Tubular scaffold is erected by specialist scaffolding companies and often requires structural calculations. Only trained and competent scaffold erectors should alter scaffolding. Access to a scaffold is usually via a tied ladder with three rungs projecting above the step off at platform level.

OUR HOUSE

You have been asked to complete a job that requires gaining access to the roof level of a two-storey building. What equipment would you choose to get access to the work area? What things would you take into consideration when choosing the equipment? Take a look at 'Our House' as a guide for working on a two-storey building.

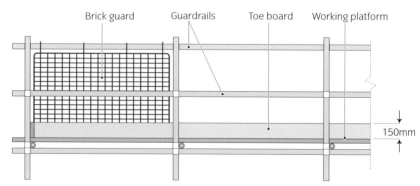

Brick guard   Guardrails   Toe board   Working platform

150mm

A safe working platform on a tubular scaffold

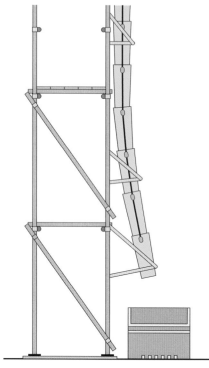

A debris chute for scaffolding

All scaffolding must:

- not have any gaps in the handrail or toe boards
- have a safe system for lifting any materials up to the working height
- have a safe system of debris removal.

Fall protection devices include:

- harnesses and lanyards
- safety netting
- air bags.

A harness and lanyard or safety netting will stop a person falling too far, leaving them suspended in the air. Air bags (commonly known as 'bouncy castles') are set up on the ground and inflated. If a person falls, they will have a soft landing. Air bags have fallen out of favour somewhat as some operatives use them as an easy way to get off the working platform – not the purpose they were intended for!

Using a scissor lift at height

LIFTING OPERATIONS AND LIFTING EQUIPMENT REGULATIONS (LOLER) 1998

The Lifting Operations and Lifting Equipment Regulations (LOLER) 1998 put responsibility upon employers to ensure that the lifting equipment provided for use at work is:

- strong and stable enough for the particular use and marked to indicate safe working loads
- positioned and installed to minimise any risks
- used safely, ie the work is planned, organised and performed by competent people
- subject to on-going thorough examination and, where appropriate, inspection by competent people.

THE CONTROL OF VIBRATION AT WORK REGULATIONS 2005

Vibration white finger or hand–arm vibration syndrome (HAVS) (see page 23) is caused by using vibrating tools such as hammer drills, vibrating pokers or hand held breakers over a long period of time. The most efficient and effective way of controlling exposure to hand-arm vibration is to look for new or alternative work methods that remove or reduce exposure to vibration.

Follow these steps to reduce the effects of HAVS:

■ Always use the right tool for each job.

■ Check tools before using them to make sure they have been properly maintained and repaired to avoid increased vibration caused by faults or general wear.

■ Make sure cutting tools are kept sharp so that they remain efficient.

■ Reduce the amount of time you use a tool in one go, by doing other jobs in between.

■ Avoid gripping or forcing a tool or workpiece more than you have to.

■ Encourage good blood circulation by:
  □ keeping warm and dry (when necessary, wear gloves, a hat, waterproofs and use heating pads if available)
  □ giving up or cutting down on smoking because smoking reduces blood flow
  □ massaging and exercising your fingers during work breaks.

Damage from HAVS can include the inability to do fine work and cold can trigger painful finger blanching attacks (when the ends of your fingers go white).

An operative taking a rest from using a power tool

Don't use power tools for longer than you need to

CONSTRUCTION SITE HAZARDS

DANGERS ON CONSTRUCTION SITES

Study the drawing of a building site. There is some demolition taking place, as well as construction. How many hazards can you find? Discuss your answers.

Dangers	Discussion points
Head protection	The operatives are not wearing safety helmets, which would prevent them from hitting their head or from falling objects.
Poor housekeeping	The site is very untidy. This can result in slips, trips and falls and can pollute the environment. An untidy site gives a poor company image. Offcuts and debris should be regularly removed and disposed of according to site policy and recycled if possible.
Fire	There is a fire near a building; this is hazardous. Fires can easily become uncontrollable and spread. There is a risk to the structure and, more importantly, a risk of operatives being burned. Fires can also pollute the environment.

Dangers	Discussion points
Trip hazards	Notice the tools and debris on the floor. The scaffold has been poorly constructed. There is a trip hazard where the scaffold boards overlap.
Chemical spills	There is a drum leaking onto the ground. This should be stored properly – upright and in a lockable metal shed or cupboard. The leak poses a risk of pollution and of chemical burns to operatives.
Falls from height	The scaffold has handrails missing. The trestle working platform has not been fitted with guard rails. None of the operatives are wearing hard hats for protection either.
Noise	An operative is using noisy machinery with other people nearby. The operative should be wearing ear PPE, as should those working nearby. Better still, they should be working elsewhere if at all possible, isolating themselves from the noise.
Electrical	Some of the wiring is 240V as there is no transformer, it's in poor repair and it's also dragging through liquid. This not only increases the risk of electrocution but is also a trip hazard.
Asbestos or other hazardous substances	Some old buildings contain **asbestos** roofing which can become a hazard when being demolished or removed. Other potential hazards include lead paint or mould spores. If a potentially hazardous material is discovered a supervisor must be notified immediately and work must stop until the hazard is dealt with appropriately.

Asbestos

A naturally occurring mineral that was commonly used for a variety of purposes including: **insulation**, fire protection, roofing and guttering. It is extremely hazardous and can cause a serious lung disease known as asbestosis

Insulation

A material that reduces or prevents the transmission of heat

FUNCTIONAL SKILLS

Using the data you collected in the Functional Skills task on page 3, produce a pie chart to show the proportion of occupational cancer that is caused by asbestosis.

Work on this activity can support FM L2.3.1 and C2.4.

Cables can be a trip hazard on site

Boiler suit

Hand cleaner

PERSONAL HYGIENE

Working in the construction industry can be very physical, and it's likely to be quite dirty at times. Therefore you should take good care with your personal hygiene. This involves washing well after work. If contaminants are present, then wearing a protective suit, such as a boiler suit, that you can take off before you go home will prevent contaminants being taken home with you.

You should also wash your hands after going to the toilet and before eating. This makes it safer to eat and more pleasant for others around you. The following steps show a safe and hygienic way to wash your hands.

STEP 1 Apply soap to hands from the dispenser.

STEP 2 Rub the soap into a lather and cover your hands with it, including between your fingers.

STEP 3 Rinse hands under a running tap removing all of the soap from your hands.

STEP 4 Dry your hands using disposable towels. Put the towels in the bin once your hands are dry.

WORKING WITH ELECTRICITY

Electricity is a very useful energy resource but it can be very dangerous. Electricity must be handled with care! Only trained, competent people can work with electrical equipment.

THE DANGERS OF USING ELECTRICAL EQUIPMENT

The main dangers of electricity are:

- shock and burns (a 230V shock can kill)

- electrical faults which could cause a fire

- an explosion where an electrical spark has ignited a flammable gas.

VOLTAGES

Generally speaking, the lower the voltage the safer it is. However, a low voltage is not necessarily suitable for some machines, so higher voltages can be found. On site, 110V (volts) is recommended and this is the voltage rating most commonly used in the building industry. This is converted from 230V by use of a transformer.

110V 1 phase – yellow

230V (commonly called 240V) domestic voltage is used on site as battery chargers usually require this voltage. Although 230V is often used in workshops, 110V is recommended.

400V (otherwise known as 3 phase) is used for large machinery, such as joinery shop equipment.

230V 1 phase – blue

Voltages are nominal, ie they can vary slightly.

BATTERY POWER

Battery power is much safer than mains power. Many power tools are now available in battery-powered versions. They are available in a wide variety of voltages from 3.6V for a small screwdriver all the way up to 36V for large masonry drills.

400V 3 phase – red

The following images are all examples of battery-powered tools you may come across in your workplace or place of training.

Battery drill   Battery-powered planer   Battery-powered jigsaw

WIRING

The wires inside a cable are made from copper, which conducts electricity. The copper is surrounded by a plastic coating that is colour coded. The three wires in a cable are the live (brown), which works with the neutral (blue) to conduct electricity, making the appliance work. The earth (green and yellow stripes) prevents electrocution if the appliance is faulty or damaged.

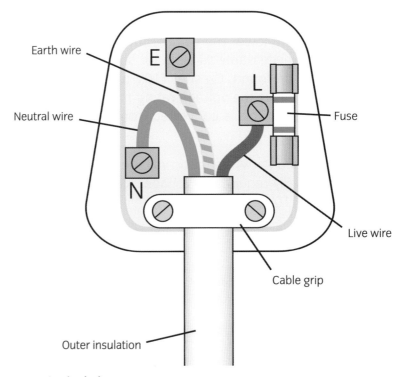

A wired plug

POWER TOOLS AND CHECKS

Power tools should always be checked before use. Always inform your supervisor if you find a fault. The tool will need to be repaired, and the tool needs to be kept out of use until then. The tool might be taken away, put in the site office and clearly labelled 'Do not use'.

Power tool checks include:

- *Look for the portable appliance testing (PAT) label*: PAT is a regular test carried out by a competent person (eg a qualified electrician) to ensure the tool is in a safe electrical condition. A sticker is placed on the tool after it has been tested. Tools that do not pass the PAT are taken out of use.

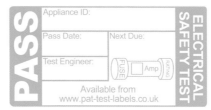

PAT testing labels

Cable protection

- *Cable*: Is it damaged? Is there a repair? Insulation tape may be hiding a damaged cable. Damaged cables must be replaced.

- *Casing*: Is the casing cracked? Plastic casings ensure the tool is double-insulated. This means the live parts inside are safely shielded from the user. A cracked casing will reduce the protection to the user and will require repair.

- *Guards and tooling*: Are guards in place? Is the tooling sharp?

- *Electricity supply leads*: Are they damaged? Are they creating a trip hazard? You need to place them in such a way that they do not make a trip hazard. Are they protected from damage? If they are lying on the floor with heavy traffic crossing them, they must be covered.

- *Use appropriate equipment for the size of the job*: For example, too many splitters can result in a web of cables.

- *Storage*: After use, power tools and equipment should be stored correctly. Tools must be returned to the boxes, including all the guards and parts. Cables need to be wound onto reels or neatly coiled as they can become tangled very easily.

Cable reel

INDUSTRY TIP

Remember, always fully unroll an extension lead before use because it could overheat and cause a fire.

FIRE

Fire needs three things to start; if just one of them is missing there will be no fire. If all are present then a fire is unavoidable:

1 *Oxygen*: A naturally occurring gas in the air that combines with flammable substances under certain circumstances.

2 *Heat*: A source of fire, such as a hot spark from a grinder or naked flame.

3 *Fuel*: Things that will burn such as acetone, timber, cardboard or paper.

The fire triangle

If you have heat, fuel and oxygen you will have a fire. Remove any of these and the fire will go out.

PREVENTING THE SPREAD OF FIRE

Being tidy will help prevent fires starting and spreading. For instance:

- Wood offcuts should not be left in big piles or standing up against a wall. Instead, useable offcuts should be stored in racks.

- Put waste into the allocated disposal bins or skips.

- Always replace the cap on unused fuel containers when you put them away. Otherwise they are a potential source of danger.

- Flammable liquids (not limited to fuel-flammable liquids) such as oil-based paint, thinners and oil must be stored in a locked metal cupboard or shed.

- Smoking around flammable substances should be avoided.

- Dust can be explosive, so when doing work that produces wood dust it is important to use some form of extraction and have good ventilation.

FIRE EXTINGUISHERS AND THEIR USES

You need to know where the fire extinguishers and blankets are located and which fire extinguishers can be used on different fires. The table below shows the different classes of fire and which extinguisher to use in each case.

Class of fire	Materials	Type of extinguisher
A	Wood, paper, hair, textiles	Water, foam, dry powder, wet chemical
B	Flammable liquids	Foam, dry powder, CO_2
C	Flammable gases	Dry powder, CO_2
D	Flammable metals	Specially formulated dry powder
E	Electrical fires	CO_2, dry powder
F	Cooking oils	Wet chemical, fire blanket

Fire blanket

INDUSTRY TIP

Remember, although all fire extinguishers are red, they each have a different coloured label to identify their contents.

CO_2 extinguisher

Dry powder extinguisher

Water extinguisher

Foam extinguisher

It is important to use the correct extinguisher for the type of fire as using the wrong one could make the danger much worse, eg using water on an electrical fire could lead to the user being electrocuted!

EMERGENCY PROCEDURES

In an emergency, people tend to panic. If an emergency were to occur, such as fire, discovering a bomb or some other security problem, would you know what to do? It is vital to be prepared in case of an emergency.

It is your responsibility to know the emergency procedures on your work site:

- If you discover a fire or other emergency you will need to raise the alarm:
  - You will need to tell a nominated person. Who is this?
  - If you are first on the scene you will have to ring the emergency services on 999.

- Be aware of the alarm signal. Is it a bell, a voice or a siren?

- Where is the assembly point? You will have to proceed to this point in an orderly way. Leave all your belongings behind: they may slow you or others down.

- At the assembly point, there will be someone who will ensure everyone is out safely and will do so by taking a count. Do you know who this person is? If during a fire you are not accounted for, a firefighter may risk their life to go into the building to look for you.

- How do you know it's safe to re-enter the building? You will be told by the appointed person. It's very important that you do not re-enter the building until you are told to do so.

Emergency procedure sign

SIGNS AND SAFETY NOTICES

The law sets out the types of safety signs needed on a construction site. Some signs warn us about danger and others tell us what to do to stay safe.

The following table describes five basic types of sign.

Type of sign	Description
Prohibition 	These signs are red and white. They are round. They signify something that must *not* be done.
Mandatory	These signs are blue. They are round. They signify something that *must* be done.

Type of sign	Description
Caution 	These signs are yellow and black. They are triangular. These give warning of hazards.
Safe condition 	These signs are green. They are usually square or rectangular. They tell you the safe way to go, or what to do in an emergency.
Supplementary 	These white signs are square or rectangular and give additional important information. They usually accompany the signs above.

Case Study: Miranda

A site has a small hut where tools are stored securely, and inside the hut there is a short bench that has some sharpening equipment including a grinding wheel.

Miranda wished to grind her plane blade, but before using it found that the grinding wheel was defective as the side of the wheel had been used, causing a deep groove.

She found another old grinding wheel beneath the bench which looked fine. She fitted it to the grinder and used it.

Afterwards, she wondered if she should have asked someone else to change the wheel for her.

- What health and safety issues are there with this scenario?
- What training could Miranda undertake?

Work through the following questions to check your learning.

1 Which one of the following **must** be filled out prior to carrying out a site task?

   a Invoice.

   b Bill of quantities.

   c Risk assessment.

   d Schedule.

2 Which one of the following signs shows you something you **must** do?

   a Green circle.

   b Yellow triangle.

   c White square.

   d Blue circle.

3 Two parts of the fire triangle are heat and fuel. What is the third?

   a Nitrogen.

   b Oxygen.

   c Carbon dioxide.

   d Hydrogen sulphite.

4 Which of the following types of fire extinguisher would **best** put out an electrical fire?

   a CO_2.

   b Powder.

   c Water.

   d Foam.

5 Which piece of health and safety legislation is designed to protect an operative from ill health and injury when using solvents and adhesives?

   a Manual Handling Operations Regulations 1992.

   b Control of Substances Hazardous to Health (COSHH) Regulations 2002.

   c Health and Safety (First Aid) Regulations 1981.

   d Lifting Operations and Lifting Equipment Regulations (LOLER) 1998.

6 What is the correct angle at which to lean a ladder against a wall?

   a 70°.

   b 80°.

   c 75°.

   d 85°.

7 Which are the **most** important pieces of PPE to use when using a disc cutter?

   a Overalls, gloves and boots.

   b Boots, head protection and overalls.

   c Glasses, hearing protection and dust mask.

   d Gloves, head protection and boots.

8 Which one of the following is **not** a lifting aid?

   a Wheelbarrow.

   b Kerb lifter.

   c Gin lift.

   d Respirator.

9 Which one of the following is a 3 phase voltage?

   a 400V.

   b 230V.

   c 240V.

   d 110V.

10 Above what noise level **must** you wear ear protection?

   a 75 dB(A).

   b 80 dB(A).

   c 85 dB(A).

   d 90 dB(A).

Chapter 2
Principles of organising, planning and pricing construction work

As you progress and gain responsibility within the industry you will be involved with interpreting a range of information. The understanding and communication of this information are crucial to the success of a building project. Mistakes made in interpreting information always prove costly in terms of labour, building materials and time. Whether reading a drawing, cross-checking information on a schedule, planning building work or working up a price, each process requires a methodical approach and should be able to be easily interpreted by those using the documentation. Much of this information is provided to you by the **contract documents**.

By reading this chapter you will know about:

1 Different types of drawn information in construction.

2 Energy efficiency and sustainable materials for construction.

3 Estimating quantities and price work for construction.

4 Planning work activities for construction.

5 Communicating effectively in the workplace.

Contract documents

These comprise of the working drawings, schedules, specifications, bill of quantities and contracts

DRAWN INFORMATION IN CONSTRUCTION

This section will discuss the different types of drawn information used and found in construction. In *The City & Guilds Textbook: Level 2 Diploma in Bricklaying* we have already looked at some of this information. Now we will recap this and look in a little more detail at how drawings are produced.

Drawings are required at every stage of building work. After the building has been designed and agreed with the client, drawings are required in order to apply for planning consent and Building Regulations approval. These drawings will show the size, position and general arrangement of the proposed construction and allow the Local Authority planning committee to decide whether approval should be given and whether the proposed building meets the current Building Regulations.

Flowchart of contracts documents

Project drawing:
1 Planning drawings: to a small scale.
2 Construction drawings: to a larger scale.

Ref:	Door size	S.O. width	S.O. height	Lintel type	FD30	Self closing	Floor level
D1	838 ×1981	900	2040	BOX	Yes	Yes	GROUND FLOOR
D2	838 ×1981	900	2040	BOX	Yes	Yes	GROUND FLOOR
D3	762 ×1981	824	2040	BOX	No	No	GROUND FLOOR
D4	838 ×1981	900	2040	N/A	Yes	No	GROUND FLOOR
D5	838 ×1981	900	2040	BOX	Yes	Yes	GROUND FLOOR
D6	762 ×1981	824	2040	BOX	Yes	Yes	FIRST FLOOR
D7	762 ×1981	824	2040	BOX	Yes	Yes	FIRST FLOOR
D8	762 ×1981	824	2040	N/A	Yes	No	FIRST FLOOR
D9	762 ×1981	824	2040	BOX	Yes	Yes	FIRST FLOOR
D10	762 ×1981	824	2040	N/A	No	No	FIRST FLOOR
D11	686 ×1981	748	2040	N/A	Yes	No	SECOND FLOOR
D12	762 ×1981	824	2040	BOX	Yes	Yes	SECOND FLOOR
D13	762 ×1981	824	2040	100 HD BOX	Yes	Yes	SECOND FLOOR
D14	686 ×1981	748	2040	N/A	No	No	SECOND FLOOR

Master Internal Door Schedule

Schedules:
Information in a table form, taken from drawings and specifications, eg a door or decoration schedule.

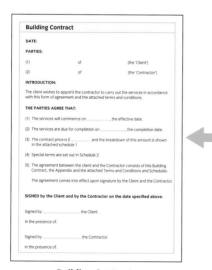

Building Contract:
Gives details of start/end dates, conditions of work, methods of payment, etc.

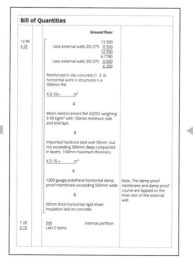

Bill of Quantities:
Written document prepared in accordance with the new Rules of Measurement Standard Information, used to provide pricing for a quote.

Specification:
Written document providing specific details of the materials and workmanship required.

TYPES OF DRAWING REQUIRED BY THE LOCAL AUTHORITY

LOCATION PLAN

This shows the proposed development in relation to its surrounding properties. It must be based on an up-to-date map and an identified standard metric **scale** (typically 1:1250 or 1:2500). The site of the proposed development needs to be outlined in red and any other land owned by the applicant that is close to or adjoining the site needs to be outlined in blue.

SITE PLAN

This shows the proposed development in relation to the property boundary (sometimes called a 'block plan'). Site plans are typically submitted at a scale of either 1:200 or 1:500 and should include the following:

- the size and position of the existing building (and any extensions proposed) in relation to the property boundary

- the position and use of any other buildings within the property boundary

- the position and width of any adjacent streets.

GENERAL ARRANGEMENT DRAWINGS

These include elevations and floor plans and sections of the proposed building. See page 49 for an example.

ELEVATIONS

These show the external appearance of each face of the building showing features such as slope of the land, doors, windows and the roof arrangement at a scale of 1:50 or 1:100 depending on the size of the project and the drawing sheet printed on.

FLOOR PLANS

These are used to identify the layout of the internal walls, doors, stairs and arrangement of bathrooms and kitchen. Again these will be drawn at a scale of 1:50 or 1:100.

SECTIONS

These are used to show vertical views through the building showing room heights and floor and roof constructions. These are generally drawn at a scale of 1:50.

Scale

A reduction in size by a given ratio, eg 1:5 means that the measurement on the scale is five times smaller than the real thing. So if the full-size object is 500mm long, the scaled-size object shown on the drawing would be 100mm long

ACTIVITY

Go to www.planningportal.gov.uk. Research what mandatory documents are required in a planning application.

Scale rule

ADDITIONAL DRAWINGS REQUIRED TO CONSTRUCT THE BUILDING
CONSTRUCTION DRAWINGS

These provide the detail required to construct a building. They provide the information required by individual trades to a larger scale and in more detail. They may include any or all of the following:

Assembly drawings

These types of drawings are used to show how various components fit together at various junctions. These are generally drawn at scales of 1:20, 1:10 and 1:5. Scale rules are used to draw scaled-down buildings on paper.

Component/range drawings

These supply information required by manufacturers producing various components for the finished building, eg purpose-made doors or kitchen units. A range drawing could also show a manufacturer's standard range of products available off the shelf, eg doors, windows and kitchen units. These are often shown at scales of 1:10 and 1:20. For instance, a range of doors or windows might be indicated on the floor plans with a simple code such as D1, D2, W1, W2, etc. These could also be shown on a door/window schedule including ironmongery and any glass requirements.

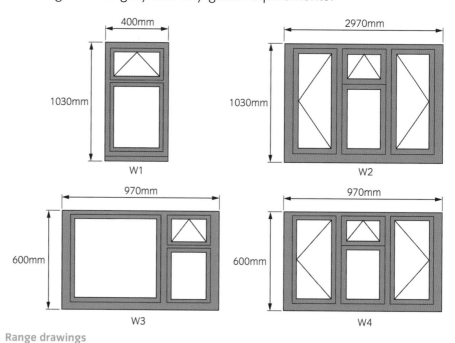

Range drawings

Detail drawings

Detail drawings show very accurate large-scale details of the construction of a particular item, eg a window or stair construction. Examples are joinery detailing and complex brickwork features. Typical scales are 1:2 and 1:5.

INDUSTRY TIP

Sketches can often be used to communicate information where a description is harder to understand. A picture paints a thousand words.

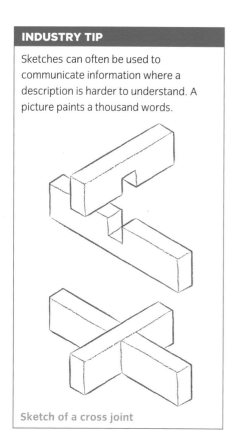

Sketch of a cross joint

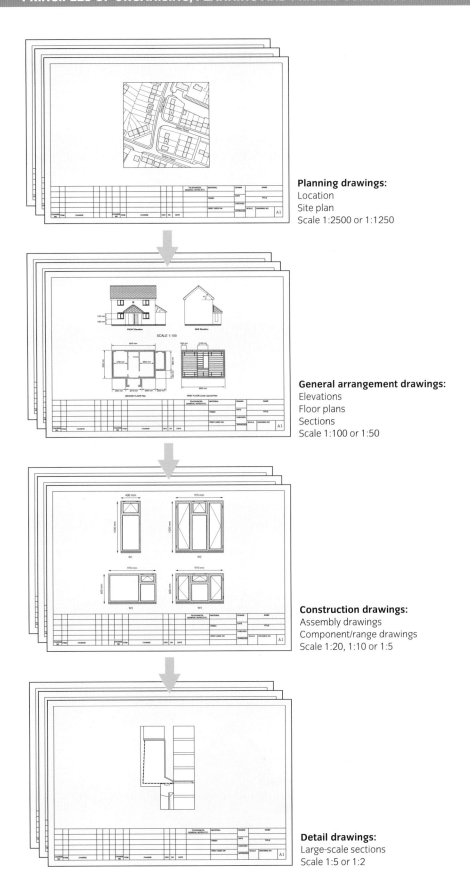

Planning drawings:
Location
Site plan
Scale 1:2500 or 1:1250

General arrangement drawings:
Elevations
Floor plans
Sections
Scale 1:100 or 1:50

Construction drawings:
Assembly drawings
Component/range drawings
Scale 1:20, 1:10 or 1:5

Detail drawings:
Large-scale sections
Scale 1:5 or 1:2

Flowchart showing progression of planning drawings

Parallel-motion drawing board

Architectural technician

A technician who is employed by an architectural practice to produce construction drawings

Walkthrough

An animated sequence as seen at the eye level of a person walking through a proposed structure

PRODUCTION OF DRAWINGS

Traditionally all drawings would have been produced by a draughtsperson by hand on large parallel-motion drawing boards. Some drawings are still produced this way but most are now produced using CAD (computer-aided design/drawing).

COMPUTER-AIDED DRAUGHTING

This is a method of producing drawings and designs using a design and drawing software package. These packages have been around for about 30 years now and have significantly improved during this period. They are now used by millions of people worldwide. Anything can be drawn using these software packages in either two or three dimensions.

All newly qualified architects or **architectural technicians** will produce drawings using CAD and this is the preferred method of production as they have a number of advantages over hand-produced drawings, which include:

- High quality drawings can be produced.
- Drawings can easily be magnified, manipulated and amended.
- Standard details can be saved and reproduced as any new contract requires.
- Drawing layers can be produced enabling particular details to be extracted.
- Objects can be viewed from any angle.
- Drawings can be archived without taking up valuable space.
- Drawings can be attached as files and sent via email.
- Three-dimensional virtual models and **walkthroughs** can be created.
- CAD can be used with other compatible software products to prepare schedules, specifications and CAM (computer-aided manufacturing) to produce products by machine.

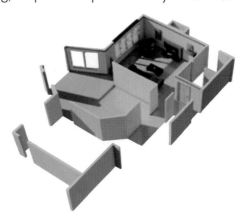

Walkthrough illustration

While the advantages of CAD are significant there are a few disadvantages:

1 The initial outlay for the equipment required, eg:

- a central processing unit (computer)
- a visual display unit (VDU)
- a keyboard
- CAD software
- a plotter or printer.

2 The training of staff to use the software.

3 The cost of updating the software.

Typical CAD workstation with large screens

There are many CAD packages that can be purchased and downloaded free. Autodesk produces a range of programs which are used universally. AutoCAD is very popular. SketchUp, formerly owned by Google, is a free-to-download program. 'Our House' was drawn using this.

SketchUp software logo

OUR HOUSE

Log on to SmartScreen and open up 'Our House'. Explore the possibilities of this package – don't worry, you can't break it!

All drawings are produced on standard-sized sheets of paper.

Drawing paper		Other sized paper	
Name	Size (mm)	Name	Size (mm)
A0	1189 × 841	A5	210 × 148
A1	841 × 594	A6	148 × 105
A2	594 × 420	A7	105 × 74
A3	420 × 297	A8	74 × 52
A4	297 × 210		

FUNCTIONAL SKILLS

Use the internet to research the standard method for folding drawings allowing the information box to be shown when folded.

Work on this activity can support FICT L2 (4).

Examples of these paper sizes are shown on the next page.

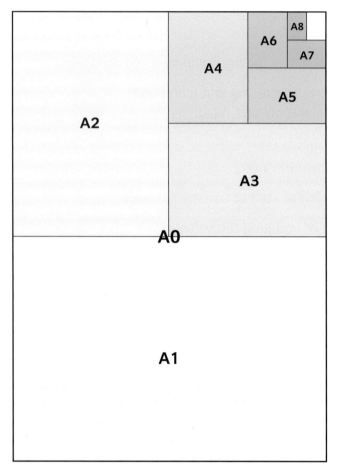

Standard drawing sheet sizes

DRAWING PROJECTION METHODS

ORTHOGRAPHIC PROJECTION

Most drawings are produced in two dimensions (length and width). These are called orthographic drawings. The only type of orthographic used in the UK is first angle orthographic. This shows how the views are projected on to a flat surface surrounding the object. The surface is then folded flat showing how the views are shown in first angle projection on a drawing.

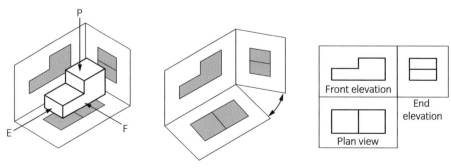

First angle orthographic

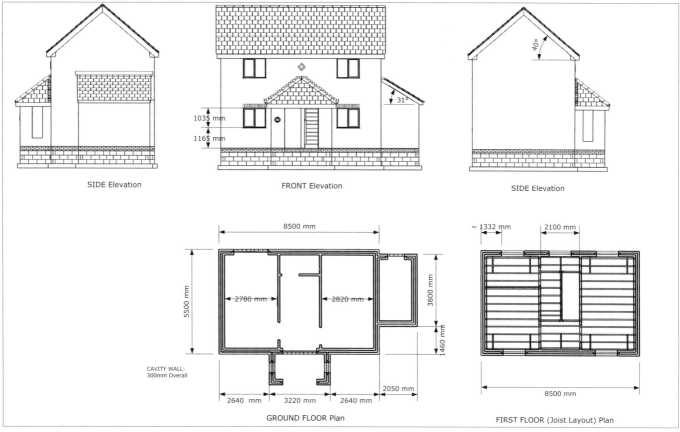

Orthographic projections of 'Our House'

In construction most of the drawings we use are produced in orthographic projection. These show a front elevation, end elevation and plan views of our proposed building. Additional views can also be added to show detail as required. In order to produce these, the following information is required:

- overall sizes of the proposed building
- room dimensions
- position and sizes of doors and windows
- position of internal walls
- brickwork, carpentry and joinery specifications
- Building Regulations requirements.

THREE-DIMENSIONAL DRAWING

This type of drawing is often termed a pictorial drawing. These drawings can add considerable clarity to a two-dimensional drawing, allowing the viewer a more life-like representation of the building. They give the client, operatives, suppliers and non-technical personnel not involved with the construction process a better understanding of how the finished product will look or give more detail of what is required of a particular part of the construction. There are many types of three-dimensional drawing, but the most common types used in construction are isometric, oblique and perspective.

ACTIVITY

Practise drawing scaled orthographic views of the project you are currently working on in your training environment.

INDUSTRY TIP

Use a 4H grade pencil for construction lines and 2H for bold lines.

INDUSTRY TIP

Before starting a drawing, calculate the widths and heights of each elevation to ensure you start the drawing in the right place, otherwise you may run out of space on your drawing sheet.

INDUSTRY TIP

Always produce all drawings in construction lines (faint lines), then draw over these in bold lines to highlight the important parts. This will make the drawing easier to understand.

Isometric projection

This is the most commonly used and produces a life-like representation of the subject. To produce these, all lines are drawn vertical or to the left or right axis at 30° (using a standard 30°/60° set square). They are often used to show an overall view of the object and form the basis of most sketches.

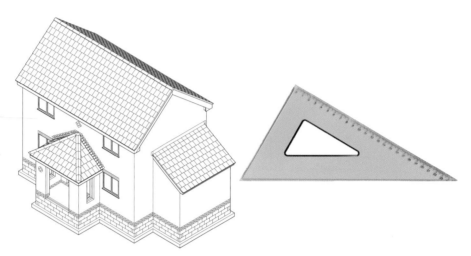

An isometric drawing of 'Our House'   30°/60° set square

ACTIVITY

Find out what type of drawing instrument can be set to draw any angle.

Oblique projection

This is a less commonly used method of showing an object in three dimensions. The front elevation is drawn to its actual size and shape, with the third dimension (or depth) shown drawn back to either the left or right as required. The lines are drawn horizontally, vertically or at 45° axis to the left or right.

45° set square   An oblique projection of 'Our House'

Perspective drawings

These give the most realistic view of a building; lines are drawn vertically or drawn back to vanishing points (VP). The vanishing points can be positioned at any height but most commonly at the viewer's eye level.

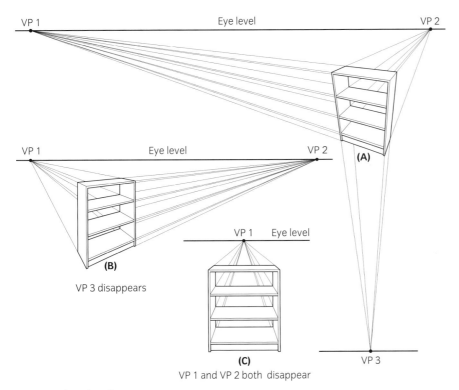

A perspective drawing

The same information is required to produce three-dimensional drawings as for two-dimensional drawings:

- dimensions
- scale required
- axis (30°, 45° or vanishing)
- position and required view of object.

An example of a perspective drawing

BASIC DRAWING SYMBOLS (HATCHINGS)

Standard symbols, also known as hatching symbols, are used on drawings as a means of passing on information simply. If all the parts of a building were labelled in writing, the drawing would soon become very crowded. Additionally, it is important to use standard symbols so that everyone can read them and they mean the same to everyone. The following images are just some of the standard symbols used.

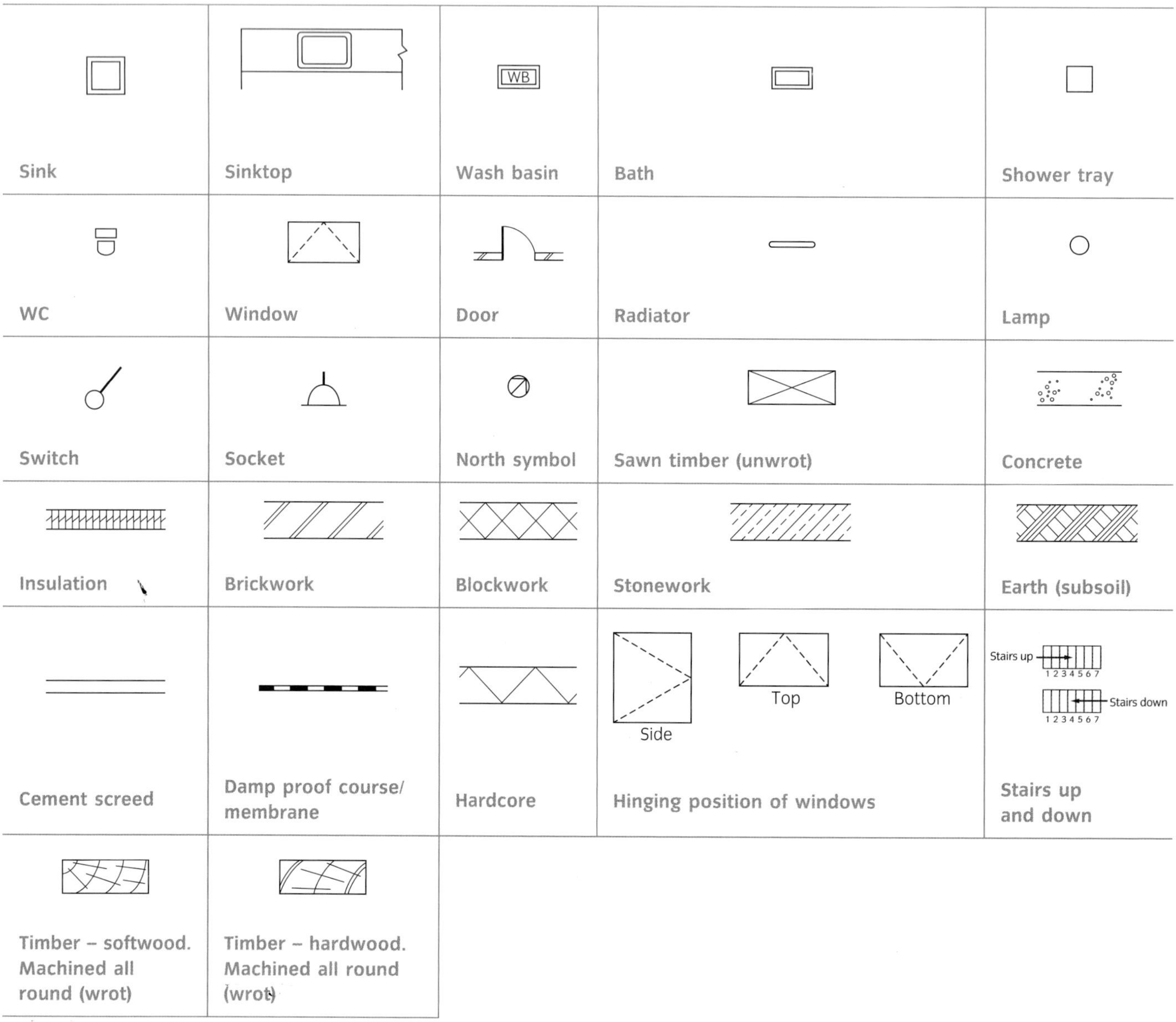

Sink
Sinktop
WB
Wash basin
Bath
Shower tray
WC
Window
Door
Radiator
Lamp
Switch
Socket
North symbol
Sawn timber (unwrot)
Concrete
Insulation
Brickwork
Blockwork
Stonework
Earth (subsoil)
Cement screed
Damp proof course/ membrane
Hardcore
Side
Top
Bottom
Hinging position of windows
Stairs up
1 2 3 4 5 6 7
Stairs down
1 2 3 4 5 6 7
Stairs up and down
Timber – softwood. Machined all round (wrot)
Timber – hardwood. Machined all round (wrot)

ESTIMATING QUANTITIES AND PRICING WORK FOR CONSTRUCTION

Before you can estimate quantities of construction materials, you need to know how to prepare a materials list using a schedule.

SCHEDULES

Drawings alone will not contain all the information required to carry out all operations efficiently. Drawings will show the positions of doors, windows, lintels and reinforcing, for example, but will not give specific detail. This information can be extracted and shown within a schedule. A schedule can be produced by a **quantity surveyor** as a part of their role. It is often read in conjunction with the range drawing. It provides an easy to handle and readable table. You can see an example on page 46.

A schedule records repeated design information that applies to a range of components or fittings, such as:

- windows
- doors
- reinforcement.

A schedule is mainly used on larger developments and contracts where there are multiples of several designs of houses, with each type having different components and fittings. It avoids a house being given the wrong component or fitting. On a typical plan, the doors and windows are labelled D1, D2, W1, W2, etc. These components would be included in the schedule, which would provide additional information about them.

To prepare a materials list to order from the schedules, you will need to use the information contained on them. This will include:

- quantity
- colour
- dimensions
- location
- installation details
- manufacturer.

Quantity surveyor

A quantity surveyor will produce the bill of quantities and later works with a client to manage costs and contracts

ACTIVITY

Produce a door schedule for at least five differently sized/types of doors within your college/training centre.

CALCULATING COSTS AND PRICING WORK

The calculation of costs for building work is carried out by an estimator. Every estimator builds their tender figures based on the information contained in the contract documents (drawings, bill of quantities and the specifications). We have already looked at drawings.

BILL OF QUANTITIES

Dimension paper

Paper with vertically ruled columns onto which building work is described, measured and costed

Taken off

Materials measured from the contract drawings

The quantity surveyor will have produced the bill of quantities. This describes on **dimension paper** each item, and the quantity of that item as '**taken off**' the drawing. The descriptions of the items that are listed for many years followed the SMM7 (Standard Method of Measurement 7) rule book. This is the standard method of measurement used in the construction of most architect-designed buildings. Through standardisation of their common features like the description of the brickwork and the quantities of that brickwork it leads to certainty of interpretation by all contractors when quoting for this work. This makes the process fair, as every contractor is pricing for the same items. The SMM7 was replaced by the NRM2 (New Rules of Measurement 2) from July 2013.

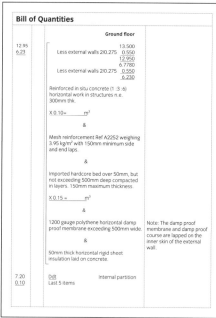

The NRM2 document is published by the Royal Institute of Charted Surveyors (RICS)

Extract from a bill of quantities

In addition to the standard items of work listed the bill of quantities will also contain information on:

Setting-up costs

These can include costs of hoarding, temporary services, temporary site accommodation, etc

- *Preliminaries:* these tend to be time-related costs that have to be built in such as management and supervision costs, **setting-up costs,** etc, rather than the costs of the actual building work.

- *Preambles:* these are statements that express the agendas and description of their accompanying documents. Therefore preambles in the bill of quantities refers to a tender letter that gives the description of the materials requirements for construction and it gives the quantity, type and characteristics of the materials required.

- *Provisional sum:* this refers to bills of work that is undefined. Such work does not have an accurate estimate and hence provisional sum best describes it. When giving a provisional sum, one has to include the type of work, how the work is to be done and the limitations likely to be encountered (such as the work hours being limited to night only).

- *Prime cost (PC):* PC sums cover work undertaken by nominated subcontractors, nominated suppliers and **statutory undertakings** and are based on quotations for items of work which can be populated into the quote.

Statutory undertakings

The various services that are brought to the site – water, electricity, etc

SPECIFICATION

This provides essential information to the contractor about the materials, finish and workmanship required for the building construction. The specification gives information that if noted on the drawing would clutter it and make it impossible to read as there would be more text than drawing. An example would be the type of brick or timber required.

On small projects the specification is written in the notes column above the information block on the drawing; on larger projects the specification is produced in a separate document.

Very often the description in the specification will be linked to the relevant British Standard.

Example of specification

THE TENDER PROCESS

The purpose of the tendering process is to provide the client with a number of estimates for the proposed work. The client will go out to known contractors, or for competitive tendering work it may be advertised and contractors come forward to quote for the work. Armed with the above contract documents the estimator will begin to build a tender cost. Each line of the bill will be costed. The figure used in the rate column will be an 'all-in rate' including:

- *Labour rates:* this will be calculated not at the operatives' hourly rate, but including costs such as annual holiday costs, employers' National Insurance contributions, pension contribution, sickness pay, bonuses, travelling time, etc.

- *Overheads:* manufacturing overheads include such things as electricity, gas, phone, insurance, plant and equipment, depreciation on equipment and buildings, factory supplies and factory personnel (other than direct labour).

- *Contingencies:* this is a small percentage added to cover unforeseen costs that may arise during the contract.

- *Profit:* the amount of money gained over and above direct costs. The rate applied will depend on a number of factors including current workload, the competition and the complexity of the project.

- *VAT* on materials.

Calculating a unit rate

Calculate a **unit rate** for brickwork using the following information.

Unit rate

The unit rate = the labour rate + the material rate

- Facing brickwork (in half-brick walling) laid in 1:5 cement mortar

- Gang of two bricklayers, one labourer

- Hourly rate: bricklayer rate £15.00, labourer £8.00

- Production rate: 50 bricks/hr (per bricklayer)

- 60 bricks/m²

- Cost of bricks/1000 = £400 (includes 5% waste)

- Pre-mixed mortar on site £30/330 litre tub (60 litre of mortar m²/brickwork)

Example

Labour rate

Labour (2 × 15) + 8 = £38 total hourly rate

$$\text{Production rate} = \frac{\text{bricks laid/hr}}{\text{bricks/m}^2} = \frac{100}{60} = 1.7\text{m}^2/\text{hr.}$$

$$\text{Labour rate} = \frac{\text{total hourly rate}}{\text{Production rate (m}^2/\text{hr)}} = \frac{38}{1.7} = £22.35/\text{m}^2$$

Materials rate

$$\text{Cost of bricks/m}^2: \frac{400}{1000} = 0.4 \text{ (each brick costs 40p)}$$

INDUSTRY TIP

Always double check all-in rates; if these are wrong the error in the costings will be considerable.

0.4 × 60 (bricks/m²) = Cost of bricks/m² = £24

Cost of mortar/m²: 1 litre/brick × 60 = 60 litres/m²

$$\frac{30}{330} \times 60 = £5.45$$

Cost of materials = 24 + 5.45 = £29.45

Unit rate

Total unit rate = labour rate + materials rate =
22.35 + 29.45 = **£51.80m²**

The total unit rate would now be used in the Rate column and would be multiplied by the Quantity column to produce the cost for this item. See the example below:

Extract from the bill of quantities

Number	Item description	Unit	Quantity	Rate	Amount	
					£	p
E.47	Half-brick walling	m²	75	51.80	3885	00

Each line of the bill should be completed and the last columns totalled to provide the final estimated cost. Once this has been completed, it should be reviewed by the management team prior to submitting to the client for consideration. It is common for a client to ask for three estimates; an estimator's job can involve a lot of work with very few contracts obtained. One in 10 jobs won would be a good success rate.

OTHER WAYS OF ESTIMATING COSTS

Traditionally many companies use a building price book, which is a complete guide for estimating, checking and forecasting building work. Again the figures in these books are established all-in rates which are updated on a yearly basis. To calculate the cost of proposed work you simply find the description of work requiring costing and use the figures in this row.

There are now also many software packages that can be purchased, which allow fixed unit rates to be used. Again enter the quantity in the correct row and the software will total up the work as you enter the work to be priced.

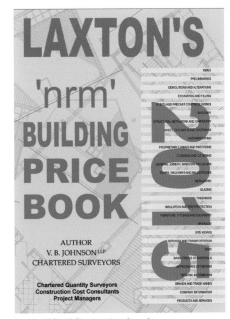

Typical building price book

Quotes

A & E BUILDERS LTD.

An estimate

WHAT IS THE DIFFERENCE BETWEEN A *QUOTATION* AND AN *ESTIMATE*?

WHAT IS A QUOTATION?

A quotation (quote) is a document submitted in formal response to a request for the cost of specific work based on the contract documents supplied. In other words, it is a promise to do work at an agreed price. It should set out exactly what work will be done for that agreed price (based on the bill of quantities). Acceptance of this quotation by the client or their representative creates a binding agreement between the two parties. Any additional work requested by the client will not form part of this quote and should be costed separately.

WHAT IS AN ESTIMATE?

An estimate is exactly that: a best guess (sometimes referred to as a guesstimate) of how much specific work will cost or how long it will take to complete. Unlike a quote, it is not an offer to carry out this work for a fixed cost. This means that the job could cost either more or less than the estimated cost. Any charges expected to be above the estimate are best flagged up as early as possible to avoid possible later disputes.

SOURCING SUPPLIERS

This is often done during the pre-contract planning phase of the contract. Local suppliers are contacted and asked for quotes and their ability to supply materials to a given schedule. It pays to set up an early working relationship with your suppliers to ensure a good service.

PREFERRED SUPPLIERS

Suppliers can become 'preferred' in a number of ways; for example, your organisation may have used them before, they may have approached you or your technical colleagues with details of their proposition, they may have made a previously unsuccessful tender, or they may have been recommended by a similar organisation. The term *preferred supplier* does not in itself guarantee a level of business, but instead should be thought of as a guide to your thinking when considering a sourcing strategy. If there are several suppliers from which you can source the same materials, the preferred supplier may be one that consistently gives the best price or the one that is most reliable and always delivers on time.

ESTIMATING QUANTITIES OF CONSTRUCTION MATERIALS

At Level 2 you will have carried out a number of calculations for quantifying materials (see Chapter 2, page 57, *The City & Guilds Textbook: Level 2 Diploma in Bricklaying*). These workings are generally based on either linear, squared or volume calculations.

UNITS OF MEASUREMENT

The construction industry uses metric units as standard; however, you will occasionally come across imperial units as these are still used in common parlance in our industry. Material sizes in particular are often still referred to in imperial units even though they are now sold in metric units. An example of this is 8ft × 4ft sheets of ply where the correct size is 2,440mm × 1,220mm.

ACTIVITY

Use an internet search engine to research:
- What imperial unit was paint purchased in?
- How many millimetres are there in a foot?
- How many inches are there in a metre?

Units for measuring	Metric units	Imperial units
Length	millimetre (mm) metre (m) kilometre (km)	inch (in) or ″ eg 6″ (6 inches) foot (ft) or ′ eg 8′ (8 ft)
Liquid	millilitre (ml) litre (l)	pint (pt)
Weight	gramme (g) kilogramme (kg) tonne (t)	pound (lb)

Units for measuring	Quantities	Example
Length	There are 1,000mm in 1m There are 1,000m in 1km	1mm × 1,000 = 1m 1m × 1,000 = 1km 6,250mm can be shown as 6.250m 6,250m can be shown as 6.250km
Liquid	There are 1,000ml in 1l	1ml × 1,000 = 1l
Weight	There are 1,000g in 1kg There are 1,000kg in 1t	1g × 1,000 = 1kg 1kg × 1,000 = 1t

INDUSTRY TIP

When ordering remember that products can be bought cheaper in large quantities.

LINEAR CALCULATIONS

Linear means how long a number of items would measure from end to end if laid in a straight line. Examples of things that are calculated in linear measurements are:

- skirting board

- lengths of timber

- foundations

- wallpaper.

We use this form of measurement when working out how much of one of the materials listed above we need, eg to find out how much skirting board is required for a room.

Here is a reminder of how to work out linear calculations:

Example

A site carpenter has been asked how many metres of skirting are required for the room below.

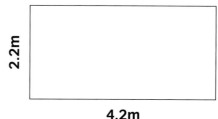

They can add all the sides together:
$$2.2 + 4.2 + 2.2 + 4.2 = 12.8m$$

Or, they can multiply each side by 2, and add them together:
$$(2.2 \times 2) + (4.2 \times 2) = 12.8m$$

Either way, **12.8m** is the correct answer.

Here is a reminder of how to obtain the area of the most common shapes:

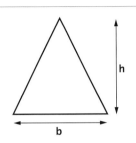	**Triangle** Area = ½ × b × h b = base h = vertical height	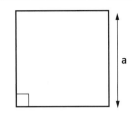	**Square** Area = a² a = length of side
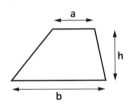	**Rectangle** Area = w × h w = width h = height	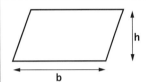	**Parallelogram** Area = b × h b = base h = vertical height
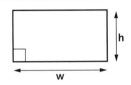	**Trapezium** Area = ½(a +b) × h h = vertical height	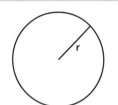	**Circle** Area = π × r² Circumference = 2 × π × r r = radius
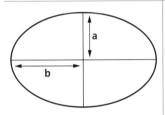	**Ellipse** Area = πab	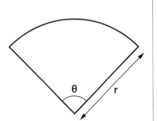	**Sector** Area = $\frac{\theta}{360}$ × πr² θ = angle in degrees r = radius

Note: a = length of side
      b = base
      h is at *right angles* to b:

ACTIVITY

What is the area of this rectangle?

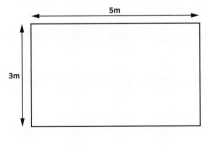

Answer: Area = 5 × 3 = 15

ACTIVITY

What is the area of this circle? Remember the value of π is 3.142.

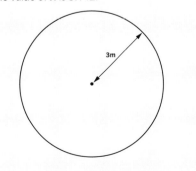

Answer: Area = π × 3² = 28.278

CALCULATING VOLUMES

Volume is measured in cubes (or cubic units). How many cubes are in this rectangular prism (cuboid)?

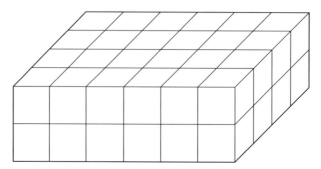

We can count the cubes although it is quicker to take the length, width and height and use multiplication. The rectangular prism above has a volume of 48 cubic units.

The volume of a rectangular prism is = length × width × height.

To calculate the volume of a rectangular prism we need to do two multiplications. We calculate the area of one face (or side) and multiply that by its height. The example below shows how the method for doing this.

FUNCTIONAL SKILLS

Calculate the cost of concrete required for a foundation 600mm wide and 300mm thick with a centre line measurement of 29.2m. Concrete costs £90m³.

Work on this activity can support FM2 (C2.7 and C2.8).

Answer: Volume: 29.2 × 0.6 × 0.3 = 5.256m³. Cost: 5.256 × 90 = £473.04

FUNCTIONAL SKILLS

Calculate the cost of timber required for 240m of 225mm × 50mm softwood floor joisting. Softwood costs £450m³.

Work on this activity can support FM2 (C2.7).

Answer: Volume: 240 × 0.225 × 0.05 = 2.7m³. Cost: 2.7 × 450 = £1,215.

Example

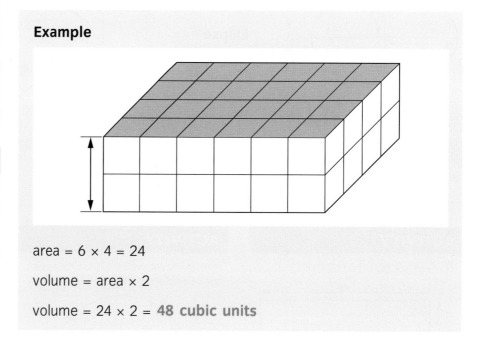

area = 6 × 4 = 24

volume = area × 2

volume = 24 × 2 = **48 cubic units**

WASTAGE

No matter how carefully we work there will be some waste involved in the construction process. There will always be offcuts of wood left over, we may not require all the mortar that has already been mixed or we may have part-full tins of paint left. Much of this unavoidable waste can be accounted for in the estimate by adding an extra allowance for this waste. Avoidable wastage can have a considerable effect on the overall profit a company makes (or loses).

How much we allow for natural waste will depend on the product. Typically we would allow an additional 5% for bricks, blocks and timber for construction purposes. This does not work for all materials; eg, if six rolls of wallpaper are required for a room, 5% would be 0.3 of a roll. As we can't order part of a roll, we would have to order a complete roll.

Calculating waste percentages

Percentage means per hundred, or *part of a hundred*.

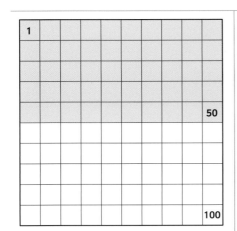

So **50%** means 50 per 100 (50% of this box is blue)

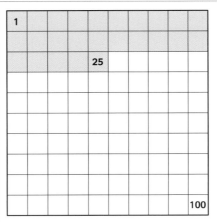

And **25%** means 25 per 100 (25% of this box is blue)

FUNCTIONAL SKILLS

Calculate the following:

a   40% of 240

b   30% of 300

c   57% of 1,140

Work on this activity can support FM2 (C2.4).

Answers: a) 96, b) 90, c) 649.8

Examples

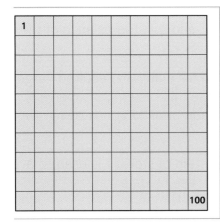

100% means all.

Example:

$$100\% \text{ of } 80 \text{ is } \frac{100}{100} \times 80 = \mathbf{80}$$

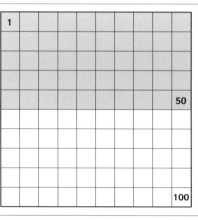

50% means half.

Example:

50% of 80 is $\dfrac{50}{100} \times 80 = \mathbf{40}$

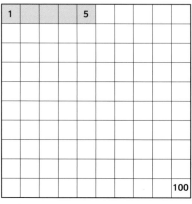

5% means 5 per 100.

Example:

5% of 80 is $\dfrac{5}{100} \times 80 = \mathbf{4}$

Waste does not just include materials but idle labour also.

Waste can be largely minimised by using common sense and having good management of **resources**.

This includes:

- ordering materials just in time for their use; this will avoid materials being damaged or stolen in storage

- having secure storage for high-value items

- using a robust method of checking in materials on delivery, recording discrepancies and reporting errors; this will ensure that goods not received are not paid for and that the balance of the goods required is timely delivered

- using an internal requisition system for materials required (large sites or joiner's shops commonly use this system); a supervisor will be required to sign the requisition prior to goods being given out by a store keeper

- ensuring a first-in first-out (FIFO) system of storage is used to avoid materials being stored beyond their use-by date.

Resources

This can include materials, personnel and equipment

INDUSTRY TIP

'To fail to plan is to plan to fail.' Plan the amount of materials you need and arrange delivery times carefully. Having no or not enough materials to work with will hold up the labour force or leave them without productive work to do.

PLANNING WORK ACTIVITIES

To ensure that work can be completed on schedule within the given budget considerable planning of the building process is required. Good management is a prerequisite for a successful build; good or bad management will determine whether a profit is made and whether the company can remain in business.

Planning should take place at various stages during the contract:

- Pre-tender planning includes deciding whether this is the type of work that can be undertaken and whether there is the capability to achieve it within the time scale, then preparation of an estimate into the tender bid.

METHOD STATEMENT			
Revision Date:	**Revision Description:**		**Approved By:**
Work Method Description	Risk Assessment	Risk Levels	Recommended Actions* (Clause No.)
1.			
2.			
3.			
4.			
RISK LEVELS: Class 1 (high) Class 2 (medium) Class 3 (low) Class 4 (very low risk)			
Engineering Details/Certificates/Work Cover Approvals:		Codes of Practice, Legislation:	
Plant/Equipment:		Maintenance Checks:	
Sign-off			
Print Name: Signature:	Print Name: Signature:	Print Name: Signature:	Print Name: Signature:
Print Name: Signature:	Print Name: Signature:	Print Name: Signature:	Print Name: Signature:

Method statement

Method statement

This is a detailed breakdown of the labour and plant required for each bill item. This allows for accurate programming

- Pre-contract planning occurs after the contract has been won and takes place prior to the build starting. The contractor may have up to six weeks to plan the commencement of work on site. During this time the following occur:
  - placing of orders for subcontractors
  - planning of the site layout in terms of temporary site buildings, storage of resources, traffic routes, position of crane, etc
  - laying on of temporary services
  - preparation of the work programme
  - production of **method statements**
  - sourcing of suppliers and labour.

- Contract planning is required to order the work in a logical sequence, determine labour, resource and plant requirements, maintain control and ensure work progresses as planned to meet the handover date.

Generally it is only the contract planning that concerns us, not in terms of producing a contract plan at this stage but in terms of knowing what one looks like and interpreting the information contained on it.

SITE LAYOUT AND ORGANISATION

As a construction worker you may be involved in the planning or setting up of the site layout. As mentioned earlier, on large sites this may form part of the pre-contract planning, while on smaller sites this would be planned by the site manager.

The purpose of site planning is to ensure that the layout, positions of the temporary site buildings, stationary plant, storage areas, any cranes, welfare facilities and site routes are placed in their most strategic and convenient positions with the overall aim of providing the best conditions for maximum economy, continuity, safety and tidiness. Site routes should be kept clear and not impede the building being constructed. Planning should include the minimal movement of materials to save double handling. Considerable additional costs can be incurred if portable buildings or materials have to be relocated. A site plan should be produced showing where all these should be positioned. Fencing or hoarding may be required. If this has to be positioned on or over a public highway a hoarding licence must be applied for from the Local Authority and displayed on the hoarding. Each authority has its own rules about how the hoarding is to be lit, decorated and formed. The hoarding must have warning signs on display at the entrance providing information to visitors and workers alike.

Safety notices

ITEMS included on SITE LAYOUT PLAN

Site security fencing.

Entrance gates.

A Welfare facilities.

B Site offices.

C Stores - lock up.

D Storage racking for finishing materials.

E Brick storage area on hardstanding.

F Formwork/reinforcement fabrication areas.

G General hardstanding area - formed up on commencement of contract.

H Area for subcontractor's accommodation and storage.

I Bagged aggregates and cement storage

● Mortar mixing area.

⊠ Position of tower crane

▱ Car parking spaces.

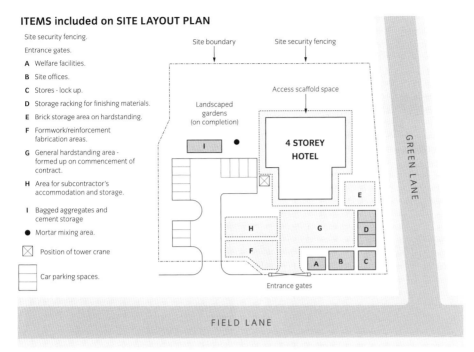

SITE LAYOUT PLAN: 4 STOREY HOTEL

Typical site layout

ACCESS AND TRAFFIC ROUTES

Consideration should also be given to how the site is accessed from the road. In some cases temporary roads have to be constructed to allow access to the site by plant and delivery lorries. Wherever possible, permanent roads should be used as the temporary roads or **hardstandings**. Permission must be obtained from the Local Authority for access over or encroachment of a public footpath. Traffic routes on site can be two way, but often there is only room for one-way traffic. Traffic routes should be clearly identified and where possible pedestrian routes provided separating them from the dangers of moving plant.

TEMPORARY SITE BUILDINGS

These are required for the management team, welfare facilities (including canteen and toilets) and storage. Where space on site is limited it can be stacked with suitable stair access. Larger sites have gatekeepers who monitor and record all comings and goings of staff, materials and visitors to the site. The buildings should be kitted out to allow each to fulfil its function effectively. Offices should be well lit with all services required laid on. The canteen should be suitably furnished, adequately heated and with facilities for drying clothes. Larger sites may have a separate facility for storing and drying clothes. The toilets should be cleaned and disinfected daily.

INDUSTRY TIP

Sometimes the Local Authority will create temporary roads if they cross the public highway, but they will, of course, charge a fee for this.

Temporary road access

Hardstanding

An area of ground where hardcore has been compacted for the placing of temporary buildings or materials

ACTIVITY

Research online types of temporary site accommodation available and the cost to hire it.

Site manager's office

Typical site accommodation set up as rest facilities

Typical clothes storage and drying facilities

Typical site accommodation set up as a meeting room facility

MATERIALS STORAGE AND HANDLING

Planning will help minimise wastage and losses arising from careless handling, poor storage, theft and double handling. Storage containers are required for high-value and fragile items, and for those that deteriorate when exposed to the weather. Open storage areas are required for bulk items such as timber, bricks, drainage pipes and roof trusses. Some items can be stored in the building as it is completed. It is essential that all materials are stored as required by the manufacturers' requirements to ensure they remain fit for purpose, do not become damaged or deteriorate.

STATIONARY PLANT

Careful planning and positioning of the site crane will allow materials to be offloaded, stored and taken to the final position efficiently and avoid double handling. The crane is often centrally positioned on the site to allow for the whole site to be encompassed by its radius. Mortar/concrete mixing or plant should of course be placed where the aggregate is stored.

Construction crane

PLANNING WORK ACTIVITIES
PROGRAMMES OF WORK

There are a number of different methods of programming building work. The most common is using some form of bar chart. Each company is likely to have its own variation of this. Traditionally this bar chart is referred to as a Gantt bar chart (as developed by Henry Gantt in the early twentieth century). Critical path analysis is another method employed to programme work but is not used much as it is more difficult to interpret the information.

Offloading construction materials from a lorry

Gantt bar chart programming

The programme of work is the key to a successful and efficiently run contract and will help the site manager and supervisors follow a set plan of action. The programme will show:

- the start date
- the sequence in which the building operations will be carried out
- an estimated time for each operation
- the labour required
- the plant required
- when materials require delivering
- the contract end date
- any public holidays.

Preparing the programme

The programme is prepared based on past experience, method statements and the measured rates from the estimate. In order to arrive at a basic programme for all trades, times are often based on a previous similar project.

Example: a small building contractor constructing a four-bedroom, brick-built house in 15 weeks.

Total time to build	100%	15 weeks (75 days)
Start to DPC	15%	11 days
DPC to watertight	45%	34 days
Internal work and finishing	40%	30 days

The programme shown above is based on the following breakdown of tasks.

Operation number	Description	Trade	Comment
1	Site preparation and setting out	Labourer, carpenter	Start to DPC
2	Excavation and concrete to foundations and drains	Labourer	
3	Brickwork to DPC	Bricklayer, labourer	
4	Back fill and ram	Labourer	
5	Hardcore and ground floor slab	Labourer, bricklayer	
6	Brickwork to first lift	Bricklayer, labourer	DPC to watertight
7	Scaffolding	Subcontractor, labourer	
8	Brickwork to first floor	Bricklayer, labourer	
9	First floor joisting	Carpenter, labourer	
10	Brickwork to eaves	Bricklayer, labourer	
11	Roof structure	Carpenter, labourer	
12	Roof tile	Subcontractor, labourer	
13	Windows fitted	Carpenter, labourer	
14	Carpentry first fix	Carpenter, labourer	Internal work and finishing
15	Plumbing first fix	Subcontractor, labourer	
16	Electrical first fix	Subcontractor, labourer	
17	Services	Subcontractor, labourer	
18	Plastering	Subcontractor, labourer	
19	Second fix carpentry	Carpenter, labourer	
20	Decoration	Subcontractor, labourer	
21	External finishing	Subcontractor, labourer	

These items of work can now be used on the bar chart.

Planned activities, labour and plant shown on a programme/progress chart

	Task	Week no	1	2	3	4	5	6	7	8	9	10	11	12	13	14	15
	Activity																
1	Site preparation and setting out		■														
2	Excavation/concrete to foundations and drains			■													
3	Brickwork to DPC			■													
4	Back fill and ram				■												
5	Hardcore and ground floor slab				■												
6	Brickwork to first lift					■											
7	Scaffolding					■							■				
8	Brickwork to first floor						■										
9	First floor joisting							■									
10	Brickwork to eaves							■									
11	Roof structure								■								
12	Roof tile									■							
13	Windows fitted									■							
14	Carpentry first fix									■	■						
15	Plumbing first fix/second fix										■			■			
16	Electrical first fix/second fix										■			■			
17	Services											■		■			
18	Plastering												■				
19	Second fix carpentry													■			
20	Decoration															■	
21	External finishing/snagging															■	■
	Labour requirements																
	Labourer		2 2	2 2	2 2	1 2	2 2	2 2	1 1	1 1	1 1	1	1	1	1	1	
	Carpenter		1 1					2		2 2	3 3	3 3	3	2 2	2 2	1 1	
	Bricklayer			2 2	2 2	4 4	4 4	4									
	Subcontractors																
	Scaffolding, roof tiler, services, plumber, electrician, plasterer, painter and decorator, landscaper.					▓				▓	▓	▓	▓	▓	▓	▓	▓
	Plant requirements																
	Ground works plant		▓													▓	
	Cement mixer			▓	▓	▓	▓	▓									
	Scaffolding				█	█	█	█	█	█	█						

Inclement weather

Weather that prevents building work from taking place, generally too cold, windy or wet, but could also be too hot

As you can see, the bar chart shows the sequence of operations and the labour and plant requirements very clearly. The second row of each activity line is used to measure progress against the planned activities. This will be shaded in another colour, usually green, as the build progresses. Careful monitoring of this progress line and early intervention of any slippage will ensure the contract remains on time. Generally, unless insufficient time has been allowed, **inclement weather**, staff sickness or lack of materials to work with will be the cause of delays. In order to 'catch up' it is likely that either extra labour is brought in, a more reliable supplier is found, or both.

Critical path analysis (CPA)

CPA is not commonly used on 'everyday contracts'. It is very specialised, and it's easy to make mistakes unless experts are used for generating it. In construction, CPA tends to be used on long, logistically difficult projects such as high-rise commercial buildings (incorporating high-tech systems) in densely populated areas (eg an international bank headquarters in any major city of the world).

CPA is used in a similar way to a bar chart to show what has to be done and by when, and what activities are critical. CPA is generally shown as a series of circles called 'event nodes'.

The key rules of a CPA

- Nodes (circles on the path) are numbered to identify each one and show the earliest start time (EST) of the activities that immediately follow the node, and the latest finish time (LFT) of the immediately preceding activities. Each node is split into three: the top shows the event/node number, the bottom left the EST, and the bottom right shows the LFT.

- The CPA must begin and end on one node.

- The nodes are joined by connecting lines which represent the task being planned. Each activity is labelled with its name, eg 'Brickwork to DPC', or it may be given a label, such as 'D', below.

- The length of the task is shown in the bottom of the node.

For example:

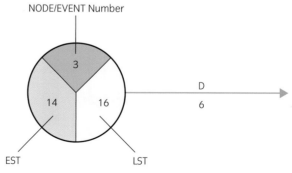

Example of CPA node

- The node is number 3.

- The EST for the following activities is 14 days.

- The LFT for the preceding activities is 16 days.

- There is 2 days' **float** in this case.

- The activity that follows the node is labelled 'D' and will take 6 days.

Every contract will have overlapping activities being carried out by various tasks, particularly large contracts. On a CPA this is shown by splitting the line.

The following CPA example shows the everyday task of making a cup of tea.

	Method statement	Secs
1	Fill kettle	10
2	Boil water	90
3	Place tea bag in cup	10
4	Add milk	10
5	Add sugar	10
6	Pour water	10
7	Let tea brew	30
8	Remove tea bag from cup	10
9	Stir tea	10
10	**Hand over to client**	160

Following this example you will see:

- Node 1 is the starting point where the kettle is filled.

- Node 2 is where the kettle is switched on for the water to boil (this node is 'critical' for the programme to work effectively and not to fall beyond time).

- Activities in nodes 3–5 can be carried out while the kettle is boiling.

- The split line shown dotted (often called a 'dummy line') links nodes 2 and 6, node 6 being pouring the boiling water onto the tea bag in the cup.

Float

In critical path analysis, the difference between the earliest start time (EST) and the latest finishing time (LFT)

Making a cup of tea

ACTIVITY

Following this example, produce a CPA for the start-to-DPC section of the bar chart on page 75.

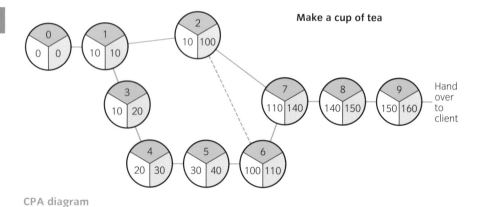

Make a cup of tea

Hand over to client

CPA diagram

Hired plant

PAT (portable appliance testing)

There is a legal requirement for all portable electrical appliances to be tested on a regular basis (dependent on use). Most companies test annually

Lead time

The delay between the initiation and execution of a process. For example, the lead time between the placement of an order for a staircase from a joinery manufacturer and its delivery may be anywhere from two to eight weeks

PLANT

During the pre-contract planning a decision will be made as to what items of plant will be purchased or hired. Some plant may already be owned by the company (cement mixers, power tools, etc) while others may not have been purchased due to their storage requirements and/or whether their initial cost can be recovered over a period of time.

HIRE OR PURCHASE?

- A crane, for example, would usually be hired as it would be cheaper to hire than purchase and does not have any storage requirements when not in use.

- Hired plant will always be supplied tested (**PAT**) and ready to use (see page 37).

- The hiring can be terminated when not in use, saving costs.

- There are no ongoing maintenance costs for hired equipment.

- Repeated use of hired equipment may be more expensive than the initial cost of outright purchase.

BUYING MATERIALS AND STOCK SYSTEMS

The programme will also allow the site manager or company buyer to plan for the advance ordering of materials (depending on the size of the company). It is good practice during the pre-contract planning to source suppliers that can provide all the materials required for the contract. In some cases these may be nominated suppliers. Most sites have limited space for storage so materials will be delivered on a regular basis. Some materials may have a long **lead time**. It is essential to know what lead times are required for all the materials needed for the project. Often discounts can be obtained if one builder's merchant is used to supply all materials required for a contract; knowledge of the reliability of this supplier is an important factor to avoid delays. As we mentioned earlier 'just-in-time' delivery methods work well to avoid wastage, damage and the need for storage space.

SITE COMMUNICATION AND ADMINISTRATION

In *The City & Guilds Textbook: Level 2 Diploma in Bricklaying*, Chapter 2, we looked at common methods of communicating information on site and the use of standard documentation to help to make communication clear, simple and accurate. Earlier in this chapter we also looked at drawings, which are one of the most informative methods of communicating information. As you will see, there is a lot of paperwork floating around every workplace and this needs careful managing. The following table shows most of the standard forms of documentation found on site or in a joiner's shop:

SITE DOCUMENTATION

Type of documentation	Description
Timesheet **Timesheet** **Employer:** CPF Building Co. / **Employee Name:** Louise Miranda / **Week starting:** 1/6/15 Date: 21/6/13 <table><tr><td>Day</td><td>Job/Job Number</td><td>Start Time</td><td>Finish Time</td><td>Total Hours</td><td>Overtime</td></tr><tr><td>Monday</td><td>Penburthy, Falmouth 0897</td><td>9am</td><td>6pm</td><td>8</td><td></td></tr><tr><td>Tuesday</td><td>Penburthy, Falmouth 0897</td><td>9am</td><td>6pm</td><td>8</td><td></td></tr><tr><td>Wednesday</td><td>Penburthy, Falmouth 0897</td><td>8.30am</td><td>5.30pm</td><td>8</td><td></td></tr><tr><td>Thursday</td><td>Trelawney, Truro 0901</td><td>11am</td><td>8pm</td><td>8</td><td>2</td></tr><tr><td>Friday</td><td>Trelawney, Truro 0901</td><td>11am</td><td>7pm</td><td>7</td><td>1</td></tr><tr><td>Saturday</td><td>Trelawney, Truro 0901</td><td>9am</td><td>1pm</td><td>4</td><td></td></tr><tr><td>Totals</td><td></td><td></td><td></td><td>43</td><td>3</td></tr></table> Employee's signature:_____ Supervisor's signature: _____	Used to record the hours completed each day, and is usually the basis on which pay is calculated. Timesheets also help to work out how much the job has cost in working hours, and can give information for future estimating work when working up a tender.

Type of documentation	Description
Job sheet	Gives details of a job to be carried out, sometimes with material requirements and hours given to complete the task.

CPF Building Co
Job sheet

Customer name: Henry Collins	Date: 9/12/15
Address: 57 Green St Kirkham London	

Work to be carried out
Finishing joint work to outer walls

Instructions
Use weather struck and half round

Variation order **Confirmation notice** **Architect's instruction**	Sometimes alterations are made to the contract which changes the work to be completed, eg a client may wish to move a door position or request a different brick finish. This usually involves a variation to the cost. This work should not be carried out until a variation order and a confirmation notice have been issued. Architect's instructions are instructions given by an architect, first verbally and then in writing to a site agent as work progresses and questions inevitably arise over details and specifications.

CPF Building Co
Variation order

Project Name: Penburthy House, Falmouth, Cornwall

Reference Number: 80475   Date: 13/11/15

From: : _____   To: _____

Reason for change:	Tick
Customer requirements	☑
Engineer requirements	☐
Revised design	☐

Instruction:
Entrance door to be made from Utile hardwood with brushed chrome finished ironmongery (changed from previous detail, softwood with brass ironmongery).

Signature _____

Type of documentation	Description
Requisition order	Filled out to order materials from a supplier or central store. These usually have to be authorised by a supervisor before they can be used.

CPF Building Co
Requisition order

Supplier Information: Construction Supplies Ltd   Date: 9/12/15

Contract Address/Delivery Address: Penburthy House, Falmouth, Cornwall

Tel number: 0207294333

Order Number: 26213263CPF

Item number	Description	Quantity	Unit/Unit Price	Total
X22433	75mm 4mm gauge countersunk brass screws slotted	100	30p	£30
YK7334	Brass cups to suit	100	5p	£5
V23879	Sadikkens water based clear varnish	1 litre	£20.00	£20.00
Total:				£55.00

Authorised by: Denzil Penburthy

Delivery note	Accompanies a delivery. Goods have to be checked for quantity and quality before the note is signed. Any discrepancies are recorded on the delivery note. Goods that are not suitable (because they are not as ordered or because they are of poor quality) can be refused and returned to the supplier.

Construction Supplies Ltd
Delivery note

Customer name and address: CPF Building Co Penburthy House Falmouth Cornwall	Delivery Date: 16/12/15 Delivery time: 9am Order number: 26213263CPF

Item number	Quantity	Description	Unit Price	Total
X22433	100	75mm 4mm gauge countersunk brass screws slotted	30p	£30
YK7334	100	Brass cups to suit	5p	£5
V23879	1 litre	Sadikkens water based clear varnish	£20	£20

Subtotal	£55.00
VAT	20%
Total	£66.00

Discrepancies: ....

Customer Signature: ....

Print name: ....

Date: ....

Type of documentation	Description
Delivery record	Every month a supplier will issue a delivery record that lists all the materials or hire used for that month.

Davids & Co
Monthly delivery record

Customer name and address:	Customer order date:
CPF Building Co Penburthy House Falmouth Cornwall	28th May 2015

Item number	Quantity	Description	Unit Price	Date Delivered
BS3647	2	1 tonne bag of building sand	£60	3/6/15
CM4324	12	25kg bags of cement	£224	17/6/15

Customer Signature: ....

Print name: ....

Date: ....

Type of documentation	Description
Invoice	Sent by a supplier. It lists the services or materials supplied along with the price the contractor is requested to pay. There will be a time limit within which to pay. Sometimes there will be a discount for quick payment or penalties for late payment.

Davids & Co
Invoice

Invoice number: 75856   Date: 2nd April 2015
PO number: 4700095685

Company name and address:	Customer name and address:
Davids & Co 228 West Retail Park Ivybridge Plymouth	CPF Building Co Penburthy House Falmouth Cornwall

VAT registration number: 663694542

For:

Item number	Quantity	Description	Unit Price
BS3647	2	1 tonne bag of building sand	£30
CM4324	12	25kg bags of cement	£224

Subtotal	£2748.00
VAT	20%
Total	£3297.60

Please make cheques payable to Davids & Co

Payment due in 30 days

Type of documentation	Description
Site diary 	This will be filled out daily. It records anything of note that happens on site such as deliveries, absences or occurrences, eg delay due to the weather.
Method statement 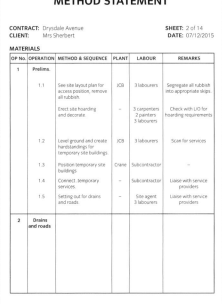	There are two types of method statement in common use. The first includes a risk assessment. This would be written during the pre-contract stage of the work programming. The second is more traditionally used to accurately estimate the time that each operation in the building process will take and is related to the bill of quantities. It is used as a guide to completing the work programme.
Risk assessment 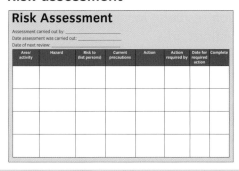	An assessment of the hazards and risks associated with an activity and the reduction and monitoring of them. Applying risk assessments and maintaining high standards of health and safety is the responsibility of everybody working in construction.

Type of documentation	Description
Permit to work **PERMIT TO WORK** 1. Area / 2. Date 3. Work to be Done / 4. Valid From / 5. Valid To 6. Company 7. Man in Charge / 8. No of Men 9. Safety Precautions 10. Safety Planning Certificate (cancelled if alarm sounds) I have inspected the above job which has been safely prepared according to requirements of a safety planning certificate Signed .... 11. Approval of Permit to Work I am satisfied that this permit is properly authorised, that safe access is provided, and that all persons affected by this job have been informed Signed .... 12. Electrical Equipment All power has been isolated/locked/tagged/tried* Circuits are live for troubleshooting only Signed .... 13. Acceptance of Permit to Work I/we* have read and understood the above precautions and will observe them. All equipment complies with relevant standards. I understand the site emergency plan. Signed .... 14. Completion of Permit to Work I/we* certify that this job is complete/incomplete*, all guards have been replaced and secured and all equipment has been removed. The job site has been left clean and tidy. Signed .... 15. Renewal of Permit to Work (same day only) Approved until .... Signed .... Approved until .... Signed .... If the alarm sounds 1. Stop Work 2. Make equipment safe 3. Leave the building by the nearest exit 4. Make your way to the main car park If you discover a fire 1. Break fire point 2. Leave the building 3. Ring 222 and give name, position, description etc 4. Report to incident controller in Main car park Do not re-enter any building until you are told it is safe	A permit to work is a documented procedure that gives authorisation for certain people to carry out specific work within a specified period of time. It sets out the precautions required to complete the work safely, based on the risk assessment.

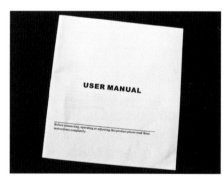

Manufacturer's instructions

MANUFACTURERS' INSTRUCTIONS

It is important to follow manufacturers' instructions for your own safety and to protect your product or purchase. Some instructions are in warning form while others are explanations of how to use that product. Warning instructions such as 'do not place in contact with fire' mean that the product is flammable, and not following this instruction could cause an explosion resulting in injuries. Other examples of manufacturers' instructions are:

- information on the back of a paint tin showing drying times, types of solvents required for cleaning brushes, etc

- information supplied as a user manual for powered equipment such as a mixer or chop saw

- information on how to use a brick cleaner safely or how to use a two-part epoxy adhesive.

Not following these instructions can lead to the invalidation of a **warranty**, lead to misuse of the product or the product not performing as expected.

Warranty

An assurance of performance and reliability given to the purchaser of a product or material. If the item should fail within the warranty period, it entitles the purchaser to a replacement

OTHER DOCUMENTS FOUND ON SITE

Minutes

Minutes of meetings normally include the following information:

- the time, date and place of the meeting

- a list of the people attending

- a list of absent members of the group

- approval of the previous meeting's minutes, and any matters arising from those minutes

- for each item in the agenda, a record of the principal points discussed and decisions taken and who is responsible for actioning any discussed items

- any other business (AOB)

- the time, date and place of the next meeting

- the name of the person taking the minutes.

Before each meeting an agenda should be drawn up, detailing the matters to be discussed. A set of minutes provides a record of what happened at the meeting, what actions are required, who is responsible for carrying them out and by what date.

Set of minutes

> **INDUSTRY TIP**
>
> Distribute (by email) the agenda before the meeting, so that members of the group have a chance to prepare for the meeting. After the meeting, follow up any action points set.

Memo

A memo, short for **memorandum,** is a brief document or note informing or reminding the receiver of something that has to be done. Memos have virtually been replaced by emails now.

Memorandum

Latin for 'to be remembered'

Memo

Hi Clare, Just to let you know that the client will be joining us at the review meeting on site next Monday.

Kind regards, Martin.

Organisational chart

An organisational chart is a diagram that shows the structure of an organisation and the relationships and relative ranks of its parts and positions/jobs.

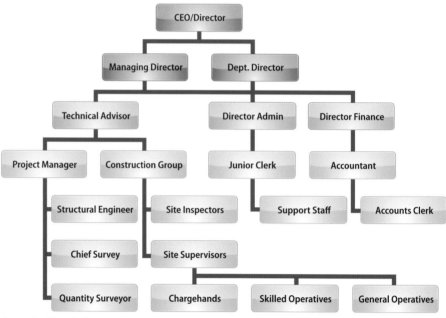

Organisational chart

SITE INDUCTION

Before starting work on site (or even when visiting) everyone should receive a site induction. This should give information on the following:

- a general introduction to the project and staff
- details of welfare facilities, site access and parking arrangements
- what to do in the case of an accident, designated first aiders, etc
- procedures for reporting accidents and near misses
- fire procedure and muster points
- notification of daily hazards such as working at height or moving vehicles
- notification about method statements and risk assessments
- site rules including those concerning drug or alcohol abuse, bullying, etc
- any other site-specific requirements.

There should be a record kept of who was inducted and when. It is in everyone's interest to ensure working on site is as safe as possible.

TOOLBOX TALKS

As mentioned in Chapter 1, a 'toolbox talk' is a short presentation given to a workforce on a single aspect of health and safety. A toolbox talk is prepared by either a company's safety officer, by a manufacturer's representative about a product being used or by a site supervisor. It should give timely safety reminders and outline any new hazards that the workforce may come into contact with. This may be new materials to be used, changes to or new risk assessments arising from an accident or near miss or working in inclement weather.

The HSE has prepared a number of standard **PowerPoint** presentations and **PDFs**; these will save companies the time and effort required in writing them from scratch.

> Health and Safety
> Executive
>
>
> **HSE**
>
> # Welfare facilities on small construction sites
>
> Presented by
>
> Martin Burdfield

HSE standard PowerPoint slide

PowerPoint

The name of a proprietary commercial presentation program developed by Windows. It presents a series of slides on an electronic whiteboard or screen

PDF (portable document format)

A file format used to represent documents in a manner independent of application software, hardware and operating system. It presents information in a fixed-layout flat document, including the text, fonts and graphics

ACTIVITY

Research what toolbox talks are available at www.hse.gov.uk.

DEFECTS SURVEY

Prior to a final certificate being written to sign off a contract, a defects survey should be carried out. Larger contracts should have a defects liability period (sometimes called a rectification period). This is a set period of time after a construction project has been completed during which a contractor is responsible for remedying defects. A typical defects liability period lasts for 12 months. A sum of money (typically 5%) is normally retained until after this period has come to an end. This will ensure that any necessary work is carried out.

To determine the work required a defects survey will need to be carried out. This could be done by a building surveyor, clerk of the works or maintenance manager. This will identify:

- poor standards of work

- poor quality of materials used

- damaged materials (damaged in construction not afterwards)

- human error.

A comprehensive list/table is collated on a schedule of outstanding works (this is commonly called a snagging list). The work identified is systematically worked through; this is often carried out by a snagging supervisor.

Schedule of outstanding works		Harold Court, Clacton-on-Sea
Location	Defect	Remedial action
Front entrance	Entry phone not working properly	Contact Door Services Ltd to send out an engineer to solve problem and recommission
First flight	Loose vinyl tiles on treads 4 and 6	Remove and bond new tiles to both steps
Front door to flat 3	Door twisted and not closing correctly	Contact Speedy Joinery to request replacement

TYPES OF COMMUNICATION

The importance of good communication cannot be overemphasised. Using effective communication skills is crucial to relationships and to success at work. There are, of course, many types of communication. We use verbal communication, non-verbal communication (eg using body language and facial expressions), written communication and many forms of each of those.

As we have said before, drawings are probably the most common means of giving information in our industry. The information shown on a drawing will communicate its contents to a speaker of any language. It is important for any communication to convey information without ambiguity.

WRITING LETTERS

Operative typing a letter

Sometimes it is necessary to put in writing a permanent record of a complaint or concern. This is often the case where it is thought that a court case could ensue. Such a letter should then be sent using a method that requires the recipient to sign to say they have received it. Letters are almost exclusively word processed now and rarely hand written. A letter should be set out with the following layout and information.

Your address

Your address, also known as the 'return address', comes first (leave this off if you're using letter-headed paper). Your return address should be on the right.

The date

Directly below your address, the date on which the letter was written should be given.

Recipient's name and address

The recipient's name and address should be positioned on the left-hand side.

The greeting

After the recipient's address, you should leave a line's space, then put 'Dear Mr Jones', 'Dear Bob' or 'Dear Sir/Madam' as appropriate.

The subject

You may want to include a subject for your letter — this is often helpful to the recipient to identify the order number or job concerned. This should be centred on the page.

The text of your letter

This should be:

- *Concise:* letters that are clear and brief can be understood quickly.

- *Authoritative:* letters that are well written and professionally presented have more credibility and are taken more seriously.

- *Factual:* accurate and informative letters enable the reader to see immediately the relevant details, dates and requirements, and to justify action to resolve the complaint.

- *Constructive:* letters with positive statements, suggesting concrete actions, encourage action and quicker decisions.

- *Friendly:* letters with a considerate, cooperative and complimentary tone are prioritised because the reader responds positively to the writer and wants to help.

The closing phrase and your name and signature

After the body of text, your letter should end with an appropriate closing phrase such as 'Yours sincerely'.

Leave several blank lines after the closing phrase (so that you can sign the letter after printing it), then type your name. You can optionally put your job title and company name on the line beneath this.

> **INDUSTRY TIP**
>
> If your letter is addressed to someone whose name you know (eg 'Dear Mrs Smith'), end it with 'Yours sincerely'.
>
> If your letter is addressed to someone whose name you don't know (eg 'Dear Sir/Madam'), end it with 'Yours faithfully'.

MB Construction
1 High Street
London
EC12 D34

01.12.2014

Mr Dawson
Scales Joinery Ltd
14 Riverside Way
Dalston Lane
Hackney E1

RE: 3 Harold Court (order 1629)

Dear Mr Dawson

Following our recent phone conversation I am writing to confirm that on further inspection the front door to 3 Harold Court is still not closing correctly. Our snagging supervisor has checked to see that the frame has been fixed plumb and informs me that it is. He has tried to ease the rebates to allow it to close more easily but the owner is still having difficulty.

Can I request that when you are delivering the paladin bin store doors you take the opportunity to have a look? If you can reliably confirm that your company will be able to rectify this problem without it re-occurring, please go ahead. Otherwise I would request that it is replaced by the 14.01.2015.

Yours sincerely,

E G Martin

E G Martin
Managing Director

ACTIVITY

Using the information on writing letters, reply to this letter, outlining what you will do as the joinery contractor to solve this issue.

FUNCTIONAL SKILLS

Draft a letter to a supplier informing them that two pallets of bricks out of 14 were badly damaged on the bottom two layers and that replacements are required.

Work on this activity can support FE2 (2.3.3a/2.3.4a).

A similar format to this letter can be used for defective or damaged materials or delivery problems.

It is important to state clearly what action you expect and in what time frame.

VERBAL COMMUNICATION

We looked at this at *Level 2* in Chapter 2. Talking face to face or on the phone is still the most common form of communication. Unfortunately there is rarely a record of these conversations; with this in mind you need to remember the following:

- Think before you speak, so that you get across exactly what you want to say.
- Be clear and concise in what you say.
- Ask for confirmation that what you have said has been understood.

BODY LANGUAGE

Body language refers to different forms of non-verbal communication. This often reveals an unspoken intention. Examples include:

- Rolling your eyes (to an onlooker meaning 'here we go again')
- Yawning   (indicating boredom)
- Hands in pocket (indicating lack of interest)
- Crossed arms   (indicating disagreement with what is being said)
- Smiling   (indicating happiness)
- Frowning   (indicating unhappiness).

Everyone has the right to be treated with respect and this will always create a productive environment. Aggression breeds aggression and leads to poor working relations and levels of production and a poor image to customers.

Types of body language

> **INDUSTRY TIP**
>
> You will always get the best from people if you treat them as you would like to be treated yourself.

ENERGY-EFFICIENT BUILDING

There is an infinite supply of materials available to us and it is our responsibility to make the most of what we have. By using sustainable products we can control the rate of consumption of resources and conserve our natural assets. Constructing buildings using these methods will pay dividends. There are a number of organisations that can help us achieve this. The Energy Saving Trust is a social enterprise with a charitable foundation. It offers impartial advice to communities and households on how to reduce carbon emissions, use water more sustainably and save money on energy bills.

They offer three main services:

- *EST endorsed products:* Setting standards for best in class for products such as boilers and glazing.
- *EST listed:* Products are listed in a directory approach and checked against quality and safety standards. Aimed at housing associations and other house builders.
- *Verified by EST:* Test reports are verified by EST to provide assurance on products' energy saving claims.

> **ACTIVITY**
>
> Go to www.energysavingtrust.org.uk to see how an existing cavity wall can be insulated.

> **INDUSTRY TIP**
>
> You can view the different EST logos at this webpage: www.energysavingtrust.org.uk/businesses/certification

HEAT LOSS FROM BUILDINGS

Heat can be lost through buildings as shown. The improved insulation of these areas will improve the energy efficiency of buildings.

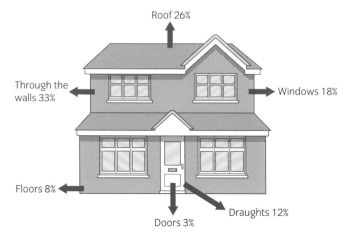

Sources of heat loss from a house

In addition heat flows from a building as shown below so it makes sense to insulate the structure to minimise this.

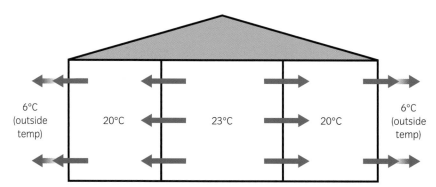

Heat flowing from a building

BUILDING REGULATIONS

Approved Document Part L, 'The Conservation of Fuel and Power in Buildings', is the standard that applies to construction projects that are new, extended, renovated, refurbished or involve a change of use.

To achieve compliance with Part L, the standard approach is to follow the guidance set out in the Government's Approved Documents, of which there are four:

- ADL1A New Dwellings

- ADL1B Existing Dwellings (extensions, renovations, change of use or energy status)

- ADL2A New Non Domestic Buildings

- ADL2B Existing Non Domestic Buildings (extensions, renovations, change of use or energy status).

Government Approved Documents

The route to compliance for new buildings and extensions is through the use of the national calculation methodology (NCM) software which calculates a dwelling or building's carbon dioxide emission rate and compares it against the target emission rate also calculated in the software for the same building. The relevant calculations are:

- SAP (Standard Assessment Procedure) for ADL1 and

- SBEM (Simplified Building Energy Model) for ADL2.

There are specialist companies that will carry out these calculations for you or you can use guidance documents that will show you how the requirements of the Approved Documents can be met.

Architects will produce their construction drawings and specifications to meet these requirements and this will be checked by the Local Authorities Building Control when a planning application and Building Regulations approval is sought.

Energy released by poorly insulated buildings can have a number of **detrimental** effects.

Highly insulated structures will help to:

- prevent heat loss

- reduce the size of heat-providing appliances

- reduce costs to the user

- help the environment

- reduce the country's energy demands.

> **INDUSTRY TIP**
>
> Invest in the best insulation affordable. It will save you money on energy in the long term.

Detrimental

Causing damage or harm

Watt

A unit of power

THERMAL TRANSMITTANCE

Thermal transmittance, also known as U-value, is the rate of transfer of heat through a structure (in **watts**) or more correctly through one square metre of a structure divided by the difference in temperature across the structure. It is expressed in watts per metre squared kelvin, or W/m²K. Well insulated parts of a building have a low thermal transmittance whereas poorly insulated parts of a building have a high thermal transmittance. The lower the U-value, the greater the insulation properties of the structure.

This chart shows the thermal conductivity of commonly used building structures.

Structure	U-value in W/m²K
Single-glazed windows, allowing for frames	4.5
Double-glazed windows, allowing for frames	3.3
Double-glazed windows with advanced coatings and frames	1.2
Triple-glazed windows, allowing for frames	1.8
Triple-glazed windows, with advanced coatings and frames	0.8
Well insulated roofs	0.15
Poorly insulated roofs	1.0
Well insulated walls	0.25
Poorly insulated walls	1.5
Well insulated floors	0.2
Poorly insulated floors	1.0

So as you will see, the Approved Documents will inform what U-value a structure must meet in order to comply with the Building Regulations.

ACTIVITY

Go to www.planningportal.gov.uk and look up the minimum U-value required for a roof.

Infra red image of heat escaping from a house

EXAMPLES OF U-VALUES REQUIRED IN A MODERN BUILDING

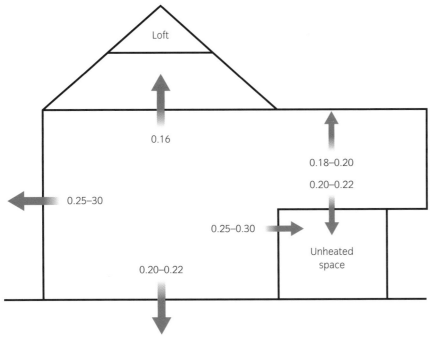

Heat lost in a typical house

The following table shows typical values of specifically constructed structures.

Element	Construction type	U-value (W/m²K)
Solid wall	Brickwork 215mm, plaster 15mm	2.3
Cavity wall	Brickwork 103mm, clear cavity 50mm, lightweight concrete block 100mm	1.6

Element	Construction type	U-value (W/m²K)
Cavity wall 	Brickwork 103mm, insulation 50mm, lightweight concrete block 100mm, lightweight plaster	0.48
Cavity wall 	Brickwork 103mm, insulation 100mm full cavity fill, lightweight concrete block 100mm, lightweight plaster	0.28
Cavity wall – timber frame 	Brickwork 103mm, clear cavity 14 studwork filled with PIR insulation	0.28
Timber frame and clad 	Tiles, render or cladding on battens	0.28

Element	Construction type	U-value (W/m²K)
Pitched roof	Tiles on battens, felt, ventilated loft airspace, 100mm mineral wool between joists, 170mm mineral wool over joists, plasterboard 13mm	0.16
Warm deck flat roof	150 PIR over joists 13mm plasterboard	0.18

ACTIVITY

Research two additional types of wall construction that will meet the current Building Regulations for a residential property.

INSULATION MATERIALS

The following types of materials are available to designers to reduce heat loss to achieve the Approved Document requirements.

Type of insulation	Description
Blue jean and lambswool 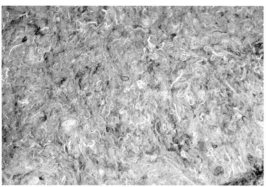	Lambswool is a natural insulator. Blue jean insulation comes from recycled denim.
Fibreglass/mineral wool 	This is made from glass, often from old recycled bottles or mineral wool. It holds a lot of air within it and therefore is an excellent insulator. It is also cheap to produce. It does however use up a fair bit of room as it takes a good thickness to comply with Building Regulations. Similar products include plastic fibre insulation made from plastic bottles and lambswool.
PIR (polyisocyanurate) 	This is a solid insulation with foil layers on the faces. It is lightweight, rigid and easy to cut and fit. It has excellent insulation properties. Polystyrene is similar to PIR. Although polystyrene is cheaper, its thermal properties are not as good.

Type of insulation	Description
Multifoil 	A modern type of insulation made up of many layers of foil and thin insulation layers. These work by reflecting heat back into the building. Usually used in conjunction with other types of insulation.
Double glazing and draught proofing measures 	The elimination of draughts and air flows reduces heat loss and improves efficiency.
Loose-fill materials (polystyrene granules) 	Expanded polystyrene beads (EPS beads) are also used as cavity wall insulation. These are pumped into the wall cavity, mixed with an adhesive which bonds the beads together to prevent them spilling out of the wall. This type of insulation can be used in narrower cavities than mineral wool insulation and can also be used in some stone-built properties.

Type of insulation	Description
Expanded polystyrene boards 	A graphite-impregnated expanded polystyrene bead board, designed to provide enhanced thermal performance. Lightweight, easy to handle, store and cut on site.
Autoclaved aerated concrete blocks 	Autoclaved aerated concrete blocks are excellent thermal insulators and are typically used to form the inner leaf of a cavity wall. They are also used in the outer leaf, where they are usually rendered.
Materials formed on site (expanded foam) 	Spray foam insulation is an alternative to traditional building insulation such as fibreglass. A two-component mixture composed of isocyanurate and polyol resin comes together at the tip of a gun and forms an expanding foam that is sprayed onto the underside of roof tiles, concrete slabs, into wall cavities, or through holes drilled into a cavity of a finished wall.
Double and triple glazing 	Single glazing has a U-value of 5, older double-glazing about 3 and new modern double glazing a U-value of 1.6, which is mainly due the fact that the cavity is gas filled improving the efficiency of the units. An added advantage of triple glazing over double is the improved reduction of external noise.

ENERGY PERFORMANCE CERTIFICATES (EPCS)

Energy performance certificates (EPCs) are needed whenever a property is:

- newly built
- placed on the market for sale
- placed on the market as a rental property.

They were introduced as a result of a European Union Directive relating to the energy performance of buildings and have been a legal requirement since 2008 for any property, whether commercial or domestic, that is to be sold or rented. Since April 2012, legislation has been set in place that makes it illegal to market a property without a valid EPC.

EPCs give potential buyers an upfront look at how energy efficient a property is, how it can be improved and how much money this could save.

The document is valid for 10 years and shows how good – or bad – the energy efficiency of your property is. It grades the property's energy efficiency from A to G, with A being the highest rating.

If you have a brand new home it's likely to have a high rating. If you have an older home it's likely to be around D or E. EPCs for a property can be obtained through specialist companies. A surveyor will be sent out and details of the structure, the heating system and even the light bulbs will be taken. A software program or tables can be used to obtain a **holistic** value to grade the property.

Holistic

Meaning overall, taking everything into account

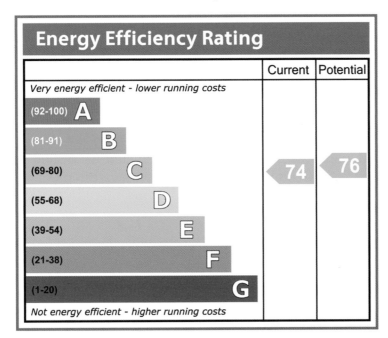

Rating chart

The chart on the previous page shows an example of a current rating and the potential improvement once changes have been made. The energy performance certificate also lists the ways to improve the rating – such as installing double glazing or loft, floor or wall insulation.

SUSTAINABLE MATERIALS

We all have to embrace reduction in energy consumption, whether this is by being limited to buying energy-efficient light bulbs or buying an 'A'-rated electrical appliance to cut down the fuel bills.

There are things we can opt into and things that are decided for us. The construction methods we use today are all decided for us, whether by codes of practice or Building Regulations. So these decide the minimum performance levels a building has to perform to for our comfort, safety and, of course, to conserve fuel and power.

Building materials can be sustainable if they are chosen carefully. For example, the process of manufacturing concrete uses considerable fuel and produces a lot of carbon dioxide (a gas that contributes to damaging the climate and our environment). On the other hand trees absorb carbon dioxide through the process of **photosynthesis** as they grow and can be grown sustainably. Trees look nice, make you feel good and they can be used to make a range of products (eg furniture, building products, paper products, medicines, cosmetics and rubbers). Some timber, however, is harvested from rainforests without thought for the surrounding environment, the life it supports or the fact that some species are close to extinction.

Brick manufacturers are doing a lot to ensure the manufacture of bricks is sustainable. They achieve this by continuous improvement of their extraction and manufacturing processes and by providing products that contribute to sustainable construction. The clay is sourced from areas local to manufacturing plants so minimising the transport required.

MANAGED TIMBER SOURCES

Managed forests where trees are replanted after harvesting provide a sustainable source of timber. The Forest Stewardship Council (FSC) is an international non-profit organisation dedicated to promoting responsible forestry. The FSC® certifies forests all over the world to ensure they meet the highest environmental and social standards.

The system has two key components:

1 Forest Management and Chain of Custody certification. This system allows consumers to identify, purchase and use timber and forest products produced from well managed forests.

An energy-efficient light bulb

Photosynthesis

The process of converting light energy to chemical energy

ACTIVITY

In pairs research three sustainable construction materials. Discuss the findings with your group.

Converted timber stamped showing certification

2 The FSC®'s 'tick tree' logo is used on product labels to indicate that the products are certified under the FSC® system. When you see the FSC® logo on a label you can buy timber and other wood products, such as paper, with the confidence that you are not contributing to the destruction of the world's forests.

FSC® logo

THE CODE FOR SUSTAINABLE HOMES™

The Code for Sustainable Homes™ is the national standard for the sustainable design and construction of new homes. It aims to reduce carbon emissions and promote higher standards of sustainable design above the current minimum standards set out by the Building Regulations. This is currently in the process of being wound down and will be replaced as a result of a Housing Standards review.

Code for Sustainable Homes™ logo

The code currently provides nine measures of sustainable design:

- energy/CO_2
- water
- materials
- surface water run-off (flooding and flood prevention)
- waste
- pollution
- health and well-being
- management
- ecology.

It uses a 1 to 6 star system to rate the overall sustainability performance of a new home against these nine categories.

The aim of this and other organisations is to prove the construction of **carbon-neutral** homes.

RECYCLED MATERIALS

A building that is sustainable must, by nature, be constructed using locally sustainable materials, ie materials that can be used without any adverse effect on the environment, and which reduce the distance travelled of those materials. When using locally sustainable materials it is essential that those materials are renewable, non-toxic and therefore safe for the environment. Ideally, they will be **recycled**, as well as recyclable.

Consideration should also be given to the extent a building material will contribute to the maintenance of the environment in years to come. Alloys and metals will be more damaging to the environment over a period of years as they are not biodegradable, and are not easily recyclable, unlike wood, for example. Also to what extent can the material be replenished? If the material is locally sourced and is likely to be found locally for the foreseeable future, travelling will be kept to a minimum, reducing harmful fuel emissions.

Examples of recycled building materials
- Crushed concrete or bricks for hardcore
- Reuse of tiles or slates
- Bricks cleaned up and reused
- Steel sections shot-blasted and refabricated
- Crushed glass recycled as sand or cement replacement or for the manufacture of kitchen worktops

Carbon neutral

Carbon neutral, or having a net zero carbon footprint, refers to achieving net zero carbon emissions by balancing a measured amount of carbon released with an equivalent amount of carbon credits (planting of trees, etc) to make up the difference. It is used in the context of carbon dioxide-releasing processes associated with transportation and energy production

Recycled

Manufactured from used or waste materials that have been reprocessed, eg bench seating made from recycled carrier bags

- Reuse of doors

- Panel products with chipped recycled timber

- Reused timber sections or floorboards

- Reuse of period architectural features

- Reuse of period fixtures and fittings (ironmongery, etc).

INDUSTRY TIP

Any number of items available can be sourced from local architectural salvage yards.

Recycled bricks

Case Study: Clare

Clare has been appointed as the site manager for the building of four two-bedroom bungalows. The company she works for has a history of constructing a range of standard dwellings. This is Clare's first role as a project manager and she is keen to show that she is capable of the task. To ensure everything runs smoothly and to time she gathers information from previous projects and plans the programme armed with this information.

There are differences, however, as this site is a little off the beaten track and the long-range weather forecast for the winter of the build is poor. Clare researches all the local suppliers and asks them to quote for the staged delivery of the materials required for the contract. She has the option of using company staff or local labour. Company staff will work out more expensive as accommodation costs will have to be paid. Local labour will be cheaper but will not have the experience of this specific work and the quality of their work will be unknown. On balance Clare chooses to use known company labour. An added advantage of using them is that if they are staying away from home they are more likely to put in extra hours if required. Finally Clare has to decide whether some of the smaller plant required should be purchased or hired. She gets quotes and decides on this occasion to use one of the material suppliers she has already contacted as she gets a preferential deal because they are also supplying the building materials.

All Clare's planning will pay dividends and should provide enough information to put together an accurate programme for the building work.

Work through the following questions to check your learning.

1 Which one of the following is an advantage of using CAD?

   a Software is not required.

   b Objects can be reproduced quickly.

   c Drawing technicians are not required.

   d Lines can be drawn quickly with set squares.

2 Which one of the following scales is used to produce a location plan?

   a 1:5.

   b 1:100.

   c 1:1250.

   d 1:4500.

3 Which one of the following is the **largest** drawing size?

   a A0.

   b A1.

   c A2.

   d A3.

4 Which one of the following three-dimensional drawings has the lines drawn back at 30°?

   a Oblique.

   b Isometric.

   c Perspective.

   d Orthographic.

5 Which one of the following is a document which promises to carry out work for a specific sum?

   a Estimate.

   b Quotation.

   c Specification.

   d Bill of quantities.

6 Which one of the following sections would setting-up costs be included in, within the bill of quantities?

   a Preambles.

   b Preliminaries.

   c Provisional sum.

   d Prime cost.

7 Who produces the bill of quantities?

   a Builder.

   b Architect.

   c Clerk of works.

   d Quantity surveyor.

8 Which one of the following **best** describes a fixed price for building work?

   a Quote.

   b Estimate.

   c Guesstimate.

   d Prime cost.

9 At what stage would the orders for subcontractors be placed?

   a Preliminary.

   b Pre-contract.

   c Programming.

   d Pre-tendering.

10 A detailed description of how a task will be carried out is called a

   a programme

   b specification

   c risk assessment

   d method statement.

11 The time it takes for a product to be manufactured is called the

   a   quote

   b   estimate

   c   lead time

   d   time delay.

12 Which one of the following is a disadvantage of hiring plant and equipment?

   a   Costs are reduced in the long term.

   b   Costs are increased in the long term.

   c   It is returned at the end of the contract.

   d   Replacement parts have to be purchased.

13 Which one of the following items will be required to document a client's request for a change to the work required?

   a   Memorandum.

   b   Specification.

   c   Variation order.

   d   Method statement.

14 The order of new-build work will be listed on a

   a   snagging list

   b   risk assessment

   c   method statement

   d   materials schedule.

15 Information on how to use a new product on site safely will be given

   a   by letter

   b   by email

   c   during an induction

   d   during a toolbox talk.

16 What is put on the top right-hand side of the page when writing a letter?

   a   Greeting.

   b   Signature.

   c   Sender's address.

   d   Recipient's address.

17 Triple-glazed windows help reduce

   a   dry rot

   b   heat loss

   c   wet rot

   d   efflorescence.

18 Which of the following construction methods has the **best** thermal value?

   a   Timber frame.

   b   Concrete wall.

   c   Brick wall.

   d   Stone wall.

19 Energy performance certificates are rated from grades

   a   A–D

   b   A–F

   c   A–G

   d   A–J.

20 Which Approved Document covers 'Conservation of fuel and power'?

   a   K.

   b   L.

   c   M.

   d   N.

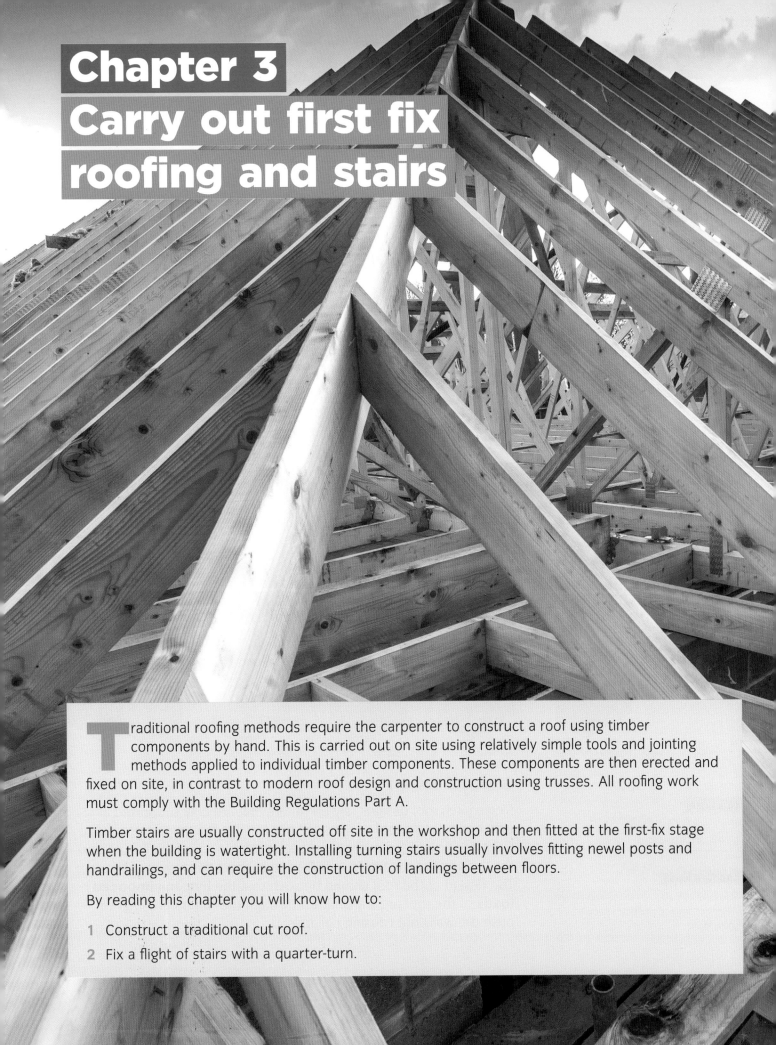

Chapter 3
Carry out first fix roofing and stairs

Traditional roofing methods require the carpenter to construct a roof using timber components by hand. This is carried out on site using relatively simple tools and jointing methods applied to individual timber components. These components are then erected and fixed on site, in contrast to modern roof design and construction using trusses. All roofing work must comply with the Building Regulations Part A.

Timber stairs are usually constructed off site in the workshop and then fitted at the first-fix stage when the building is watertight. Installing turning stairs usually involves fitting newel posts and handrailings, and can require the construction of landings between floors.

By reading this chapter you will know how to:

1 Construct a traditional cut roof.
2 Fix a flight of stairs with a quarter-turn.

TRADITIONAL CUT ROOFS

TYPES OF ROOF CONSTRUCTION

Traditional cut timber pitched roofs can be divided into three types, the choice of which usually depends on the span required:

- single roof
- double roof
- triple roof.

Cut roofs can be further categorised by their shape. The common ones are:

- flat
- lean-to
- mono-pitch
- gable-end
- hipped-end

- mansard
- gambrel
- jerkin-head
- valley.

Cut roof

A construction technique that builds up the roof without the use of trusses

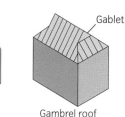

Flat roof

Lean-to roof

Mono-pitch roof

Gable-end roof

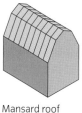

Hipped-end roof

Mansard roof

Gablet

Gambrel roof

Jerkin-head roof

Valley roof

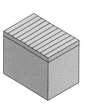
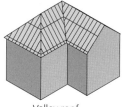
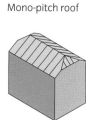

SINGLE ROOF

Single pitch roof designs rely on rafters that are only supported at either end. Single roof designs are not suitable for spans in excess of 5.5m because of the exceptional section size in length and width and the thickness of the timber required to support the weight of the roof. The design of the roof should be such that the **live** and **imposed loads** of the roof are transmitted through the rafters and do not push the walls out causing collapse.

Live loads

The weight of the structure. In this case the timber and tiles/slates

Imposed loads

This is a live load such as the additional weight of snow. Sometimes known as a snow load

Single roofs can be further divided into the following three types.

Couple roof

This type of roof has a pair of rafters, each of which typically sits at its lower end on a wall plate that sits on top of the inner wall, and at its highest point in the centre of the roof against another piece of timber called a ridge board. (See the next section of this chapter for more information about the components used within traditional cut roofs.)

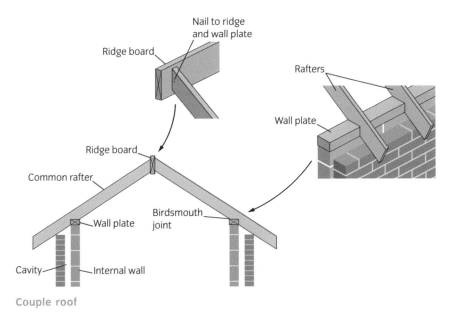

Couple roof

Close-couple roof

This type of roof is very similar to the couple roof, except the ceiling joists are fixed to the ends of the rafters at wall plate height. Vertical timbers called 'hangers' can be fixed from the common rafters to the ceiling joists; this helps stop the ceiling joists from sagging. This type of roof construction can help reduce the risk of the roof spreading.

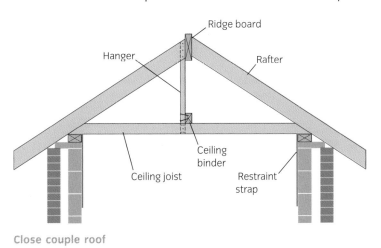

Close couple roof

Collared roof

This design of roof has a tie that is positioned part way up the rafters, allowing an increased span to be used. The collar tie should not be positioned up the rafters more than one-third of the rise. This type of roof is also known as 'collar-tie roof' and 'collar roof'.

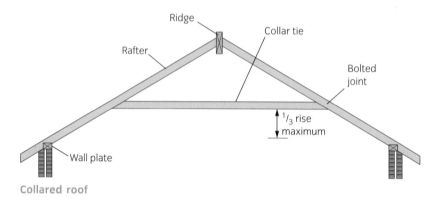

Collared roof

DOUBLE ROOF

Double roof designs allow greater spans to be covered. They require longer rafters to be used but without having to use excessively large sectional sizes of timbers. This is achieved by providing intermediate support along the rafters, called 'purlins', which are usually positioned midway along the rafter.

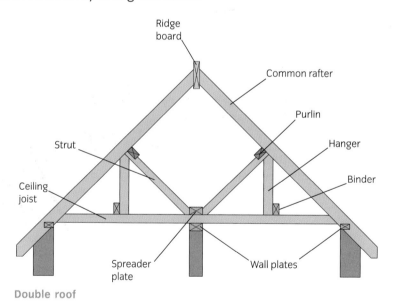

Double roof

TRIPLE ROOF

This type of roof was traditionally used for very large spans. It is nowadays mainly found in renovation work, heritage buildings, barn conversions and oak-framed buildings. This roof design includes a truss, typically a 'king post truss', which is fixed at intervals along the roof to provide intermediate support for the purlins, which in turn provide support for the common rafters.

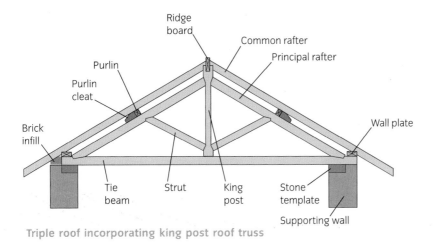

Triple roof incorporating king post roof truss

COMPONENTS WITHIN A TRADITIONAL CUT ROOF

The traditional cut roof can be divided up into several clear areas:

- *gable:* the upper portion of the end wall that is triangular and carries the slope of the roof; the 'gable end' refers to the whole end wall
- *verge:* the part of the roof that overhangs the gable
- *eaves:* the bottom end of the rafter where it meets the walls, normally projecting beyond the walls to protect the upper part of the wall from water entry
- *valley:* the part of the roof where two roof surfaces meet at an internal corner (producing a join shaped like a letter 'V')
- *hips:* the part of the roof where two external roof surfaces meet on an external corner
- *gablet:* a small gable at the top of a hipped roof.

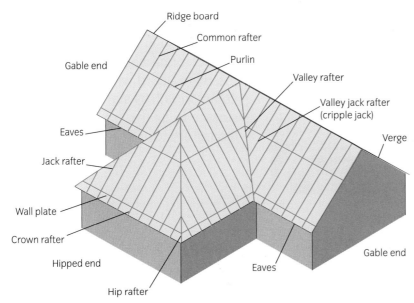

Components within a traditional cut roof

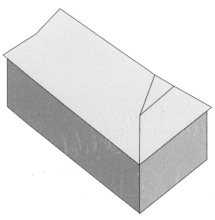

Gablet on a roof

The main components that make up a cut roof are shown in the following table.

Component	Description
Wall plate 	The piece of timber bedded on top of the internal walls, which are held in place with metal restraining straps to prevent movement. The wall plate helps spread the load of the roof and provides a fixing for the bottom of the rafter.
Ridge board 	The horizontal board at the apex (top) of the roof against which common rafters are fixed, providing longitudinal support for the roof.
Common rafters 	The main load-bearing components of the roof, these are fixed at the lower end by a birdsmouth joint to the wall plate, pass over a purlin (if fitted), and are then fixed at the top of the roof to the ridge board.
Hip rafters 	The substantial piece of timber running from an external corner of the roof to the ridge board (usually up against a saddle board that is fixed up against the end of the ridge).
Saddle board 	A piece of timber fixed to the end of the ridge board where the first set of common rafters start, used as a surface to fix the crown rafter and hip rafters to.

Component	Description
Crown rafter 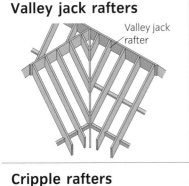 Crown rafter	The longest rafter used at the centre of the hipped end, running from the wall plate up to the centre of the ridge board (usually up against a saddle board fixed up against the end of the ridge).
Valley rafter Valley rafter	The timber running from the wall plate to the ridge at an internal corner.
Hip jack rafters Jack rafters / Jack rafters	The timbers running from the wall plate to the hip rafter at an external corner.
Valley jack rafters Valley jack rafter	The timbers running from the ridge board to the valley rafter at an internal corner.
Cripple rafters Cripple jack rafter	Cripple rafters touch neither the ridge board nor the wall plate.

INDUSTRY TIP

Valley jack rafters are nowadays often called 'cripple jack rafters'.

INDUSTRY TIP

Cripple rafters are also known as 'flying jacks' or 'flying rafters'.

Component	Description
Purlins 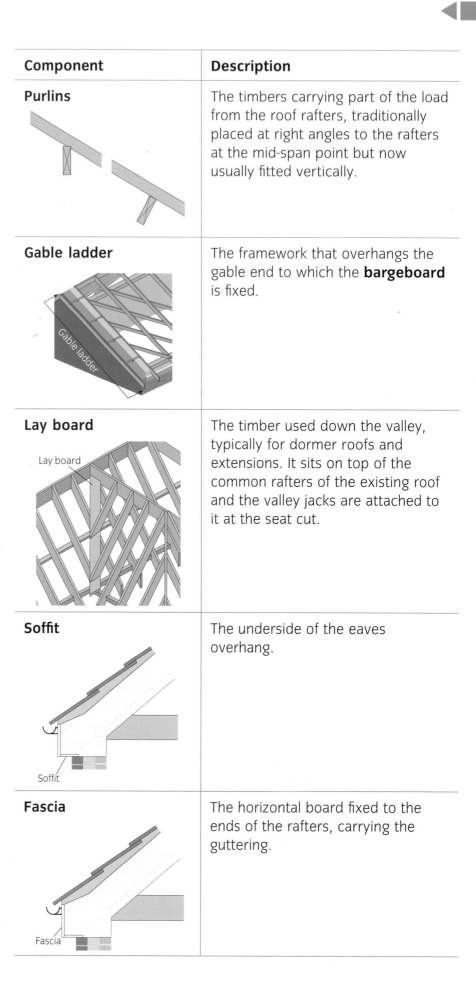	The timbers carrying part of the load from the roof rafters, traditionally placed at right angles to the rafters at the mid-span point but now usually fitted vertically.
Gable ladder	The framework that overhangs the gable end to which the **bargeboard** is fixed.
Lay board	The timber used down the valley, typically for dormer roofs and extensions. It sits on top of the common rafters of the existing roof and the valley jacks are attached to it at the seat cut.
Soffit	The underside of the eaves overhang.
Fascia	The horizontal board fixed to the ends of the rafters, carrying the guttering.

Component	Description
Collar ties 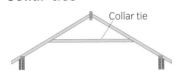 Collar tie	The timbers attached to the common rafters to help reduce the spread of the roof.
Restraining straps 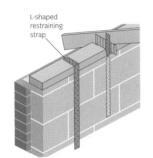 L-shaped restraining strap	Required to tie the rafters to the gable wall to prevent lateral movement of rafters and to fix down the wall plate. Restraining straps should not be more than 2m apart.
Struts 	Used to prevent the purlin from deflecting, these should not be positioned at an angle greater than 45° to the vertical, and should be used in conjunction with anti-slip blocks fitted at the **ceiling joists**.
Angle tie 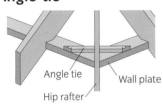 Angle tie Wall plate Hip rafter	Used with hipped roofs at the corner of the wall plate, providing additional support for the seating area of the hip rafter.
Dragon tie and beam 	Used to provide a decorative support (often carved) for the hip rafter, typically in heritage work and new timber-framed roofing.

Ceiling joists

Usually fixed at wall plate level to both the wall plate and rafter. The underside of the ceiling joist provides the fixing for the ceiling plasterboard

CALCULATING ROOFING COMPONENTS

INFORMATION REQUIREMENTS

To correctly calculate the true lengths and angles of roofing components, the following four pieces of information are usually available, of which two are needed: either the rise and the run or span, or the pitch and the run or span.

- *span:* the distance measured from the outside of one wall plate to the outside of the other wall plate

- *run:* the distance from the outside of one wall plate to the centre of the ridge board, measured along a horizontal plane and equal to half the span

- *pitch:* the angle at which the roof rises

- *rise:* the distance from the wall plate to the apex of the roof.

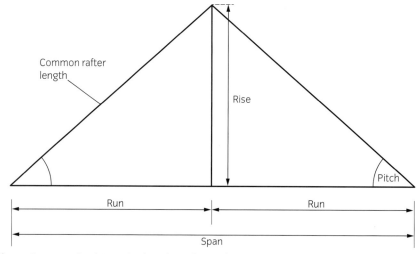

Information required to calculate lengths and angles

METHODS USED TO DETERMINE RAFTER LENGTHS AND BEVELS

To enable the correct angles and lengths of roof components to be calculated, you could use any of the following five common methods:

- full-size setting out

- scale drawings

- a roofing square

- rise and run

- a ready reckoner.

FULL-SIZE SETTING OUT

This is only really suitable for small-scale roofing that requires a short span with a small rise, and that can be set out on a sheet of plywood or similar. However, by joining several pieces together it can be used for larger roofs. The wall plate could also be used to set out the full-size rafter provided they are fixed at right angles to each other. If the wall plate is not at a right angle on the corner it will not only give incorrect angles but also incorrect true lengths.

Producing true lengths and angles for common rafters

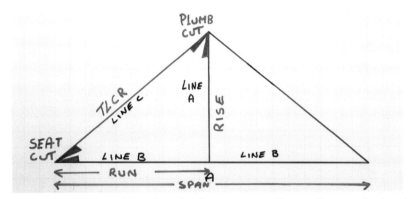

Producing true lengths and angles for common rafters using a scale drawing

1 On a sheet of paper, mark out a right angle. Label the vertical line 'A' and the horizontal line 'B'.

2 Along line A, mark the required rise.

3 Along line B, mark the required run.

4 Join the ends of lines A and B to create line C – this is the length of the common rafter.

5 The angle that is formed by lines B and C is the pitch. It gives the required angle for the seat cut, which will sit on the wall plate.

6 The angle that is formed by lines A and C is the plumb cut, which will sit up against the ridge board.

7 Set up sliding bevels for each of these angles.

ACTIVITY

Set out a roof that has a rise of 1,000mm and a run of 1,600mm. Measure the length of the **hypotenuse**. Is this the same length as if you were to use Pythagoras' theorem? Using a protractor measure both the plumb and seat cut angles.

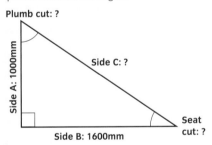

Hypotenuse

The side opposite the right angle and the longest side of a right-angled triangle

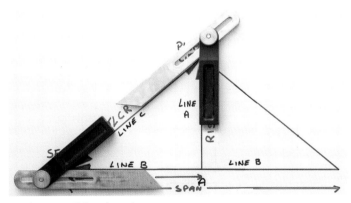

Setting up sliding bevels

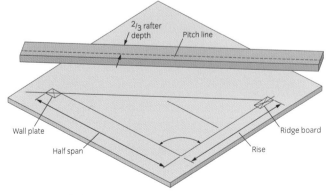

Setting out full size

SCALE DRAWINGS

Scale drawings can also be used to produce roof components, their lengths and their angles. The accuracy of these lengths and angles will depend on the skill of the person who produced the drawing, as well as the scale of the drawings. The larger the scale the better, but in practicality the size of the paper will determine the scale.

Drawings produced by **CAD** will be more accurate but often small in size to take information from, although the information might be printed as part of the drawing.

Using scale drawings can seem daunting and confusing compared with using a roofing square or a ready reckoner (methods covered later in this section). However, once you understand how scale drawings are produced, they can be created quickly and easily with only a few simple tools. Unlike a CAD program, which also calculates the exact lengths and angles of components, scale drawings depend totally on the skill and accuracy of the draftsperson to obtain these exact lengths and angles.

To keep the scale drawing as clear as possible, abbreviated labels should be used.

CAD

Short for 'computer-aided design', used to create drawings when setting out

Table of abbreviated labels for roofing components

Abbreviation	Definition	Abbreviation	Definition
TL	True length	SCVR	Seat cut valley rafter
PC	Plumb cut	SCJR	Seat cut jack rafter
SC	Seat cut	SCCrR	Seat cut cripple rafter
EC	Edge cut	PCCR	Plumb cut common rafter
TLCR	True length common rafter	PCHR	Plumb cut hip rafter
TLHR	True length hip rafter	PCVR	Plumb cut valley rafter
TLVR	True length valley rafter	PCJR	Plumb cut jack rafter
TLJR	True length jack rafter	ECHR	Edge cut hip rafter

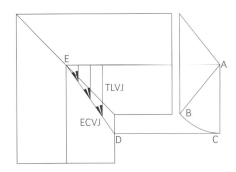

Table of abbreviated labels for roofing components

Abbreviation	Definition	Abbreviation	Definition
TLCrR	True length cripple rafter	ECVR	Edge cut valley rafter
SCCR	Seat cut common rafter	ECCrR	Edge cut cripple rafter
SCHR	Seat cut hip rafter	SCP	Side cut purlin
ECP	Edge cut purlin	SCVJ	Seat cut valley jack
ECVJ	Edge cut valley jack	ECJR	Edge cut jack rafter

Equipment required to produce scale drawings

To produce scale drawings, you will need the following:

- scale rule
- tee square
- set square
- protractor
- good-quality compass
- a 2H pencil
- paper
- drawing board
- tape to attach the paper to the board.

Producing a scaled drawing for a hipped-end roof

The following steps show you how to produce the scaled drawing shown to the right.

1. Decide on a suitable scale to enable the developed roof to fit on the paper.
2. To find the true lengths and angles of the hipped ends rafter, produce a rectangle to the required size using corners A^1, A^2, B^2 and B^1.
3. Draw in line D^1–D^2, which represents the ridge and is equal to the run, which is half the span.
4. Draw a plan view of hip A^2–E and B^2–E. Most hipped ends are set at 90° so point E is equal to an angle of 45° from points A^2 and B^2.

ACTIVITY

Using the equipment list to the side and following the steps below, develop the true lengths and angles of roof components.

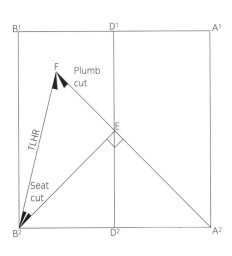

5 Extend line A²–E from point E. Measure a distance equal to the rise at point F.

6 Draw line B²–F. This is the true length of the hip rafter. Plumb cut is at point F, being the angle EFB². The seat cut is at point B², being the angle between lines EB²F.

7 Mark the angles, label, set sliding bevels to the angles and transfer to a pitch board for use in the workshop.

The dihedral angle or backing bevel to the hip rafter

The **dihedral angle** is the angle at which the two sloping roof surfaces meet against the hip rafter. This angle provides a level surface for the tile battens to be fixed to the hip rafter. The dihedral angle has dropped out of fashion and is seldom used nowadays, mainly due to increased pressure to produce roof structures quickly. (The opposite of a dihedral angle is an 'anhedral angle'. This is the 'V' shape applied to the top edge of valley rafters to allow for the tile lathe to be easily fixed and follow the flat surface of the roof.)

Dihedral angle

The angle formed on the top edge of the hip rafter running along its length from the eaves to the ridge

The following steps show you how to produce a scale drawing for a dihedral angle.

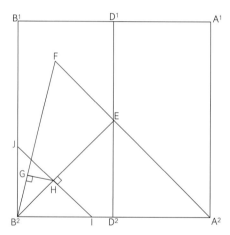

1 Using the previous scale drawing, mark point G on line B²–F. It can be at any distance along this line, but if it is too close to B² then the angles will be so small that it will be difficult to see; too close to point F and it will become large and overcomplicated. About half way along the line will give an acceptable size to set the sliding bevel.

2 From point G, mark a line that is at a right angle to line B²–F until it strikes line B²–E at point H.

3 At point H, mark a line that is at a right angle to line B²–E until it strikes the wall plate at points I and J.

4 With your compass point on H and its radius set at G–H, draw an arc at point K.

5 Draw in lines I–K and J–K.

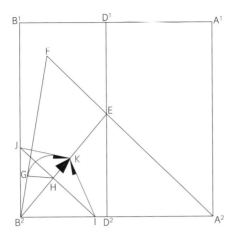

6 The dihedral angle for the hip rafter is the angle between B²–K and J–K. Mark the angles and label them.

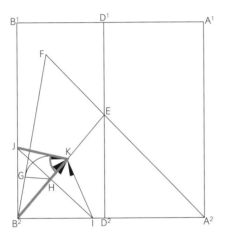

Edge cut for the hip

The following steps show you how to produce the scaled drawing shown.

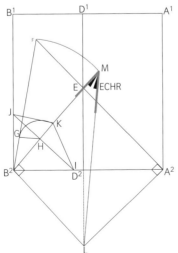

1 Repeat the steps for the previous scale drawing.

2 Extend line D^1–D^2 down until it meets the lines taken at right angles from line B^2–E and A^2–E to create point L.

3 With your compass at point B^2 and the radius set at B^2–F, draw an arc to meet extended line B^2–E at point M.

4 Mark the intersections of the lines M–L and B^2–M, which is the edge cut to the hip rafter at point M.

Lengths and angles of the jack rafters

The following steps show you how to produce the scaled drawing shown.

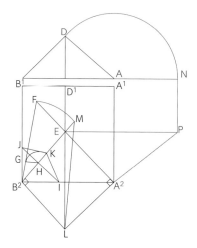

1 Repeat Steps 1 to 4 for 'Producing a scaled drawing for a hipped-end roof'.

2 With your compass centre at A and the radius set as A–D, draw an arc to point N.

3 From point N, draw down a line to point P (which is horizontally across from point E, the position of the first common rafter).

4 From point P draw a line to A^2 and draw in horizontal lines for jack rafters to the correct **centres**.

Centres

Here, the distance from the centre of one point to the centre of the next, ie used with rafters, floor joists and studwork

5 Mark on the edge cut angle and true length.

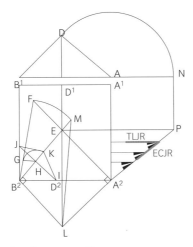

Edge and side cuts for purlins

The following steps show you how to produce a scaled drawing for edge and side cuts for purlins.

1 Produce a common rafter at the required pitch (40°) and a hip rafter plan view (45°).

2 Draw to a suitable scale on any portion of the common rafter the purlin points ABCD.

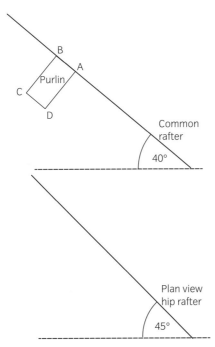

3 Draw a horizontal line through point B.

4 Using a compass centred at B and with a radius set as A–B, draw an arc to the right of B to point E, on the horizontal line through point B.

5 Drop a vertical line from point A to intercept the hip rafter at point F.

6 Take a horizontal line from point F to meet the vertical line from point E, creating point G.

7 Drop a vertical line from point B to meet the hip rafter to achieve point H.

8 Join points G and H to achieve the edge cut angle for the purlin.

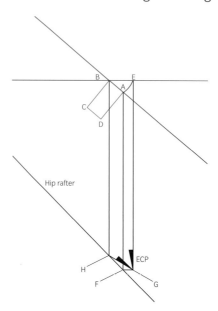

9 Drop a vertical line from point C to meet the hip rafter at point J.

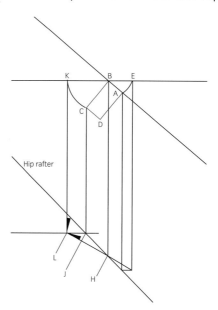

10 Using a compass centred at point B and with a radius set as B–C, draw an arc to point K.

11 Draw a horizontal line from point J to intercept the vertical line from K, creating point L.

12 Join points H and L to achieve the side cut for the purlin.

A 'lip cut' is the part of the purlin that sits under the hip or valley rafter and touches the other purlin.

Relationship between the edge cut and side cut of a purlin and the hip rafter

For the development of valley rafters, valley jack rafter lengths and angles, follow the same principles as those used for hip rafters and jack rafters.

1 Produce a plan view of the roof, showing the valley.

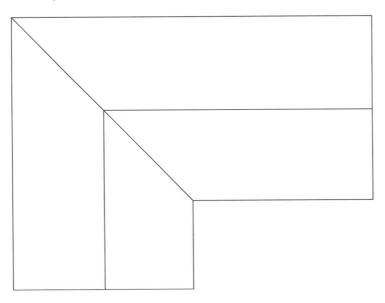

2 Draw the rise at point A, which is 90° from the plan view of the valley rafter position A–C, to give point B.

3 Draw a line from point B to the wall plate in the valley to point C.

4 Mark on the plumb cut angle ABC.

5 Mark the seat cut angle between ACB.

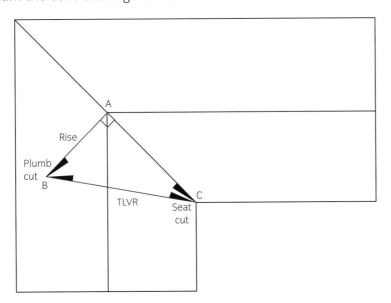

The edge cut to the valley rafter can be produced using the following method.

1 Repeat the steps for the previous scale drawing.

2 At point C, draw a line at 90° to line A–C to point D, on the ridge line.

3 Using a compass centred at point C and with a radius set to C–B, mark an arc to point E.

4 Draw line D–E, and mark the edge cut angle CED.

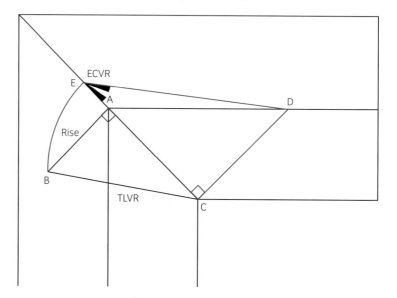

The valley jack and cripple rafter lengths and angles can be produced using the following method.

1 Draw a plan view of the roof, with the elevation of the roof showing the correct pitch/rise.

2 Using the radius of A–B with a centre on point A, draw an arc to the vertical line taken from point A at point C.

3 Take a horizontal line from point C to point D.

4 Take a line from point D to point E, on the ridge line.

5 Mark the valley jack rafter at correct centres, and draw the edge cut angle and true lengths.

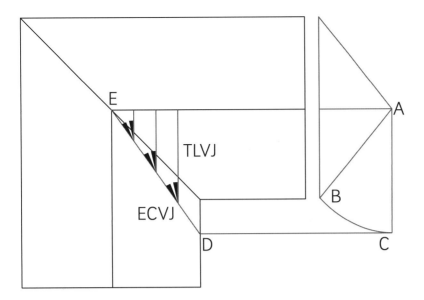

ROOFING SQUARE

The roofing square is a valuable addition to the carpenter's tool bag, as well as an aid to setting out right angles on a large scale such as when working on stud walling. Many other activities fall within its scope, including stairs and roofing.

The roofing square has a wide side called the blade and a narrow side called the tongue. Both the tongue and the blade have millimetres marked along their edges. They can also have pitch degrees marked along them and a chart giving rafter lengths per metre run for standard pitches.

By calculating the appropriate metre run for the required pitch, the true length of rafters can be worked out. The plumb cut is marked off along the blade and the seat cut along the tongue.

The true lengths of rafters are always measured along the **pitch line**. This line is from the outside edge of the wall plate to the centre of the ridge board. The true length of jack rafters is measured from the outside edge of the wall plate to the centre line of the hip rafter, while for valley jack rafters it is from the outside edge of the wall plate to the centre of the valley rafter.

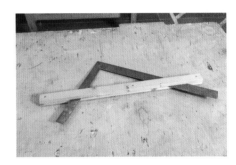

Roofing square

Pitch line

A line measuring two-thirds of the way down from the top of the common rafter and which is used to set out the lengths of the rafter. Regionally known as the setting out line

INDUSTRY TIP

The birdsmouth on a rafter incorporates the seat cut and plumb cut. Combined, they always form a 90° triangle that sits on the wall plate.

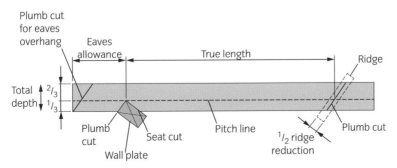

Example of common rafter that is set out

Producing the common rafter

The following steps demonstrate the use of a roofing square to produce common and hip rafter cuts. In this example, the specification for the roof is:

- span of 3.6m

- 40° pitch

- eaves overhang of 300mm

- ridge board thickness of 25mm.

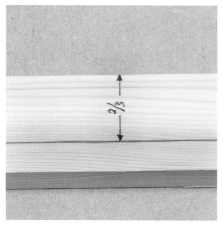

STEP 1 Mark along the common rafter at two-thirds of its depth: this is the pitch line.

STEP 2 Position a roofing square at the required point on the common rafter with the required pitch marks on the pitch line.

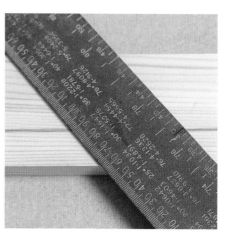

STEP 3 Use the roofing square to mark the plumb cut on the pitch line.

STEP 4 Use the roofing square to mark the common rafter seat cut marks on pitch line.

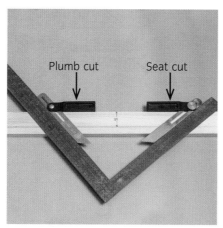

STEP 5 Set sliding bevels to the seat cut and plumb cut angles.

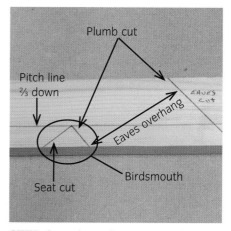

STEP 6 Mark out the eaves overhang and birdsmouth cut.

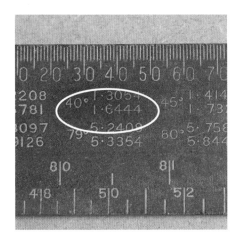

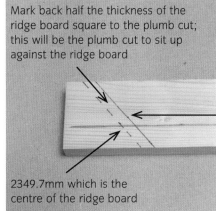

Mark back half the thickness of the ridge board square to the plumb cut; this will be the plumb cut to sit up against the ridge board

Plumb cut sitting against ridge board

2349.7mm which is the centre of the ridge board

STEP 7 Calculate the true length of the common rafter: in this example, it is 1.3054 × 1,800mm = 2,349.9mm or 2.349m. Remember that the top number is for the common rafter and the bottom number is for the hip or valley rafter.

STEP 8 Measure 2.349m along the pitch line. This is the measurement from the birdsmouth to the centre of the ridge board. Measure back half the thickness of the ridge board square to the original plumb cut to allow the face of the common rafter to sit against the ridge board.

ACTIVITY

Set out a common rafter using a roofing square from the specification given on the previous page.

Producing the hip rafter without the dihedral angle

1 Measure and mark the pitch line from the top surface of the hip rafter equal to that of the common rafter pitch line.

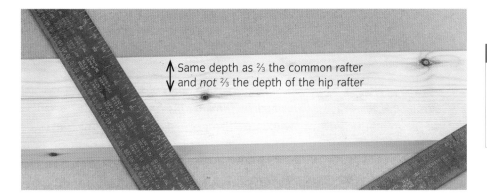

Same depth as ⅔ the common rafter and *not* ⅔ the depth of the hip rafter

INDUSTRY TIP

Once the setting out has been completed, the first rafter can be cut (apart from the eaves overhang cut, which is usually done after fitting). The rafter can then be used as a pattern to mark out the remaining rafters.

2 Position the roofing square on the pitch line to the required pitch (40°) and mark point B on the square, which is the hip rafter mark.

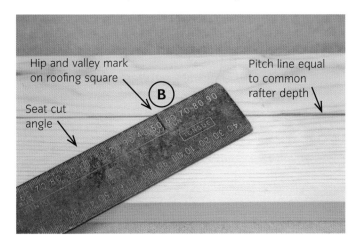

Hip and valley mark on roofing square

Pitch line equal to common rafter depth

Seat cut angle

B

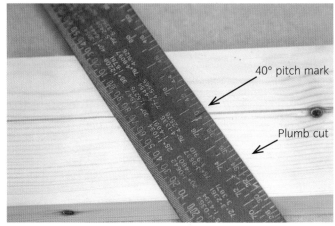

40° pitch mark

Plumb cut

3 Set sliding bevels to the correct plumb and seat cuts.

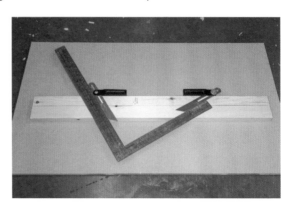

4 Mark the eaves overhang and the birdsmouth. To find out the eaves overhang length, multiply the eaves overhang (300mm) by the required pitch calculation for the hip rafter (1.6444, see photo under Step 5). So 300 × 1.6444 = 493.32mm eaves overhang.

5 Calculate the length of the hip rafter: 1.6444 × 1,800mm = 2,960mm (rounded up). This is to the centre of the ridge board along the pitch line and will require a reduction. The reduction is equal to the (rounded up) diagonal distance from the face of the saddle board to the centre of the ridge board (on plan) if you were to continue the line of the hip rafter. This measurement only works if all other factors with the roof are correct; for example if the hip is exactly 90°, the pitch is exactly 40°, the run is exactly 1.8m and the crown rafter is exactly 40°. Considering all the variables, it is better to take a working measurement from the roof, as outlined in the following steps.

6 Measure across the gap where the hip fits between the top of the crown rafter, saddle board and first common rafter. This mark is the point from which to start the measurement.

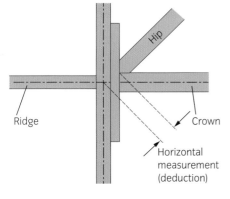

Ridge

Hip

Crown

Horizontal measurement (deduction)

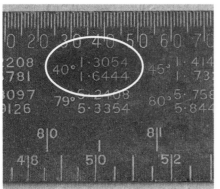

40° pitch measurements per metre run. Top measurement for common rafters. Bottom measurement for hip rafters.

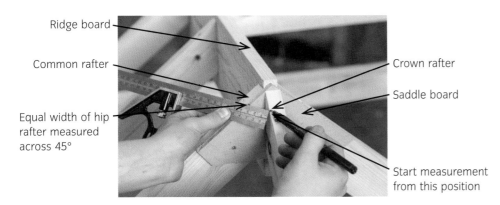

Ridge board

Common rafter

Equal width of hip rafter measured across 45°

Crown rafter

Saddle board

Start measurement from this position

7 After cutting back the wall plate, the overall measurement can be taken.

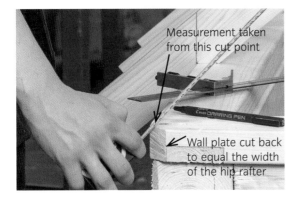

Measurement taken from this cut point

Wall plate cut back to equal the width of the hip rafter

8 Mark the edge cut to the top of the hip rafter for the **compound cut**.

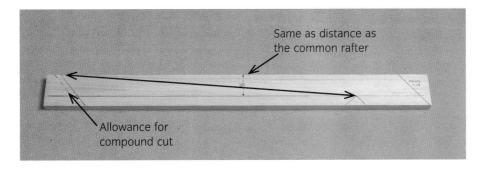

Same as distance as the common rafter

Allowance for compound cut

Compound cut

A cut that consists of two angles: the bevelled angle from the canted saw blade and the mitre angle (or crosscut angle) from the fence

INDUSTRY TIP

In practice you could take a physical measurement from the part-constructed roof using a length of tile lath for the length of the hip rafter.

9 Cut the seat and plumb cuts and fit the hip rafter. The top edge of the hip rafter should line through with the common rafters.

Producing the hip rafter when the dihedral angle is required

STEP 1 Having set the sliding bevel to the dihedral angle, mark this angle on the hip rafter from the centre of the rafter and continue the line down the rafter's length.

STEP 2 Take the vertical pitch line measurement from the common rafter.

STEP 3 Mark the plumb cut position on the hip rafter ensuring you allow enough material for the eaves overhang.

STEP 4 Measure down the plumb line using the measurement taken in Step 2. Secondly, the measurement must start from the dihedral line and will end at the starting point for the seat cut.

STEP 5 Mark off the seat cut.

STEP 6 Measure across the gap produced at the top of the crown rafter, saddle board and first common rafter. This mark is the point from which to start the measurement (refer to labelled image on previous page).

STEP 7 After cutting back the wall plate, the overall measurement can be taken (marked on the photo with an asterisk).

STEP 8 Measure from the seat cut to the dihedral angle line, applying the measurement taken in Step 7.

STEP 9 Mark the edge cut to the top of the hip rafter, forming a compound cut. Plane the dihedral angle down the hip rafter. Cut both the seat cut and compound plumb cut. Fit the hip rafter, ensuring that the dihedral angle follows the horizontal line from the common rafters.

Dihedral angle on hip rafter lining through with common rafters, showing setting out is correct

Valley rafters are produced in a similar manner to hip rafters, but ensuring the valley rafter's birdsmouth is cut to fit into the corner of the wall plate, in order to achieve the best possible solid and secure fit.

RISE AND RUN

The rise and run method is a good way of finding the pitch on an existing roof. The horizontal distance of the roof is measured as a fraction of the vertical distance, typically expressed with the rise first and run second.

Typically the roofing square is used at 'one-tenth' scale, ie if the run is 3,000mm and the rise is 1,500mm, this would be applied to the square as 300mm and 150mm. This is then stepped along the rafter 10 times to achieve the full length. If 300mm is used for the run and the rise equals 300mm, then the pitch of the roof will be 45°.

There are several online systems and smartphone apps available to convert the rise into degrees of pitch.

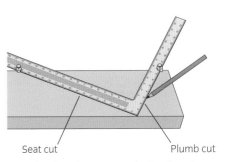

Seat cut   Plumb cut

Transferring the seat and plumb cuts

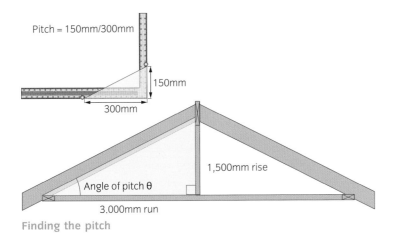

Pitch = 150mm/300mm

150mm

300mm

1,500mm rise

Angle of pitch θ

3,000mm run

Finding the pitch

Rise= 150mm
Run= 300mm

Plumb cut

Slope of roof

Seat cut

150mm

300mm

Rise and run method for producing seat and plumb cuts

READY RECKONERS

Ready reckoner books consist of a series of tables that provide the required angles for the seat and plumb cuts for common rafters, hip rafters, valley rafters, jack rafters and the purlin. Also provided are the diminish lengths for the jack rafters. The 'diminish' is the amount that each jack rafter is shorter than the last as it steps down the hip.

To allow the true lengths of the common and hip/valley rafters to be calculated, also included is a table stating the required length of the rafter per metre run or part thereof. See the example of the chart for a roof with a 40° pitch. Using the example shown, we can work out the true length of the common rafter for the roof.

RISE OF COMMON RAFTER 0.839m **PER METRE OF RUN**								**PITCH** 40°	

	BEVELS:	COMMON RAFTER	– SEAT	40
		" "	– RIDGE	50
		HIP OR VALLEY	– SEAT	30.5
		" "	– RIDGE	59.5
		JACK RAFTER	– EDGE	37.5
		PURLIN	– EDGE	52.5
		"	– SIDE	57.5

JACK RAFTERS	333mm	**CENTRES DECREASE**			435	(in mm to 999 and
	400 "		"	"	522	thereafter in m)
	500 "		"	"	653	
	600 "		"	"	783	

Run of rafter	0.1	0.2	0.3	**0.4**	0.5	0.6	0.7	0.8	0.9	**1.0**
Length of rafter	0.131	0.261	0.392	**0.522**	0.653	0.783	0.914	1.044	1.175	**1.305**
Length of hip	0.164	0.329	0.493	0.658	0.822	0.987	1.151	1.316	1.48	1.644

Extract from a ready reckoner/rafter table for 40° pitch. Measurements in bold are used in the following example

Example

A roof has a span of 4.8m and an eaves overhang of 400mm.

Step 1

Divide the span by 2 to arrive at the run.

4.8m ÷ 2 = 2.4m

Step 2

From the example chart, we can see that for every 1m in the run the rafter length 1.305m. Therefore we multiply the full metre lengths by 1.305m

2 × 1.305 = 2.61m

Step 3

The chart shows that for every 400mm of run the rafter length is 522mm. Now we need to add on the remaining 0.4m of the 2.4m run from Step 1.

2.61m + 0.522m = 3.132m

Step 4

The chart shows that for every 400mm the rafter length is 522mm. Now we need to add the eaves overhang, again 0.4m of the run.

3.132 + 0.522 = **3.654m**

The length of the rafter along the pitch from the edge of the fascia to the centre of the ridge board is 3.654m. Half the thickness of the ridge board must be deducted from this length to arrive at the required cutting length. Using the above information, the rafter can now be set out and cut.

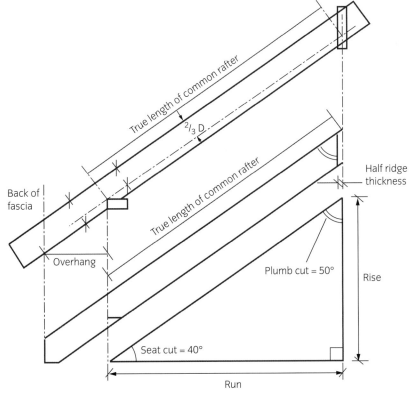

Rise of common rafter

LENGTHENING ROOFING TIMBERS

The method used to join structural timber components along their length is known as 'scarfing'. This technique enables the joint to be produced with a smooth or flush appearance on all its faces. In the case of scarfing joints that are used in roofing (usually found in the purlins or ridge boards), the length of the scarfing joint should be between two-and-a-half and four times the depth of the component being jointed. They incorporate hardwood wedges that are driven into the joint that force together and lock all of the surfaces.

Most scarfing joints are 'fished jointed'. A fished jointed beam also incorporates a steel or timber plate fastened to either side of the scarfing joint and then bolted together.

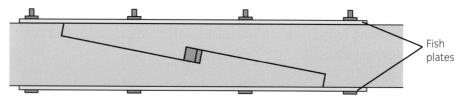

Fish plates

Structural scarfing joint that is fished jointed

ACCESS EQUIPMENT

While working on roof structures you will inevitably be working at height, regardless of whether you are working on new build, refurbishment or alteration, and consideration must be given to safe working practices. A detailed risk assessment and method statement will need producing, paying particular attention to the type of access equipment that is required.

Access equipment used could be a full, independent scaffold that is erected by trained and competent scaffolders (typically needed with full roof construction), platforms/podiums, or a simple mobile tower scaffold to gain access to the eaves area during repair and maintenance.

When selecting the type of access equipment to use, consideration should be given to:

- how long the work is likely to take

- the type of work that is to be done

- how many people will need to use the platform

- the type of access and egress to the platform (such as ladders)

ACTIVITY

Research different methods of preventing falling from height while working on a roof.

- what the ground and weather conditions are like

- whether there are likely to be passers-by/pedestrians in the vicinity

- the risks from dust and failing debris.

Whatever the type of access equipment being used, never adjust or alter it in any way, not even to enable you to work more efficiently. Any adjustments must only be carried out after approval from the installers or other suitably trained personnel, and even then it can only be performed by those who have received suitable training on the erection and adjustment of the access equipment.

For more information on health and safety issues, including risk assessments and access equipment, refer back to Chapter 1.

OUR HOUSE

You have been asked to help select suitable access equipment that would be acceptable for use on our two-storey building. You are required to replace the existing roofing members and have been given a time frame of three weeks. Identify the type of access equipment required, stating your reasons, and outline any variable considerations that could affect the length of time expected to complete the work.

FIXING AND POSITIONING RAFTERS

Before fixing and positioning rafters, ensure that the wall plate has been levelled and is correctly positioned to ensure the span is correct and consistent along its length. The wall plate can be fixed down using restraining straps fixed at centres of no more than 2m apart. The wall plate can now be marked out. Before starting work on any roof, always double check it has suitable working platforms that are correctly guarded and properly installed.

GABLE-ENDED ROOF

Below is a method for marking out the wall plate and fixing the rafters with a gable-ended roof.

1 Mark out the first set of rafter positions, 50mm from each gable wall.

2 Following the roof specification, mark out the remaining rafter positions. These will usually be at 400mm centres.

3 Mark out the ridge board, ensuring sufficient length to allow the ridge to carry on through the gable end to form the part of the gable ladder, if required.

4 Temporarily fit the ceiling joists and board out to act as a working platform.

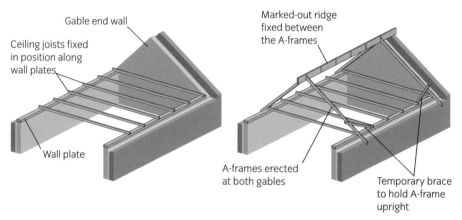

Gable end wall

Ceiling joists fixed in position along wall plates

Wall plate

Marked-out ridge fixed between the A-frames

A-frames erected at both gables

Temporary brace to hold A-frame upright

Gable roof erection

5 Fit the two outer pairs of rafters onto the wall plate. This is usually done by skew nailing through the birdsmouth on each side of the rafter into the wall plate and/or using metal fixing plates attached to the wall plate and through the rafter. Temporarily brace each as you work.

6 Temporarily attach a length of tile lath to the top edge of the ridge board to provide a stop to which the common rafters will sit and to prevent the ridge board from falling while fixing.

7 Push the ridge board up through the gap at the top of the rafters, allowing the plumb cut of the rafters to sit up against the ridge board and tile lath. It usually takes several operatives to achieve this safely and efficiently.

8 Correctly position the ridge board and fix the rafters by skew nailing.

9 If a purlin is required, it is usually positioned now to aid the fixing and positioning of the remaining rafters and help overcome

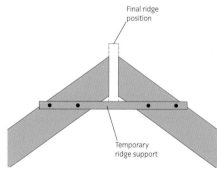

Final ridge position

Temporary ridge support

Temporary ridge support

potential problems with sagging rafters. However, this can only be done if the gable wall has been constructed. The purlins are either built into the gable walls or the wall is corbelled out (see the illustration) to support the purlin.

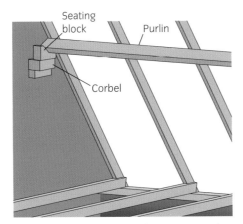

Close-up of the corbelled support to the purlin

10 Position and fix the remaining rafters.

11 Correctly position and fix the ceiling joists against the sides of the rafters.

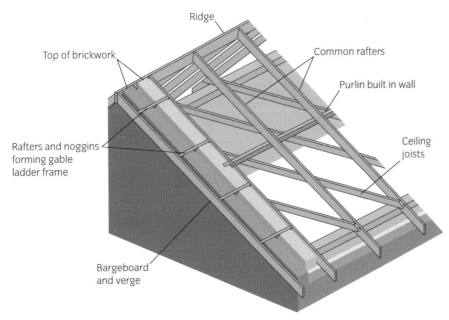

Gable end roof

12 Allow for the gable wall to be finished, if not already done. Fit noggins for the gable ladder.

13 Fit purlins, if not previously done.

14 Fit diagonal struts for the purlins to prevent any deflection.

15 Finish the roof at the verge and eaves with the gable ladder, bargeboard, fascia and soffit, as required.

HIPPED-END ROOF

Below is a method for constructing a hipped-end roof.

1 On a correctly laid and fixed wall plate, mark the position of the crown rafter. The run of the roof will be the centre of the crown rafter.

2 The same measurement as the run is marked up on both sides of the roof. This is the centre position of the first set of common rafters. Repeat on the other hipped end.

3 Mark the position of the remaining rafters to the correct centres as specified, usually 400mm.

4 Fit the first set of rafters as described in Steps 1 to 8 in the previous section on fixing the rafters in a gable-ended roof.

5 Fit the saddle board to the end of the rafters, forming the hip.

6 Fit the crown rafter, remembering to allow for saddle board deduction.

7 Cut the corner of the wall plate at 45° to equal the thickness of the hip rafter.

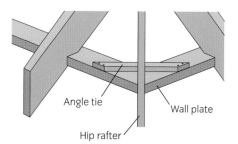

Angle tie

Wall plate

Hip rafter

Angle ties used with a hip rafter

8 Fit any angle ties across the corner of the wall plate as required. These provide extra supporting surfaces for the hip rafter and helps to prevent the wall plates spreading.

9 Check the length of the hip rafter before cutting. If the roof is even only slightly out of square, plumb or pitch, the hip rafter length could alter considerably.

10 Cut and fit any purlins required, along with strutting.

11 Position and fit the remaining common rafters.

12 Fit the jack rafters, allowing for the diminish in lengths.

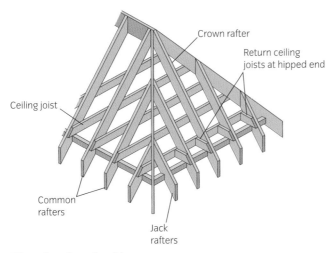

Crown rafter

Return ceiling
joists at hipped end

Ceiling joist

Common
rafters

Jack
rafters

Hipped-end roof gable

13 Correctly position and fit the ceiling joists, including the short return joists to the feet of the jack and crown rafters on the hipped end.

Fitted rafters

> **INDUSTRY TIP**
>
> A 'dragon tie' is a method of providing extra support for the hip rafter at its junction with the wall plate and angle tie. It is often found in traditional building methods and timber-framed construction. See page 117 for an example.

14 Fit any collars and hangers, as required.

15 Finish the roof at the eaves overhang, as required.

TRIMMING ROOFING STRUCTURES FOR OPENINGS

Dormer

An opening in the roof surface where a secondary roof projects away from the plane of the existing roof. Typically used to give greater head room within the primary roof and add light to the roof space/loft

When openings are required in roof structures, for features such as chimneys, roof lights or **dormer** windows, the roof structure requires altering so that it fits around them.

To allow for the increased gap to accommodate these features, the rafters of the roof will require cutting and strengthening. This will involve cutting out parts of the rafter and putting in extra support to carry the weight of the roof over the missing rafters.

TRIMMING AROUND A CHIMNEY

To comply with building regulations, when constructing the timberwork around a chimney, the rafters need to be placed with a 50mm gap between them and the brickwork of the chimney to prevent the passage of damp from the wall to the timber. They also need to be 200mm from the inside face of the flue to prevent the risk of combustion. Additional rafters may also need to be fitted in order to comply with your specification's maximum spacings.

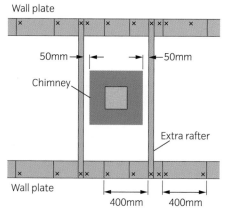

Trimming around a chimney

The names given to the rafters surrounding an opening are:

- *trimming rafter:* the rafter that runs from the ridge board to the wall plate and alongside the opening, with the trimmer rafter fixed to its side

- *trimmer rafter:* the rafter that runs between the two trimming rafters and will have the trimmed rafter fitting up against it

- *trimmed rafter:* the rafter that is shortened at either the plumb cut or the seat cut end and instead fits up against the trimmer.

As both the trimming and trimmed rafter take increased loads due to the gap created by the chimney, each of these rafters will need increasing in thickness and strength. Alternatively each of these rafters is doubled up and bolted together for increased strength.

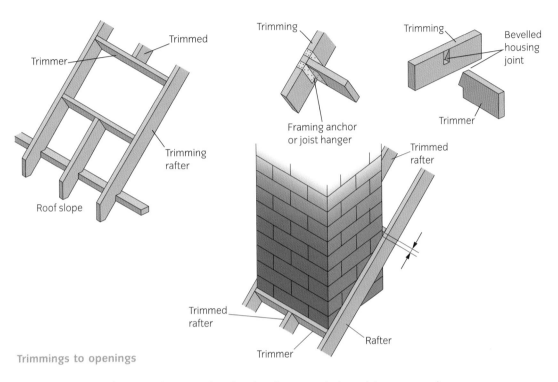

Trimmings to openings

To enable the roof surface to be finished around the chimney and prevent water penetration, a back gutter, front apron and side flashing need to be constructed – see the following illustration.

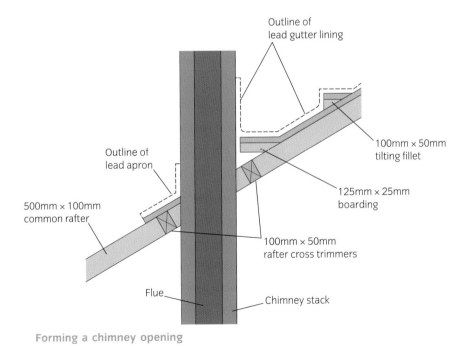

Forming a chimney opening

CONSTRUCTING DORMER WINDOWS AND ROOF LIGHTS

Dormer windows are usually referred to by their roof shape or design, typically falling into one of four categories:

- *flat:* although these are flat along their horizontal plane, they usually have a fall from the main roof

- *segmental or arched:* with a curved roof surface and a horizontal plane running back towards the main roof

- *pitched:* usually having the same pitch as the main roof

- *eyebrow:* one of the most difficult types of roof to construct, consisting of a double-curvature roof line.

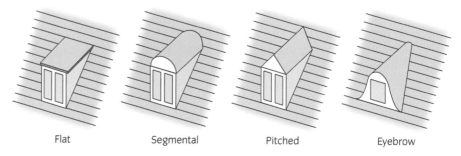

| Flat | Segmental | Pitched | Eyebrow |

Dormer roof styles

TYPICAL FRAMING ARRANGEMENTS FOR DORMER ROOFING

As with chimney openings, double rafters need erecting either side of the dormer opening to carry the increased loads effected by the opening.

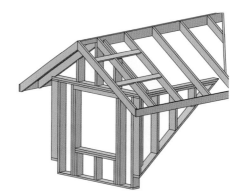

Ptched dormer framing arrangement

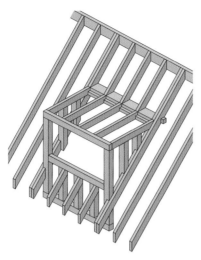

Flat roof dormer construction

Segmental dormer construction

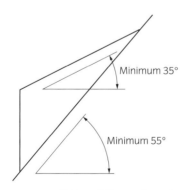

Minimum 35°

Minimum 55°

Minimum roof pitches to ensure sufficient fall to take rain water away without the use of soakers

The dormer cheeks will normally be of timber stud construction following the same principles as stud walling and timber-framed buildings. Suitable section sizes need to be used to allow a sufficient thickness of insulation to be placed between the studs to achieve good thermal insulation.

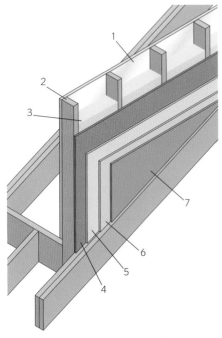

Components in a dormer installation

TYPICAL CONSTRUCTION LAYERS OF A DORMER CHEEK

The following list describes the components that are labelled in the illustration.

1 Internal plasterboard, 15mm thick with a foil back vapour barrier or 2 × 9mm thick boards cross jointed.

2 Timber stud work, typically 50mm × 100mm but larger sections may be required if a specific U-value is required.

3 Insulation of required thickness to achieve the required U-value in accordance with Part L of the Building Regulations. (See page 94 for more about U-values.)

4 Structural **sheathing**, eg exterior grade plywood or OSB board.

5 Breather membrane.

6 External cladding or vertical tiles hung on tile battens.

7 Decorative finish, if required.

Sheathing

The covering of the top of the rafters/roof surface with timber-based material

INDUSTRY TIP

Vapour barriers are placed on the warm side of the wall, while moisture barriers are fixed to the external sheathing on the outer edge.

INSTALLING ROOF WINDOWS INTO EXISTING ROOFS

Often roof windows are fitted into roofs that have already been constructed and finished. The procedure required to fit roof windows in these circumstances is different compared with the procedure used when they are being built into new roofs. In either case, always ensure that suitable working platforms are provided.

The following guide takes you through the installation procedure for roof windows in existing roofs.

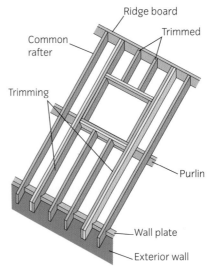

A roof trimmed out to take a roof window

1 Strip back the tiles or slates from an area sufficiently large to uncover around the proposed site of the window.

2 Remove tile battens and felt from the area.

3 Mark out and cut the roof rafters to suit the size of the new roof light, allowing room for trimming out. Temporary supports may be needed if a large part of the roof structure is being removed.

4 Install trimmer and trimmed rafters.

5 Reinforce the existing rafters by installing additional rafters along either side.

6 Install the roof window and secure it in place.

7 Place new felt, ensuring sufficient overlap with the existing felt.

8 Fit a flashing kit according to the manufacturer's instructions.

9 Fit a new tile lath and refit the roof tiles, etc.

PREVENTING HEAT LOSS: INSULATION USED IN DORMER ROOFS

All roof spaces should be insulated to make sure they comply with current Building Regulations Approved Document L. Not only does it make sense from the point of view of comfort, but with increased prices of fuel bills it makes both economic sense and environmental sense to insulate homes.

Insulation is incorporated into the building mainly for thermal purposes, but it can also be used for fire and sound protection. Many insulation materials can be used to achieve more than one of these roles and this is often the factor that determines which is used.

The density and thickness of the material affect its ability to prevent heat loss. A low-density material that traps air within its structure is a good insulator. Typical examples of insulation materials that contain lots of air pockets are:

■ *Rockwool®, rock fibre or mineral wool/fibre:* available as quilted rolls for use in floors, lofts and timber wall construction, as sheets or 'batts' for use in cavity wall insulation, and as loose fill for use in floors, lofts and secondary insulation of cavity walls (by being blown into the cavity of older homes)

■ *polystyrene:* available in sheet form and loose fill, its uses are the same as above and also for between rafters

Mineral wool

■ *polyurethane:* available in large rigid sheets and used for floor and wall insulation, often combined with plasterboard for use in insulating older properties with no cavity in the walls

Sheep wool

Vapour barriers

Breather membrane

- *recycled materials:* these include plastic bags and sheep wool, typically available in rolls for loft insulation and timber wall construction

- *reflective film:* typically foil backing to plasterboard, polystyrene or polyurethane sheets, acting as a heat-reflecting vapour barrier, but also used to encase Rockwool® quilt rolls.

In addition to insulation, the timber construction must incorporate the use of vapour and moisture barriers.

Vapour barriers

Vapour barriers are used on the inside of the dormer cheeks, usually situated between the plasterboard and the insulation. The purpose of a vapour barrier is to help prevent any warm, moist air from passing through the wall from the interior of the building and condensing in the colder areas of the dormer cheek.

Moisture barriers/breather membranes

Moisture barriers or breather membranes are usually a type of lightweight building felt that is both waterproof and breathable. This allows any water vapour to pass through the barrier to the outside. These are fixed on top of any sheathing materials used for the dormer cheeks and also act as a secondary line of defence from water penetration.

EAVES FINISHING

The projecting foot of the rafter after they join the wall at the birdsmouth is known as the 'eaves overhang'. The rafters are usually finished with fascia board and soffit, providing a closure of the roof. Eaves can be finished in four different ways:

- flush eaves

- open eaves

- closed eaves

- sprocketed eaves.

FLUSH EAVES

With flush eaves, the end of the rafters are cut just past the outer walling by about 10 to 15mm. The rafter end then has a fascia board attached which in turn carries the guttering. The gap between the wall and the back edge of the fascia board creates a ventilation gap for the roof.

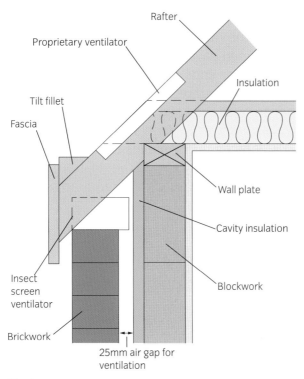

Flush eaves

OPEN EAVES

With open eaves, the rafter ends project past the outer wall to provide extra weather protection. The rafter ends are usually boarded on their top edge to provide a decorative finish when viewed from ground level. Fascia board, to which guttering can be attached, is then fixed to the rafter ends.

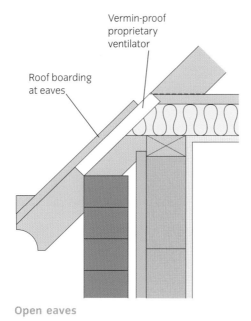

Open eaves

CLOSED EAVES

With closed eaves, the rafter ends project past the outer wall in the same way as with open eaves but, instead of boarding the top edge of the rafters, the rafter ends are enclosed with fascia and soffit. To hold the soffit in place, brackets are fixed to the sides of the rafters and used either to push the soffit down onto the outer wall or hold it flush against the outer wall. Ventilation needs to be provided in the soffit; this is done either by means of a continuous ventilation grill fixed into the soffit, or by circular ventilation grills fixed at intervals along the soffit to provide the required amount of ventilation.

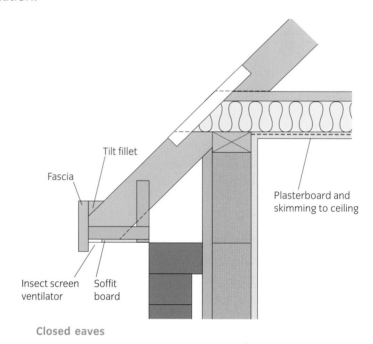

Tilt fillet

Fascia

Plasterboard and skimming to ceiling

Insect screen ventilator

Soffit board

Closed eaves

SPROCKETED EAVES

On steeply pitched roofs, the slope of the roof is reduced at the eaves; if it is not reduced, rain water can have a tendency to overshoot the guttering. To enable this reducing of the roof's slope, a sprocket (short rafter) is fixed to the sides of the rafters at the eaves. This then provides the fixings for the fascia and soffit, as with the closed eaves finish.

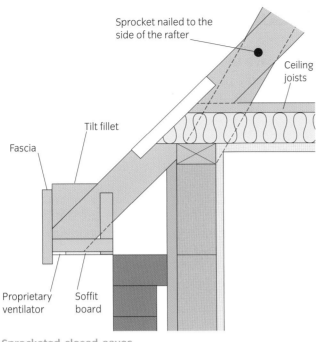

Sprocket nailed to the side of the rafter

Ceiling joists

Tilt fillet

Fascia

Proprietary ventilator

Soffit board

Sprocketed closed eaves

HOW TO FIX TURNING STAIRCASES

Stairs vary considerably in their design: from simple straight flights typically found in domestic dwellings (either built in between walls or with one side up against a wall with newel posts and balustrade), to more elaborate turning stairs featuring in more expensive domestic dwellings and public buildings.

The design, manufacture and installation of stairs is governed by the Building Regulations Part K. Fitting and fixing stairs can be one of the most difficult tasks involved in first fixing operations because of the number of components that make up a staircase.

TYPES OF TURNING STAIRCASE

Types of stairs are discussed in Chapter 7, pages 360–361.

COMPONENT PARTS ASSOCIATED WITH TURNING STAIRCASES

Parts of staircases are covered in Chapter 7, pages 365–369.

ACTIVITY

Using Part K of the Building Regulations, find the minimum going of tread that turning staircases can have at the narrow end.

Answer: 50mm.

INDUSTRY TIP

One type of winder is the 'kite winder', which is a trapezium-shaped tread like a kite.

STAIR REGULATIONS

Site carpenters have to identify and understand the building regulations that apply to stairs. The construction and design of stairs is controlled under the Building Regulations Approved Document (AD) Part K. The categories of staircases covered under the regulations are:

- *private*: flights that are installed within one dwelling (for example a two-storey house)

- *general access*: flights installed where a substantial number of people will gather (for example a public building)

- *utility*: flights in all other buildings; they are also known as 'common' stairs as they are used by more than one dwelling, for example flats).

Different requirements are necessary for the different categories of stairs. The following table outlines some of the requirements that apply to them.

	Private stairs	General access stairs	Utility stairs
Minimum rise	150mm	150mm	150mm
Maximum rise	220mm	170mm	190mm
Minimum going	220mm	250mm	250mm
Maximum going	300mm	400mm	400mm
Application of rise and going	For dwellings, external tapered steps and stairs that are part of the building, the going of each step should be a minimum of 280mm. The maximum pitch for a private stair is 42°. For school buildings, the preferred going is 280mm and the rise is 150mm. For existing buildings, the dimensions shown above should be followed unless, due to the sizes on site, they are not possible. An alternative method should be discussed with the Local Authority's building control department, with particular attention paid to disabled access (Approved Document Part M).		
Handrail heights for flights and landings	Between 900mm and 1,100mm, measured from pitch line or floor. Handrails can form the top of the guarding if the heights can be matched. If the stairs are 1,000mm or wider then a handrail must be provided on both sides. If buildings other than dwellings, then continuous handrails must always be provided on both sides of flights and landings.		

	Private stairs	General access stairs	Utility stairs
Guarding of stairs – in the form of wall, screen or balustrade	When there is a drop of more than 600mm. Guarding height for landings = between 900mm and 1,100mm.	When there are more than two risers. Guarding height for landings = 1,100mm.	When there are more than two risers. Guarding height for landings = 1,100mm.
	Guarding should be provided in all buildings. In buildings likely to be used by children under 5 years old, the guarding should be designed so that a 100mm sphere cannot pass through any opening in the guarding. Children should not readily be able to climb the guarding.		
Minimum headroom height	2m	2m	2m
Widths of stairs	No minimum width unless it is used as a fire escape or for disabled access, when it must be at least 1m wide.		
Landings Drop newel post fitted at quarter landing	Landings need to be provided at the top and bottom of every flight of stairs, with the width of the landing being at least as wide as the narrowest flight. Each landing should be clear of permanent obstructions. A door can open onto a landing at the bottom of a flight provided it will leave a clear space of at least 400mm across the full width of the flight. Relation between doors and stairs on landings		
Tapered treads	To comply with the required going for tapered staircases of less than 1m wide, measure in the middle of the staircase. For staircases exceeding 1m in width, measure 270mm from the side. The going of tapered treads should measure at least 50mm at the narrow end.		

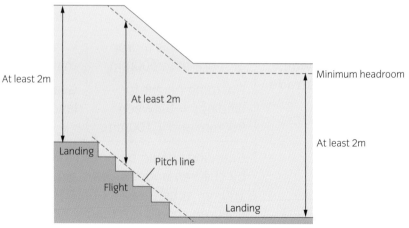

Rise and going of a flight of stairs

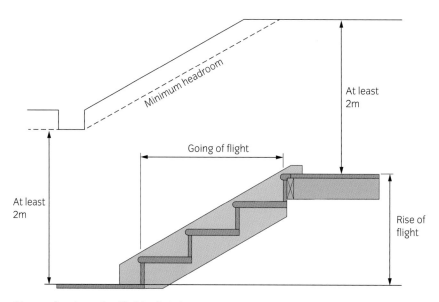

Rise and going of a flight of stairs

Building Regulations Approved Document Part K states that steps for straight flights should have the same rise and going, and that twice the rise plus the going (2R + G) should be between 550mm and 700mm.

The minimum headroom height can be relaxed in situations such as loft conversions, where a clear headroom height can be deemed to be satisfactory if it measures 1.9m at the centre of the stair width, reducing to 1.8m at the side of the stair. Part K also states that all flights of stairs should have a handrail on at least one side if they are less than 1m wide, but on both sides if they are wider.

Although there are no recommendations given by the building regulations for minimum stair widths for dwellings, off-the-shelf flights are available at 860mm widths for domestic installations.

For flights under 1m wide the going is measured along the centre of the flight, while a minimum of 50mm going is required at the narrowest side of the tread.

DESIGN PRINCIPLES ASSOCIATED WITH STAIRS TURNING THROUGH 90°

To enable a staircase to turn through 90°, it can be built:

- using a landing known as a 'quarter landing'

- with tapered or winding treads

- as a **geometrical staircase**.

As the name implies, this type of design allows the staircase to turn through 90° at any position through its rise, turning either to the right or left as required. The newel post used for turning stairs may terminate at landing level, carry on through and act as a dropped newel, or rest on the finished ground level.

Geometrical staircase

A staircase that turns and rises without the use of landings – the stair strings are formed to rise and turn and are also known as 'wreathed strings'

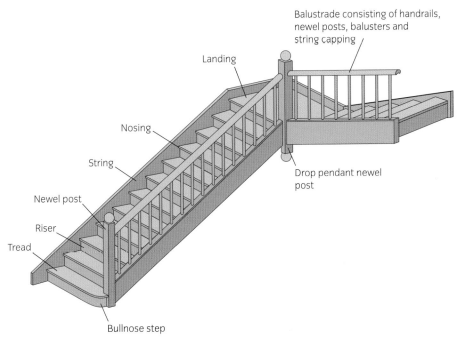

Drop newel post fitted at quarter landing

LANDINGS

Possibly the most simple method to adopt for turning staircases, this consists of two sets of straight flights of the required lengths with the construction of a landing at the required height on which to fit each flight. The landing acts as the ground floor level for the upper flight and the bulkhead for the lower flight. Each flight must have the same rise and going but can have a different number of steps.

Remember when constructing landings to ensure that you have at least 2,000mm head room when using the stairs in order to comply with the building regulations. And if a doorway opens onto a landing at the bottom of a flight of stairs then you must have an unobstructed area measuring at least 400mm square to comply with the building regulations.

Aside from that, the design and construction of the landing will depend on which type of newel post is to be used and usually falls into one of the following three methods. Newel posts need to be securely fixed to any landing. This usually involves notching the newel around the landing joists and either bolting or screwing the newel to the trimmer.

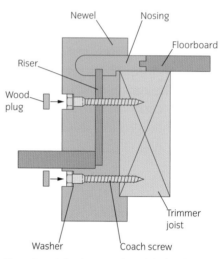

Newel notched around and bolted to trimmer joist

Newel post terminating at quarter landing floor level

This is where the landing is constructed without the newel passing through the floor. It is typically built with the trimmer built into or onto two walls, the materials used being similar to those used in the construction of the upper floor, or with the use of a support framework that can be boxed in or turned into a cupboard, known as a 'spandrel framing'.

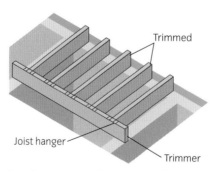

Landing support from two walls

Framework below stairs for a quarter landing

Newel post passing through the quarter landing but not continuing to the ground floor

This is where the newel post passes through the quarter landing floor and terminates just below the ceiling level, usually forming a decorative feature in the ceiling. It is known as a 'drop pendant newel'.

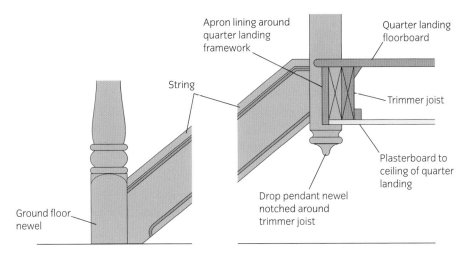

Apron lining around quarter landing framework

Quarter landing floorboard

String

Quarter landing floorboard

Trimmer joist

Plasterboard to ceiling of quarter landing

Drop pendant newel notched around trimmer joist

Ground floor newel

Drop pendant newel

Newel post terminating at ground floor level

This is where the newel post continues through the quarter landing, terminating at ground floor level and acting as part of the structural support for the staircase.

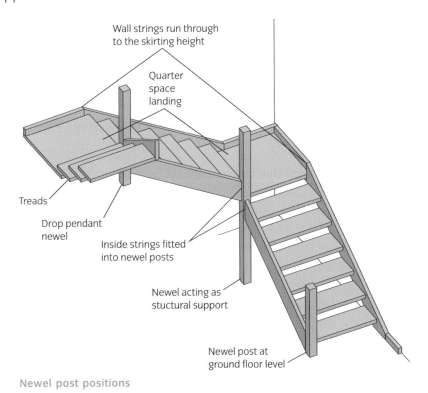

Wall strings run through to the skirting height

Quarter space landing

Treads

Drop pendant newel

Inside strings fitted into newel posts

Newel acting as stuctural support

Newel post at ground floor level

Newel post positions

DECORATIVE STEPS

More elegant staircases are finished with some form of decorative step, rather than the more common straight riser and tread. These fall into three main design classifications, these being bullnose, curtail and angled/splayed. These are discussed in detail in Chapter 7, pages 377–380.

FIXING A QUARTER-TURN FLIGHT OF STAIRS WITH WINDER TREADS

INDUSTRY TIP

In some cases, if the staircase is constructed from hardwood or was to receive a clear finish, it would be fixed after plastering and as late as possible. This is to reduce any likelihood of damage.

When the flight of stairs is delivered to site it usually arrives in its completed form as far as possible. But for ease of handling and access, its newels, winder treads, handrail, balustrade, the shaped bottom step, nosing and the top riser are usually supplied loose, ready for on-site completion.

It is usual to fix the stairs just before plastering is to take place, allowing the staircase to be fixed to the bare wall. The staircase then requires protection from all the other trades that will use the flight during first and second-fix stages. This can be achieved by covering each step with plywood or hardboard, and by covering the strings, handrailing and balustrade with cardboard.

METHOD FOR FIXING A QUARTER-TURN STAIRCASE

1 Before carrying out any installation of a flight of stairs, make sure a suitable risk assessment has been carried out. Always ensure you comply with all the requirements of the risk assessment. This should include making sure you meet relevant regulations, such as manual handling regulations, and wear appropriate PPE (see Chapter 1, pages 20–23).

2 Check that the total rise between floor levels is correct and that the floor surfaces are level. Use the specification to check the height of the skirting that is to be used, and check that the upper flooring material is the same thickness as the thickness of the treads – it may be necessary to rebate the underside of the top nosing to fit over the trimmer and ensure that the finished floor level and the level of the top tread are the same. Clear the stairwell and ensure that other trades do not try to access the upper floor via the stairwell until it is safe to do so.

3 Carefully lift the flight and position it with the wall string placed on saw benches. It is normal for the wall string to be left 'long' to protect it from damage and also to allow for site conditions (eg incorrect levels or any other problem that might present itself during the fixing operations).

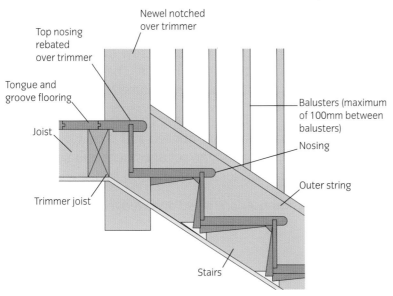

Stair components positioned at trimmer

4 To fit the staircase you will first need to measure your joist thickness. Mark this thickness on your newel post to allow it to hang on your trimmer joist. To achieve the correct horizontal measurement, your wall string and newel post will require cutting right back to the riser, underneath the nosing. The top riser will be held in place with glue and screws.

5 The newel posts will have been dry fitted in the workshop and should go straight on. Once you are happy that the newel fits correctly, apply glue to the tenon and dowel the newel into place. The newels will be draw bored onto the strings and will pull the newel into place. Now chisel the tops of your dowels off. Do not forget to fit the handrails at this time, if they are tenoned into the newel posts.

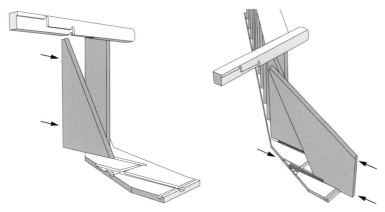

Positioning the first two kite winders

6 Once your newel post is on, you can slide the winder tread into place, starting with the top winder tread. Make sure the winders are fully down in the string. You can screw through the wall string into the treads to pull them home. Then drive home the glued wedges. Fit the riser and drive home the glued wedges, then screw the riser to the tread.

7 The wall string has its seat cut applied and is now glued and screwed together, ensuring correct alignment and square. This joint is usually a half lap joint or a grooved housing joint.

8 The last winder and riser are glued and wedged into place as before.

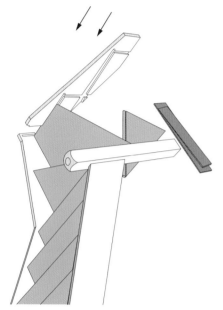

Positioning the lower string onto the winder tread and against the other wall string

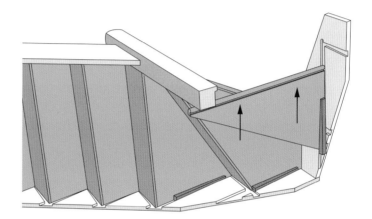

Fitting final winder tread into newel and lower wall string

9 Apply generous amounts of glue on the glue blocks and push them into place. You can pin these into the treads while the glue sets.

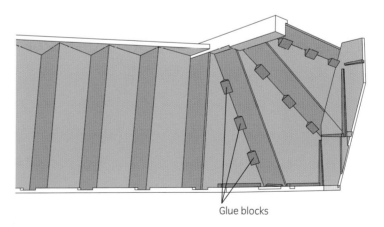

Glue blocks

Positioning glue blocks to underside of winder steps

10 If you have a bullnose or curtail step, you need to glue and screw it to the newel post as well as wedging it into the string trenching. Before you fix this, make sure the newel post is square to the tread.

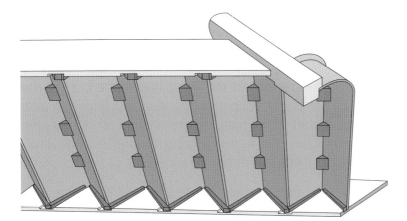

Bullnose step positioned into newel

11 You can now mark and cut the skirting board position.

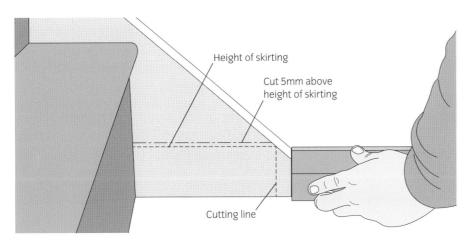

Marking skirting cutting height

12 The flight can now be lifted and offered to the trimmer. This will require a team lift to ensure no one is hurt and to protect the flight from being damaged. A temporary strut can be used to help keep the flight in place. Check the treads for level with a boat level, and ensure the flight is fully located against the trimmer and landing (the floorboarding should have been cut back to allow for the nosing to fit snugly against it). Check the bottom newel post for plumb.

13 Now drill and screw/bolt through the wall string into the wall, ensuring you apply at least five or six fixings through the underside of the string to ensure the screw heads are not seen. You will also need to bolt or screw the top newel post to the trimmer at landing level to ensure a strong fix.

INDUSTRY TIP

If using contemporary handrail fixings, ensure the handrail is fitted between 900mm and 1,100mm above the pitch line in accordance with the building regulations.

FUNCTIONAL SKILLS

Calculate how many balusters are required for a landing that has a clear gap from newel post to newel post of 2,135mm, each baluster being 32mm square. You are required to comply with Part K of the Building Regulations.

Work on this activity can support FM2 (C2.2).

Answer: 16 balusters.

INDUSTRY TIP

Each tread will usually have two balusters, one just back from the nosing and the other positioned mid-way between this position and its equal on the next step, giving two balusters per step.

14 The handrail and balusters can now be fitted, ensuring a gap of 100mm is not exceeded between each baluster in order to comply with the building regulations. It is easy to be a little out with the correct gap so regularly check spacing and the vertical line of the spindles and adjust as required. The balusters fit in a groove in the underside of the handrail and the groove in the top of the string capping.

15 Starting at the bottom, fit the first baluster the correct horizontal distance from the bottom newel, ensuring it is plumb and correctly seated in the handrail groove and string capping. Pin and glue the baluster into the string capping and handrail, ensuring the fixing will not show.

16 Now the infill spacer, which has been cut at the same angle as the pitch of the stairs on one end and can be square on the other, is fitted. The square end is located against the baluster and the angle cut the correct way up to receive the next baluster. Glue and pin both the top and bottom spacers in the groove in the handrail and capping, using veneer pins.

17 Repeat this process, constantly checking the balusters to ensure accuracy is maintained. It only takes 0.5mm difference on a raking cut on the infill strips to make the spacing wrong. This then will accumulate as the work proceeds.

18 Any landing adjacent to the flight of stairs will also require handrailing, nosing, capping and balusters. Use the same method for the landing as with the stair balustrade.

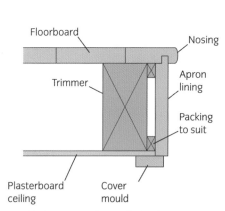

Vertical section through landing

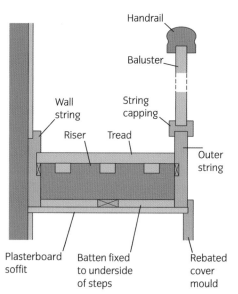

Vertical section through stairs

19 Complete the opening with an apron lining to complement the stair material.

20 To finish the handrail against the wall, it is good practice to use a half-newel.

OUR HOUSE

You have been asked for suggestions on changing the balustrade in 'Our House' from square spindles to turned spindles, thereby creating a more elegant staircase. What considerations and equipment would you require to achieve this? State the particular requirement of the building regulations in reference to handrail heights and baluster positioning, and list the stages involved in completing this task.

FUNCTIONAL SKILLS

Calculate the rise of each step for a staircase with a total rise of 2,600mm. Suggest a suitable going to complete the step, ensuring you comply with the current Building Regulations Approved Document K.

Work on this activity can support FM2 (C2.5).

Case Study: Daniel

Daniel has been asked by his supervisor to carry out two tasks on a local building project. The first task is to construct a cut roof. Daniel has been asked how the edge of the roof is to be finished and whether anything can be done to help stop the rain missing the guttering during heavy rain storms.

- What type of eaves finish would you recommend and why?

- Make a sketch of and label the finished eaves, detailing with particular emphasis placed on reducing overspill of the guttering.

The second task refers to options about the different types of turning staircases available.

- Describe two types of turning staircase with the advantages and disadvantages of each one compared with the more common straight flight staircase.

- List the safety concerns that should be considered when installing turning staircases.

Work through the following questions to check your learning.

1 What part of the Building Regulations covers roofing?

   a Part L.

   b Part K.

   c Part A.

   d Part D.

2 Which is **not** a specific requirement for setting out rafter lengths?

   a Type of eaves overhang.

   b Height of wall plate.

   c Pitch of roof.

   d Span.

3 On which one of the following rafters would a dihedral angle be found?

   a Hip.

   b Jack.

   c Cripple.

   d Crown.

4 The purpose of an angle tie in a roof is to provide additional support for the

   a hip rafter

   b purlin

   c valley rafter

   d ridge board.

5 The verge is the part of the roof

   a under the eaves

   b along the hip

   c along the valley

   d at the gable end.

6 What is the standard abbreviation for the hip rafter's true length?

   a HRTL.

   b CRTL.

   c TLHR.

   d THRL.

7 The type of joint used to lengthen a ridge board is called a

   a dovetail joint

   b scarfing joint

   c bridle joint

   d housing joint.

8 A lay board would be found

   a along the hip rafter

   b along the valley of a dormer window

   c next to the restraining straps

   d under the purlin.

9 The vapour barrier in a dormer cheek would be found on the

   a cold side of the wall

   b side of the outer skin of the wall and the insulation

   c warm side of the wall behind the plasterboard

   d warm side of the wall before the plasterboard.

10 The current building regulations require roof insulation to be a **minimum** thickness of

   a 350mm

   b 270mm

   c 200mm

   d 150mm.

11 An apron lining would be found

   a   attached to the bullnosed step

   b   sitting on top of the stair string

   c   fitted between the handrail and the string capping

   d   fitted around the floor joists.

12 The term 'bulkhead' refers to the

   a   joists forming the landing on which the strings sit

   b   collection of balusters

   c   combination of one tread and one riser

   d   joint used to attach the newel post.

13 The continuous strings used to turn a staircase **without** the need for newel posts are called

   a   dog-leg

   b   quarter

   c   wreathed

   d   winder.

14 For domestic stairs, Building Regulations Approved Document Part K specifies that the rise and the going should fall between

   a   550–750mm

   b   500–750mm

   c   550–700mm

   d   550–850mm.

15 According to the building regulations, a domestic staircase would be allowed a **maximum** pitch of

   a   42°

   b   41°

   c   40°

   d   39°.

16 The **minimum** width for a tapered tread is

   a   25mm

   b   50mm

   c   55mm

   d   75mm.

17 Winder treads are found where the

   a   staircase turns and rises at the same time

   b   treads join the landing

   c   strings are cut to fit along the trimmer joist

   d   bottom tread joins the wall string.

18 The shaped end of a bullnose step is

   a   quarter round

   b   tapered

   c   squared

   d   semi-circular.

19 In order to comply with building regulations, what is the **largest** sphere size allowed to pass between balusters?

   a   70mm.

   b   80mm.

   c   90mm.

   d   100mm.

20 For domestic staircases the building regulations state that twice the rise plus the going (2R + G) should be between

   a   500mm and 700mm

   b   550mm and 750mm

   c   500mm and 750mm

   d   550mm and 700mm.

Chapter 4
Carry out second fixing double doors and mouldings

Installing double doors and mouldings can greatly enhance any room.

Double doors allow the movement of larger numbers of people through an opening without the need to construct a single door out of over-sized materials. The use of double doors is common practice in shops and public buildings, as well as for industrial buildings and garages.

The skills and knowledge required to install double doors are the same as those required for fixing single doors, but a greater emphasis is required on the setting and marking out skills, fitting the frames, door hanging skills and ironmongery fitting. Even small discrepancies and fitting faults will be greatly magnified in the installation of double doors.

Installing curved and raking mouldings can test the carpenter's setting out, mitre cutting and skills.

By reading this chapter you will know how to:

1 Install double doors and ironmongery.
2 Install curved and raking mouldings.

DOUBLE DOORS

The main function of double door openings is to provide access to and **egress** from buildings or between rooms. Doors are designed, manufactured and installed to meet performance requirements depending on the situation and location of the door. When choosing which type of double doors to fit, the following considerations should be taken into account:

Egress

A means of exiting a room or building

- door design and construction methods
- weather performance: watertightness and wind resistance
- mechanical performance
- operating forces
- resistance to distortion
- resistance to impact
- performance in use (opening and closing)
- effect of temperature and humidity variation.

These considerations are explored in more depth in the following sections.

PERFORMANCE

Particularly with external doors, greater emphasis is now placed on:

- weather proofing
- security
- thermal performance
- sound insulation.

While the door itself may be designed and constructed to perform well against these criteria, it relies on the correct fitting of the door to the frame, along with its associated ironmongery, in order to achieve the best performance.

External doors are usually fitted with a 3mm clearance around all edges to allow moisture movement. Weather seals should be positioned so that where possible they are not interrupted or damaged by door locks or hinges. Multi-point locking systems using **espagnolette bolts** help to prevent distortion of the door leaf and provide enhanced security and weather resistance.

Espagnolette bolt

A multi-point locking device that is normally mounted vertically on the locking side of the door

Matchboarded doors are unlikely to meet modern thermal performance requirements for heated buildings. Exposure of end grain should be avoided where possible, with any end grain fully sealed before application of the surface finish.

WEATHER PROTECTION

Weather protection is an important aspect. How exposed are the doors to elements such as rain, snow, wind, direct sunlight or extreme temperature change?

SECURITY

Are the doors intended for internal or external use? Are the materials to be used in the construction of the door, the door's construction method, and its types and quantities of ironmongery all suitable?

SOUND AND THERMAL INSULATION

External doors need to prevent reasonable levels of cold air and penetrating winds from entering the building. A correctly designed door incorporating double glazed units, solid door construction and door seals will help with both sound and thermal insulation.

External doors and frames are included when calculating the thermal performance of the external wall, or taken into account in whole building performance calculations. They do not come under the building regulations relating to the passage of sound from the outside of a building to the inside, but planning restrictions may specify particular acoustic performance requirements on exposed sites. These sites could include those near flight paths, motorways and railway lines.

Manufacturers must be able to provide evidence of the **U-value** achieved by their doors, and these considerations will affect the design and choice of the external door.

TIMBER CHOICE

Timber is **hygroscopic**, so the door will either shrink or expand depending on the temperature and humidity of the surrounding air until it stabilises to **equilibrium moisture content**. It swells when it absorbs moisture and shrinks as it loses it, a response known as 'movement'. Timber species differ in their movement characteristics.

External doors subjected to both internal and external conditions, in addition to seasonal variations, must be designed and fitted to accommodate this movement. Timber that is known to have a 'small movement' is preferred. Poor timber selection can negate good door design. Low-grade timber is not suitable for use in external doors.

Matchboarded doors

A series of timber boards jointed together (usually by a tongue and groove arrangement) in their width

U-value

Also known as 'thermal transmittance', this is the rate of transfer of heat through a structure (in watts) or more correctly through one square metre of a structure divided by the difference in temperature across the structure. It is expressed in watts per metre squared kelvin, or W/m²K

Hygroscopic

Able to absorb or lose moisture

Equilibrium moisture content

The moisture content at which the wood is neither gaining nor losing moisture

The timber used for external doors must be of a high quality to stand up to the weather and offer durability, with good sound and thermal properties. Hardwoods such as oak are extremely durable but also extremely expensive. For this reason, other timbers are used by joinery workshops to manufacture doors. Idigbo, Utile and European redwood are common alternatives.

Large-scale joinery manufacturers make sure they use timber approved by the **Forest Stewardship Council (FSC®)**, indicating that the forests where the timber is sourced are managed responsibly and are sustainable. For increased durability, the timber is often **impregnated** with preservative, often referred to as 'tantalised' or 'double vac' (short for 'vacuumed').

Forest Stewardship Council (FSC®)

An international not-for-profit organisation established in 1993 to promote responsible management of the world's forests

Impregnated timber

Timber that has had preservative forced into its cells, typically in a vacuum tank

PRIVACY

Unglazed doors with one-way door viewers prevent unwanted viewing into rooms or buildings, while still allowing the occupant to see out. Carefully choosing the hanging side of a door can help restrict direct viewing into a room.

A typical door viewer in use

GLAZING REQUIREMENTS

- Does the door require full glazing, as might be the case in shops, or partial glazing to allow **borrowed light** from an area?

- Does the glazing need to meet fire-resisting requirements?

- Is there a requirement to use toughened glass in vulnerable situations?

Borrowed light

Light that is allowed to enter a room or corridor from an adjoining room through a glazed opening, usually above a door

FIRE RESISTANCE

Although all doors give a certain level of fire resistance, only approved fire doors can be used at specifically required fire door locations. Provision for an escape route (as specified in the building regulations) may govern your choice of:

■ door size

■ locks

■ hardware

■ whether the door opens inwards or outwards.

Escape routes impact on your choice of door

ACCESS AND EASE OF OPERATION

When designing entrances to buildings or specifying and fitting doors, you should consider those building regulations that cover access and ease of operation. These will affect your thinking on issues such as:

■ widths of door openings for access by wheelchair users

■ types of thresholds

■ the means by which the door is opened and closed

■ how large and heavy the door will be

■ who is likely to use the door.

All of these will help determine the type of ironmongery selected, and whether any door opening and closing assistance will be required.

APPEARANCE AND SURFACE FINISHES

Is the door intended to make a grand statement while providing all the above functions, or is it simply to be practical and functional?

The surface finishing treatments of external timber doors are not just for decorative purposes but play a vital role in the long-term performance and protection of the door. **High-build exterior stains** and good-quality exterior paints are among the most suitable types of finishes. If finishing on site, apply at least two coats of protective finish to a door as soon as it is delivered or is removed from its wrapping. Leaving a door unprotected for days or weeks means that it will react to its environment more rapidly and to a greater extent than a fully sealed door.

High-build exterior stains

A microporous multi-layer timber coating that resists surface mould/algae and protects against UV light, resulting in a humidity-controlling timber finish

STORAGE AND INSTALLATION

Poor site storage can lead to damage and/or distortion of the doors. Keep any shrink wrapping or other protection on pre-finished doors in place until the doors are ready for installation, and then keep the wrapping in place for as long as possible after installation to protect against wet trades and construction damage.

Store doors flat on a level surface, fully supported along their full length and across their full width, and on at least three bearers. Do not store doors by leaning them against a wall: that would encourage twisting.

If possible, do not install the doors until the building has dried out.

Installation usually involves on-site modification of the doors including letter plate apertures, glazing, door viewers, and locks and hinges. On-site installation and fitting of doors should not be carried out without considering the effect this may have on the performance of the doors. The considerations include:

- leakage of water into the building

- locks and keeps failing to close the door correctly or coinciding with construction joints in the door framing, such as mortice and tenons from mid rails, reducing the strength and stability of the doors

- the absence, or incorrect design and fitting, of weatherboards.

HEALTH AND SAFETY IMPLICATIONS

A suitable risk assessment should be carried out before installation of double doors and mouldings. All relevant safety considerations should be taken into account when producing the risk assessment. Beside specific job-related risks, other safety-related issues that must be considered are covered by the following pieces of legislation:

- Provision and Use of Work Equipment Regulations (PUWER) 1998

- Personal Protective Equipment (PPE) at Work Regulations 1992

- Control of Substances Hazardous to Health (COSHH) Regulations 2002

- Control of Noise at Work Regulations 2005

- Control of Vibration at Work Regulations 2005.

Further information about risk assessments and health and safety information can be found in Chapter 1.

TYPES OF DOUBLE DOORS

Doors are typically divided into five classifications:

1 Matchboarded doors

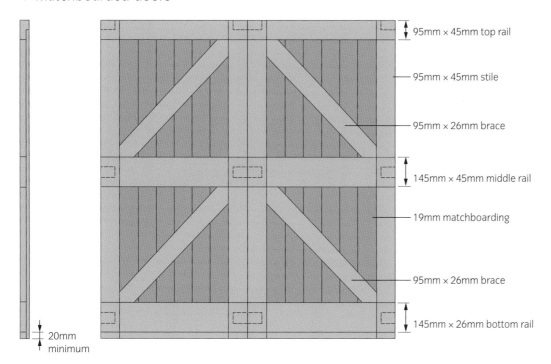

95mm × 45mm top rail

95mm × 45mm stile

95mm × 26mm brace

145mm × 45mm middle rail

19mm matchboarding

95mm × 26mm brace

145mm × 26mm bottom rail

20mm minimum

2 Flush doors

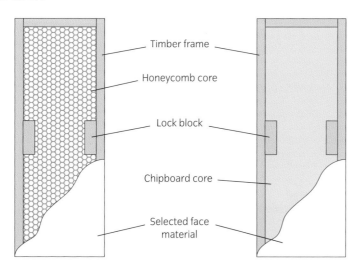

Timber frame

Honeycomb core

Lock block

Chipboard core

Selected face material

3 Panelled doors

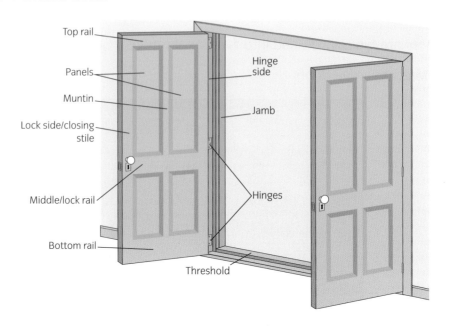

4 Glazed or part-glazed doors

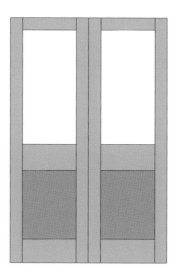

5 Fire doors.

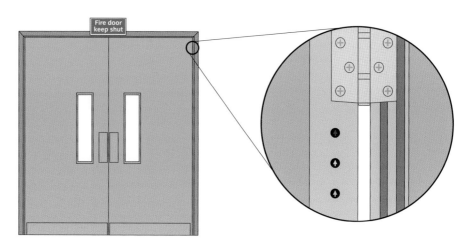

These are then generally further defined as external or internal, or by operation.

EXTERNAL DOORS

External double doors provide access to and egress from buildings. They are designed and constructed from materials that will help prevent the elements from entering the building or having an undue effect on either the door or the materials used. External doors are normally constructed from thicker materials and in ways that make them more secure and stronger than internal doors.

INTERNAL DOORS

Internal double doors are used for access to and egress from rooms within a building and to provide privacy for a room. They are not normally constructed for strength and are typically made from thinner material.

A range of external doors

METHODS OF OPERATION

The method of operation is a way of describing how the double doors are to be used or operated. Doors can be:

- side hung, opening in or out

- swinging

- sliding

- folding

- up and over.

The chosen method will depend on several factors, typically:

- *the location:* whether external, internal or a bathroom

- *the method used to construct the door:* matchboarded, flush or panelled (see later in this chapter for more information on the different types of door)

- *the required performance of the door:* such as high security, easy access, fire resistance or heavy usage.

Once the door's location, construction method and performance requirements are known, the hanging method can be selected.

Modern sliding doors

Master and slave door set: the master door opens on the right

Slave door

With a rebated pair of doors, the door that is bolted shut is called the 'slave door' or 'dead door', while the main opening door leaf is called the 'master'

A door closer

INDUSTRY TIP

Fixing timber frames and linings for single doors was covered in *The City & Guilds Textbook: Level 2 Diploma in Site Carpentry and Bench Joinery*. Refer back to Chapter 7, Unit 208 of that book to review information such as:
- jointing arrangements for door linings
- assembling linings
- fixing to studwork
- fixing to masonry
- built-in frames
- temporary frames
- fixed-in frames.

Where greater access is required, it is usual to fit a wider door frame or lining with two or more doors. In some public buildings, such as hotels, hospitals and schools, unequal pairs of doors are often used to give more accessibility for moving trolleys or equipment, or for when high volumes of people are anticipated. These usually consist of a full-width door and a half-width door. The full-width door (often termed the 'master' or 'live' door) is used for general access purposes, whereas the half-width door (often termed the 'slave door' or 'dead' door) is bolted shut most of the time, allowing the master door to be shut and secured against it.

In public buildings, double doors can be the norm and they are often fitted with power-assisted operating systems that help both open and close the door.

In a lot of cases the doors are designed and hung in such a way that the doors can swing in both directions, allowing easier movement of people or goods. For example, swinging double doors are often installed in hospitals to help with moving trolleys.

With double fire doors, the doors are often fitted with an automatic door release system that enables the doors to remain open when in general use, but the door release system is activated when the fire alarm is activated. The doors then close, fulfilling their function as a smoke and fire barrier.

In most cases it is usual to fit some form of door closer that controls the closing speed of the door. This is particularly important with heavier doors and in locations where young, elderly or infirm people are expected to use the doors. The door closers allow sufficient time for people to pass through the door while ensuring that the doors remain closed when not in use; this can help prevent the passage of sound or heat from one side of the door to the other, thereby allowing easier control of the local environment.

INSTALLING A DOUBLE DOOR FRAME

It is vital when fixing double door frames within openings that you make sure the frame is not twisted, bowed, out of square or distorted in any way. As with the fixing and positioning of single door frames, poor positioning and fixing techniques will affect the hanging and performance of the double doors. But the need for care and accuracy is even more vital with double door frames and linings, as any discrepancies will be magnified when the pair of doors are hung and come to meet at the meeting stiles.

The following guide should ensure that any problems are reduced to a minimum.

1   Position the frame within the opening and temporally pack it out so there is an even gap all the way around. This will hold the frame in place while further positioning adjustments are made.

2   Check the frame is plumb using a plumb bob or a 1.8m spirit level. Square the frame diagonally from corner to corner; the two measurements should be the same. Sight across the frame to ensure it is not twisted. Ensure the frame is not bowed by measuring horizontally across the top, middle and bottom; these measurements should be the same. It is essential the frame is not bowed or distorted in any way or this will affect the correct operation of the doors.

3   Drill through the frame using an HSS drill, making sure that holes are around 100mm from each corner with at least another two evenly spaced down the frame and along the head and cill. If fixing into brickwork the holes should be lined up so the fixing enters the brick and not the mortar course; this ensures good fixings as some mortar joints are not full with voids behind the surface, allowing the plug to pull free.

4   Using a TCT masonry bit, drill into the brickwork or similar masonry to the correct depth. Make sure you drill deep enough to accept the wall plug and then clean out the hole. Remember not to catch the chuck of the drill on the frame. If fixing into studwork or similar timber surrounds, a suitable HSS drill will be required instead of the TCT drill bit.

5   Insert flat packers (or **folding wedges**) either side of the fixing hole and insert the plastic anchor fixing into the hole. Screw it home with suitable screws, making sure that you do not bow or distort the frame when tightening the screw. The packers should be nipped tight. Check the frame width to ensure it has not been over tightened, causing the frame to become bowed.

Folding wedges

A pair of wedges used 'back to back' to create a pair of parallel faces. By sliding one wedge against the other, the total parallel thickness of the folding wedges can be adjusted to pack out the required gap

It is sometimes advisable not to drill and fix the lower fixing on one side of the frame and the cill until the doors have been hung. This will allow for a small amount of movement at the lower end of the frame to compensate for any slight twist within the doors and ensure the meeting stiles meet flush through their entire height. When the doors are hung and correctly operating, these lower fixings can then be made, again ensuring no distortion takes place.

1   Hang the doors with suitable ironmongery as required.

2   Drill and fix the lower remaining fixings for the frame, if not previously done in Steps 4 and 5.

3   Use expanding foam to fill all the gaps around the frame, making sure not to over fill.

4 Cut away the excess foam once it has cured using a suitable knife, and in the case of external frames apply silicon all around the frame. Internal frames will usually require architrave fitting.

As an alternative method, some operatives prefer to fix the frame with expanding foam before drilling and screwing home. Then, when the foam has dried, they proceed from Step 3. However, this alternative method could cause the frame to distort as the foam expands and will not allow for any adjustments.

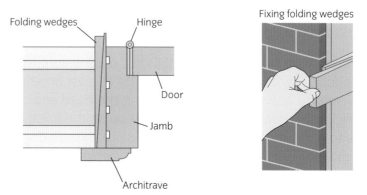

Section through an internal frame showing fitting folding wedges

MATCHBOARDED DOUBLE DOORS

This group of doors involves a relatively simple form of construction. They are suitable for internal and external use, but are commonly used for sheds, gates and older period properties, as well as industrial and agricultural buildings. These types of double doors are typically side hung but larger industrial doors are often sliding.

LEDGED DOORS

The simplest form of this type of door is the ledged and matchboarded door. This consists of a set of (usually three) horizontal ledges fixed to the back of the vertical matchboarding, with one fixed at each end and one centrally. Traditionally, boards are then nailed onto the ledges through the face using lost head or oval nails that are then 'clenched' over on the back to prevent the boards from pulling away. Modern practice usually involves screws being used to secure the ledges to the boards instead, which are then counter-bored and filled with timber pellets to conceal the fixings.

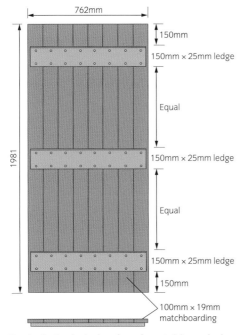

Typical single ledged and matchboarded door

Cutting the timber pellet flush

LEDGED AND BRACED DOORS

Ledged and braced doors are constructed in the same way as ledged and matchboarded doors, but have diagonal braces added to prevent the door from dropping. Bracing must be fitted in the correct direction to support the side of the door opposite the hinges; this means pointing upwards, away from the hinge side. If they are fitted wrongly, the door can drop and fail to function properly. Braces are fitted as close as possible to 45° to the ledges; if they are fitted below this angle, they will not give the door adequate support.

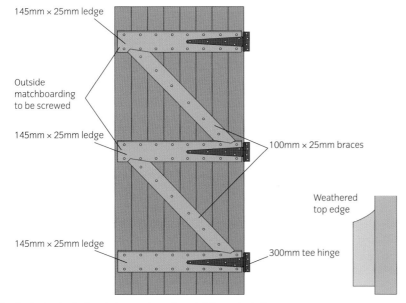

Typical single ledged, braced and matchboarded door

FRAMED, LEDGED AND BRACED

Framed, ledged and braced (FLB) doors are an improvement on ledged and braced doors as the matchboarding is fixed into a framed door, increasing the door's strength. These are jointed with mortice and tenon joints. The matchboards are either tongued into a groove that runs around the frame or more typically fitted into a rebate that runs around the frame.

To allow the matchboards to pass over the face of the middle and bottom ledges, the ledges are reduced in thickness. This means the jointing arrangement for the middle and bottom ledge requires the use of a **barefaced tenon**.

Barefaced tenon

A tenon that is shouldered on only one face of the joint

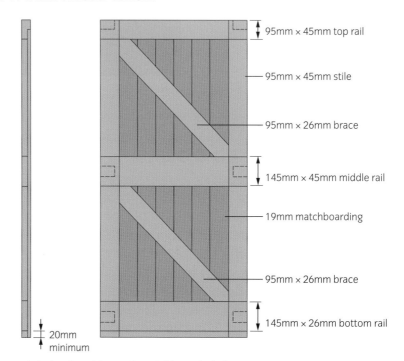

- 95mm × 45mm top rail
- 95mm × 45mm stile
- 95mm × 26mm brace
- 145mm × 45mm middle rail
- 19mm matchboarding
- 95mm × 26mm brace
- 145mm × 26mm bottom rail
- 20mm minimum

A framed, ledged and braced matchboarded door

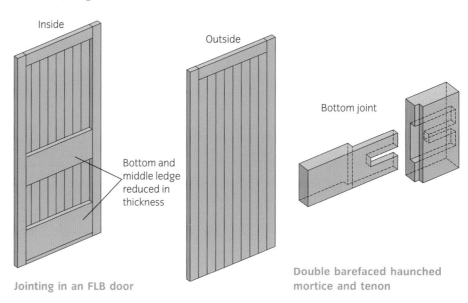

Inside

Outside

Bottom and middle ledge reduced in thickness

Bottom joint

Jointing in an FLB door

Double barefaced haunched mortice and tenon

NON-STANDARD MATCHBOARDED DOORS

Matchboarded doors are manufactured to suit many types of opening sizes, as well as for internal and external use. Typical locations in which matchboarded double doors are found include:

- domestic garages

- industrial garaging

- industrial locations (workshops, factories and warehousing).

Matchboarded garage doors

These oversized doors typically come as double door sets that are either sliding, hinged or a combination of both.

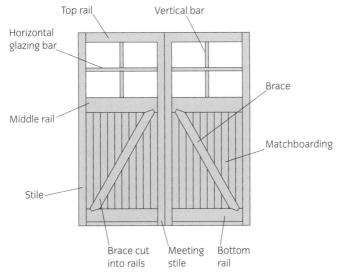

Double FLB part-glazed matchboarded doors (inside view)

Double matchboarded double door sets usually have the braces fitted, whereas single doors usually supply the brace loose to be fitted on site (which allows the single doors to be used for either hand). Larger door sets may have four braces, all pointing from the outer corners to the centre of the middle rail. This not only stiffens up the door but also allows more fixing points for the boarding and ensures an angle as near to 45° as possible is maintained for the braces, thereby reducing the possibility of the door dropping on the side opposite the hinges.

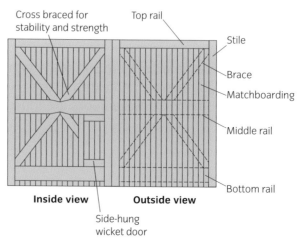

A pair of sliding industrial-type matchboarded doors with wicket door

Weathering

A means of preventing water from gaining entry

Where double doors meet in the middle, some form of **weathering** is required. The meeting stiles are rebated over each other to allow the doors to finish flush. It is vital that the rebate is applied to the correct side of the door, thus ensuring that the correct door is allowed to open first. An alternative method is to fit a planted cover bead to the door, again making sure you fit it to the door that is to open first.

ACTIVITY

Draw and label a full-size section through the meeting stiles of a pair of bead and butt framed, ledge and braced doors.

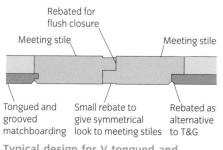

Typical design for V tongued and grooved matchboarding

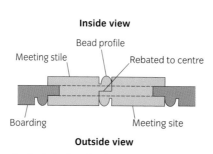

Typical design for bead and butt matchboarding

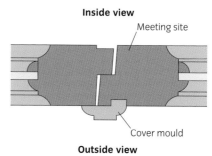

Combination of rebate to the inside and a cover mould bead to the outside of glazed doors

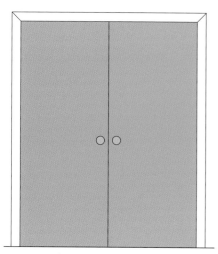

Honeycomb flush doors

FLUSH DOUBLE DOORS

Flush double doors are constructed with either a hollow or a solid core, faced with sheets of hardboard, plywood or moulded plastic. Joinery workshops rarely manufacture flush doors; they are usually produced on a mass-production scale by larger joinery manufacturers. They are lightweight, cheap and simple to manufacture. The hollow core variety is the most common: a stapled softwood frame in-filled with (usually) cardboard, set in a honeycomb or lattice pattern that is then covered with the facing material (see illustration on page 175).

Because they are hollow, a section of solid timber called a 'lock block' is sandwiched into the core of both sides of these doors to enable a lock or latch and the handles to be securely fixed. The position of the lock blocks are clearly identified by the manufacturer within the instructions that accompany the door, or more commonly a key symbol or the word 'lock' is printed on the top of the door on the side that contains the lock block.

Lock block identifed by manufacturer

Moulded panel doors are a variation on flush doors, using a hollow door construction with the application of facings that are pressed or moulded to give the impression of a traditional panelled door. They often have a textured grain finish to simulate real timber.

Flush doors can be designed to incorporate vision panels. These provide light and a view through to the other side. Apertures cut into hollow doors require additional framing to accommodate the glazing and the glazing beads. Below are examples of typical profiles that could appear around glazed areas within doors to hold the glass in.

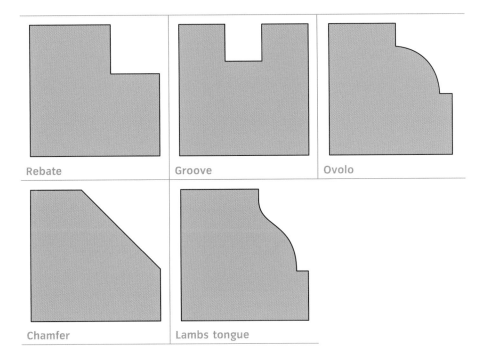

Rebate   Groove   Ovolo

Chamfer   Lambs tongue

Rebated glazing beads (called 'bolection beads') are the best way to hold the glass in place. They can be secured with countersunk screws and recessed cups to the main framework, allowing easy removal if the glass needs replacing, or simply nailed in place.

These types of doors are typically side hung but can be sliding or folding, particularly where rooms are divided up and where there is restricted space for doors to swing.

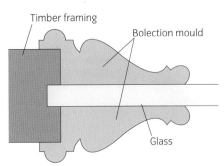

Timber framing

Bolection mould

Glass

Profile of beading holding glass in place

PANELLED AND GLAZED DOUBLE DOORS

Panelled and glazed double doors have a similar construction technique. Both consist of a frame that has a groove or rebate around its inside edges to receive panels or glazing. The framing members for these doors vary in the number and arrangement of the panels. The door construction consists of horizontal members (rails) and vertical members (stiles or muntins). Rails are named according to their position in the door, such as 'top rail', 'intermediate rail', 'middle rail' and 'bottom rail'. When an optional upper intermediate rail (positioned below the top rail) is used, it is often referred to as the 'frieze rail'. The two vertical members on the outside of the door are called stiles, while all intermediate vertical members in panelled doors are called muntins.

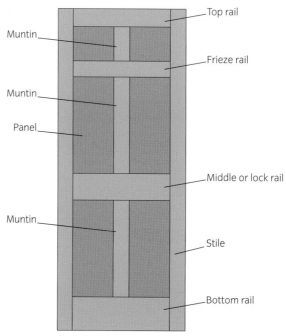

Panelled door components

French doors

Doors that are fully glazed or contain glazing bars, typically known as 'French doors', are very popular in domestic housing and are usually situated in a rear living room, leading on to the garden. They have an advantage over solid doors in that they are able to flood the room with light and bring the garden into the room. Because they are mainly used at the rear of the property, privacy tends not to be a problem. French doors are also widely used to divide up larger rooms without obscuring the view into the other room.

JOINTS

In good construction, panelled and glazed doors are usually constructed using mortice and tenon joints, while in poorer, mass-produced construction they are dowel jointed.

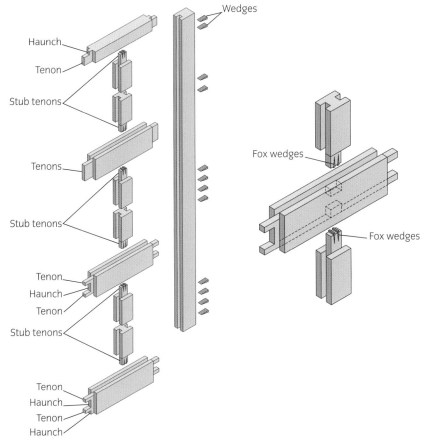

Panelled door mortice and tenon joints

Glazed doors may have glazing bars to replace the middle and intermediate rails and the muntins. In the case of half-glazed doors with panels to the bottom of the door, the middle rail should be constructed using a diminished shoulder construction, often referred to as 'gunstock shoulders'. This form of construction involves a considerable amount of time and difficulty compared with parallel shoulder construction; as a result this type of joint is seldom used on cheaper doors.

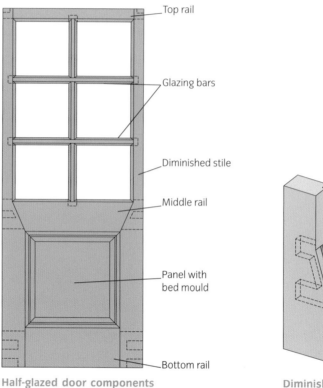

Top rail

Glazing bars

Diminished stile

Middle rail

Panel with
bed mould

Bottom rail

Half-glazed door components

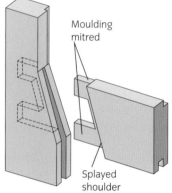

Moulding
mitred

Splayed
shoulder

Diminished stile joint detail

DECORATIVE PANELS

While panelled doors can have plain flat panels, such as the plywood that is used in cheaper doors, more decorative and bespoke doors will have panels that give the door a more elegant appearance. To achieve this, decorative panels are used.

Panels can be divided up in to three parts. The centre part of the panel, and usually the largest part, is called the 'field'. Around this is the margin; this part of the panel is next to the door components. The part of the panel that fits into either the groove or rebate is called the 'flat'.

The main types of panels can be described as:

- *raised panel:* having a flat and bevelled section

- *raised and fielded panel:* having a flat, and a margin that is bevelled leading to a raised field

- *raised, sunk and fielded panel:* having a flat, and a margin that is bevelled but sunk below a raised field

- *raised, sunk and raised fielded panel:* having a flat, and a margin that is bevelled but sunk below a further raised and bevelled field.

Section

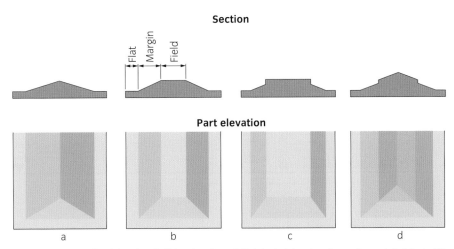

Part elevation

a   b   c   d

Decorative panels: (a) raised; (b) raised and fielded; (c) raised, sunk and fielded; (d) raised, sunk and raised fielded

GLAZING

Safety glass must be used in areas as described in the Building Regulations Approved Document Part N. For example, a fully glazed door would require safety glazing, as would a door that has glazing at a height below 800mm. The regulations state that the glazing in these areas must:

- break safely, in a way that does not produce pointed shards with razor sharp edges

- be robust and of adequate thickness to reduce the likelihood of breakage

- be permanently protected by screens or similar devices.

To meet these requirements, **toughened** or **laminated glass** is used in larger areas. For smaller glass panes, standard **annealed glass** can be used provided:

- it is as least 6mm thick

- it is not larger than 0.5m²

- it has a maximum width of 250mm.

Toughened glass

Standard annealed glass that is heated and rapidly cooled to give added strength

Laminated glass

Two pieces of annealed glass with a thin plastic interlayer that acts as an adhesive holding the glass together in the event of breakage, typically used in car windscreens

Annealed glass

The standard type of glass readily available

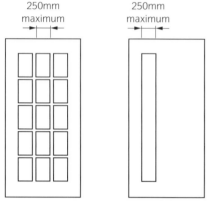

Maximum area of single pane not to exceed 0.5m², small panes of annealed glass should not be less than 6mm thick

Dimensions and areas of small panes

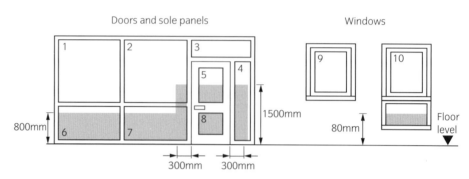

Shaded areas show critical locations to which requirement N1 applies (ie glazing in areas numbered 2, 4, 5, 6, 7, 8, 11)

Critical locations in internal and external walls

DOUBLE FIRE DOORS

Double fire doors (also known as fire-resisting doors) are mainly manufactured using a solid core construction method. The core consists of laminated fire-retarding material faced as a flush door or as a moulded panel door. The main function of a fire-resisting door is to delay the spread of flames, smoke and gases into escape routes for a set period of time in order to give occupants of the building enough time to evacuate in the event of a fire. These types of doors are side hung.

The performance requirements of fire-resisting doors are dictated by the building regulations. They have to undergo rigorous testing to ensure they meet the British Standards before being certified with a fire rating. The rating is determined by the fire door's performance in resisting penetration by flames or smoke through the core of the door, and is given the prefix FD. A fire door with the rating FD30 has the ability to resist fire for 30 minutes. If it then has the suffix S, rated as FD30S, then the fire door also has the ability to resist smoke. Fire doors can be identified by a label attached to the top of the door, giving the rating. They also use a colour-coded plug system in which a plug is inserted into the edge of the door to identify the rating.

Colour-coded plugs identifying rating of door

Time fire door certification scheme

Fire doors are available in standard sizes and various thicknesses, depending on their rating. The most common ones are FD30 (which are 44mm thick) and FD60 (which are 54mm thick). Fire doors are tested as a complete assembly and this includes not only the door but also the door frame, the **intumescent seals** and the ironmongery, including the hinges, latches or locks and door closers. Intumescent seals are necessary to prevent fire and smoke getting round the edges of the door. They are available in sizes of 10mm × 4mm (to suit FD30 fire-resisting doors) and 20mm × 4mm (to suit FD60 fire-resisting doors). These strips are inserted into a groove machined into the edges of the frame or the door.

Both types of fire door can also have a plastic smoke seal strip and brush pile added, to further help prevent smoke from passing through the gap between the frame and the door.

Intumescent seal

A seal that swells as a result of exposure to heat, closing the gap between the frame and the door

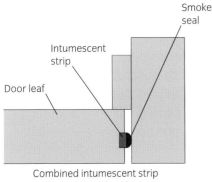

Combined intumescent strip and smoke seal (brush type) fitted in door leaf

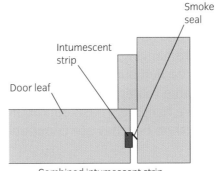

Combined intumescent strip and smoke seal (flexible blade) fitted in door leaf

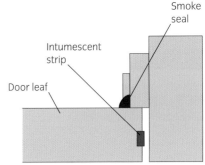

Perimeter smoke seal fitted to door stop – intumescent strip fitted in door leaf

Typical methods for fitting intumescent seals

Georgian wired glass

The name given to glass that has a thin wire mesh embedded into it, used in fire doors and where added safety is required

Fire doors can also have vision panels or be glazed, provided that the glass used is either fire rated or wired glass (**Georgian wired glass**). The glass must be bedded in an intumescent glazing gasket and held by glazing beads. These are protected by being manufactured from non-combustible material or coated in intumescent paint. The beads should be bolection beads bedded on an intumescent seal.

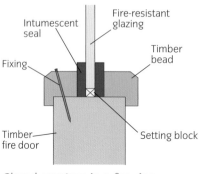

Glazed aperture in a fire door

PREPARING DOUBLE DOORS FOR HANGING

Georgian wire glass used in double fire doors

External double doors can be hung before plastering takes place in order to help secure the building and make it watertight. Internal double doors are hung after plastering has been completed and after the building has completely dried out. It is normal for internal doors to be hung before the architrave and skirting are fixed.

All doors need to acclimatise to the conditions that they are to be hung in. They should be stacked flat in a dry, well ventilated room, supported on three evenly spaced bearers to allow air flow underneath them and ensure even weight distribution.

External doors may require the bottom rail to be rebated so as to fit over a water bar that is integral to the threshold or the cill design. This is best done with the aid of a portable router. The rebate should then be coated with a suitable sealer and primer and allowed to dry before hanging. They may also require a letter plate fitting, which is best carried out with a letter plate jig and the use of a jigsaw and portable router. Again, the aperture can then be sealed and primed before hanging. These operations can be carried out in the joiner's workshop but they are commonly carried out on site.

IRONMONGERY

Door ironmongery is also known as door hardware or door furniture. It refers to a wide range of fittings and fixtures for doors that are available to accommodate any weight of door or required operating method.

The ironmongery or hardware required for double doors can be divided into sections according to their use, as shown in the table.

Type of ironmongery	Typical use
Hinges	Hanging the door
Locks and latches	Door security and restraint of the door
Closers	Self-closing mechanisms
Bolts and security devices	Securing the door
Door furniture	Door handles and attachments
Transom and floor springs	Hanging the door with unobtrusive closing mechanisms (typically glass doors)

Other items that can be installed alongside ironmongery are intumescent strips and weather seals.

On larger contracts there will be an ironmongery schedule giving precise details of the ironmongery to be used with each door. These schedules are very useful on site as an accurate check of what is required. The type of door and its intended location will dictate the types of ironmongery that will be used with the doors.

OUR HOUSE

While discussing with your supervisor the suitable door designs, construction methods and material choices for the various locations within 'Our House', list the considerations that should be taken into account in each location. For each location, decide the best door type along with its construction method and suitable material.

Ferrous

Containing iron

Non-ferrous

Not containing iron

Brittle

Liable to break or crack without bending

MATERIALS USED FOR IRONMONGERY

Ironmongery is usually made from either **ferrous** or **non-ferrous** metal.

Ferrous metal

The most common material used for butt hinges is mild steel, which is also the cheapest. The appearance of mild steel can be enhanced by electro-brass or zinc plating. Cast iron is suitable for heavyweight doors but can be **brittle**. Stainless steel resists staining and corrosion and is therefore used for hardwood doors and external uses, as well as on fire doors.

Non-ferrous metals

Brass, which is mostly used for ornamental purposes, is extremely resistant to rusting and staining and is suitable for both internal and external use. It is particularly good for use with oak.

However, brass screws are soft and can be easily broken while screwing into hard materials.

The use of iron and steel ironmongery needs careful consideration as these metals rust when they come into contact with moist air or with acidic timbers such as oak. This reaction can cause unsightly staining to the joinery.

Information on screw types can be found in Chapter 3 of *The City & Guilds Textbook: Level 2 Diploma in Site Carpentry and Bench Joinery*.

TYPES OF HINGE

Type of hinge	Description
Butt hinge 	Butt hinges consist of two leaves joined by a pin that passes through a knuckle formed on the edge of both leaves. They are fitted so that the leaves are recessed equally into the door and frame, with a margin between so that the door operates without binding. The knuckle is usually set to project past the face of the door to give increased clearance. There are normally an odd number of knuckles, and the leaf with the greater number of knuckles is positioned on the door frame. Butt hinges are available in a range of sizes from 25mm up to 150mm. Brass butt hinges are susceptible to wear on the knuckles, so are available with stainless steel or phosphor bronze washers fitted between the knuckles to prevent this. The washers also reduce squeaking.

Type of hinge	Description
Ball race butt hinge 	High-performance ball bearing or ball race butt hinges give a much smoother action, and are much more durable on heavy doors. They are available in sizes from 75mm to 150mm.
Loose pin butt hinge 	Loose pin butt hinges enable easy removal of the door by removing the pin from the hinge knuckle. Lift-off butt hinges enable the door to be lifted off when it is in the open position. The hinges are handed and incorporate one long and one short pin hinge. The long pin hinge is the lower hinge and the short pin hinge is the upper one, to enable easier repositioning of the door.
Rising butt hinge 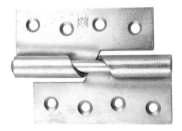	Rising butt hinges have a spiral-shaped knuckle, allowing the door to rise as they open. These are particularly useful for clearing uneven floors, mats and rugs. The shape of the knuckle also gives the door a self-closing action. The top leading edge of the door must be eased to prevent the door fouling the head of the door frame as it opens and closes. These hinges are handed.
Parliament hinge 	Parliament hinges have knuckles that protrude from the door so that when the door is fully open it stands away from the wall.
Flush hinge 	Flush hinges are only suitable for lightweight doors. They are quick to install as they do not need to be cut in to the door or frame.

Type of hinge	Description
Single- and double-action spring hinges 	Adjustable spring hinges are designed to make a door self-closing. They have large cylindrical knuckles that can be tensioned to the required closing action. Single-action spring hinges have two leaves, are fitted the same way as normal hinges, and can open up to 180°. Double-action spring hinges have three leaves and open 360°. To correctly install double-action spring hinges, a planted strip of timber that is the same thickness as the door should be fitted to the hanging edge of the frame to allow free movement of the cylindrical knuckles. Double-action spring hinges are frequently used in corridors in public buildings such as schools and hospitals, and between kitchen and dining areas in restaurants and public houses.
Hawgood swing door hinge 	Hawgood swing door hinges are suitable for heavy-duty and industrial double swing doors. The cylinders enclosing the springs are morticed and recessed into the door frame and the moving shoes are cut and fit around both sides of the door. This type of hinge design allows the door leaf to swing away from the frame due to the design of the shoe that fits around the door; this means there is only a small gap between the door leaf and the frame when the door is closed.
Concealed hinge 	Concealed hinges are suitable for flush doors with timber, steel and aluminium frames. A common type is Tectus, which is fully adjustable to ± 3mm and is designed for heavy-duty doors. Concealed hinges are commonly used in shop fitting. Soss hinges are used for light-duty applications.
Floor and transom springs 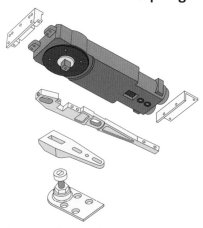 Transom door spring	Floor springs are used mostly for heavy-duty entrance doors. They can control both single-action and double-action doors. Doors are pivoted via the floor spring at the bottom and a pivot centre at the top. The floor spring is housed in a box that is set into the floor. The bottom of the door is recessed to accept a shoe, which is fitted to the floor spring. The pivot centre is attached to the underside of the head of the door frame and a socket is housed into the top of the door. Floor springs also act as door closers. The transom overhead door closer has a similar function to a floor spring but is designed with the spring concealed within the transom header bar of an aluminium door frame, where a load arm pivot is attached to the top rail of the door and a floor pivot is fixed to the bottom rail. It provides controlled closing of a single- or double-action aluminium door or doors.

Type of hinge	Description
Strap hinge	Strap hinges are surface fixed, being screwed or bolted directly to the door and frame. The most common type of strap hinge is the tee hinge, which is made from thin-gauge steel and is usually **black japanned** or galvanised. Strap hinges are used predominantly with ledged and braced doors for sheds and gates.
Cranked, hook and band strap hinge	Heavy-duty hook and band hinges are made from stronger galvanised or stainless steel. They are used for heavier industrial or garage doors, framed, ledged and braced doors, and farm gates.

Hinge selection and positioning

Selecting the correct hinges for hanging double doors is essential to ensure the performance of the door. If the doors do not hang perfectly within the frame, the lock or latch may fail to operate and the doors will not shut correctly.

Lightweight internal doors (hollow core flush doors) usually only require one pair of 75mm hinges per door (although bathroom and en-suite doors may be specified to be hung on one-and-a-half pairs of 75mm hinges, ie three hinges). For 35mm-thick timber panelled and glazed internal doors, use one-and-a-half pairs of 75mm hinges. It is best practice to hang all 44mm-thick doors, whether internal or external, on one-and-a-half pairs of 100mm hinges. Fire doors also need to be hung on one-and-a-half pairs of fire-rated 100mm hinges.

Hinge positions for doors have regional variations but the standard positions are 150mm down from the top of the door and 225mm up from the bottom, while the centre hinge is positioned an equal distance between the top and bottom hinges. On heavier doors the middle hinge is often moved up to 200mm below the top hinge.

Black japanned

Finished with a black enamel lacquer, originally associated with products from Japan

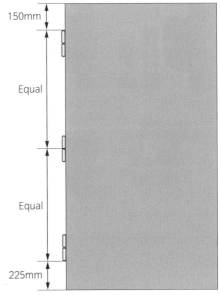

Standard positioning for hinges

LOCKS AND LATCHES

Most locks and latches are morticed into the closing edge of the door although a few, like rim locks, sit on the face of the door. There is an increasingly wide variety of locks and latches available but most fall into five main categories, these being:

- mortice deadlock
- mortice locks/latches
- mortice rebate locks
- mortice latches
- rim locks and latches.

Locks are mostly produced from steel, with a variety of different finishes that can be matched to the finish of other door furniture. Most locks and latches bought today are supplied with fitting and positioning instructions included.

Type of lock or latch	Description
Mortice deadlock  Deadlock Euro pattern deadlock Hook lock for use with sliding doors Euro lock barrel	Mortice deadlocks are simple key-operated locks that require no handle to operate them. They are commonly used for storerooms or as additional security for external doors. They are available with a claw deadbolt or hook deadbolt for use with sliding doors. They are morticed and recessed into the edge of the door. The keyhole is covered by an escutcheon. The lock is controlled by levers: the more levers the lock has, the more secure it is as it is more difficult to pick. Three-lever and five-lever deadlocks are the most frequently used types. The deadbolt engages into a box striking plate, which is morticed and recessed into the door jamb. European (Euro) pattern cylinder deadlocks, which use a barrel lock to operate the deadlock, are also available.

Type of lock or latch	Description
Mortice lock/latches 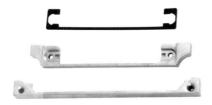 Mortice lock/latch	Mortice lock/latches (also known as 'mortice sash locks') are a combination of a mortice deadlock and mortice latch. They are available as a vertical mortice lock/latch or a horizontal mortice lock/latch. Vertical mortice lock/latches are suitable for fitting in most types of door and come in various sizes, the most common being 75mm deep with a 57mm **backset**, and 65mm deep with a 44mm backset. Horizontal types are only suitable for fitting in doors with wide stiles, solid flush or set in the middle of the lock rail, as the length of the lock/latch would protrude past the width of a standard door stile. A typical horizontal mortice lock/latch is 150mm deep, with a spindle backset at 127mm and keyhole backset at 51mm. They are typically used with doorknobs and escutcheons. Both types have reversible latches to enable them to be used with left-hand (LH) and right-hand (RH) opening doors. The most commonly used are three-lever and five-lever mortice lock/latches; three-lever types engage into a striking plate, whereas five-lever types engage into a box striking plate. Mortice lock/latches are available as Euro pattern cylinder lock/latches. Bathroom lock/latches have an additional spindle hole rather than a keyhole.
Mortice rebate locks 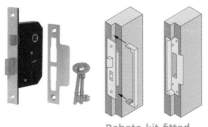 Rebate kit fitted	Rebated mortice lock/latches are available for double doors where the meeting stiles are rebated into each door. Rebate kits are available to convert standard lock/latches to suit rebated doors with a standard rebate of 12mm.
Mortice latches/tubular latch 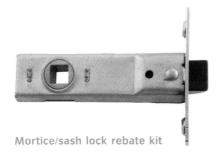 Mortice/sash lock rebate kit	Mortice latches are mainly used for internal doors that do not need to be locked. The most common type is the tubular mortice latch. The latch engages into a striking plate on the lining or frame, which keeps the door shut. It is operated from either side of the door by a pair of lever handles or door knobs. Mortice latches are reversible, so can be used on LH or RH opening doors. They are available in different lengths and backset depths to suit different applications. Rebated tubular mortice latches are used for rebated doors; alternatively rebate kits are available to enable standard mortice latches to be converted for rebated doors.

Backset

The distance from the face of the latch to the centre of the spindle

Type of lock or latch	Description
Rim deadlocks and rim lock/latches  Rim deadlock 	Rim deadlocks and rim lock/latches are face-fixed and offer a cheaper alternative to locks that require the additional labour of being morticed in. They are usually operated by a pair of door knobs. They are often used with ledged and braced doors, as these doors are not thick enough to receive mortice locks. They are screwed to the inside of the door in conjunction with an escutcheon on the outside. Doors fitted with rim deadlocks usually have a pull handle on the outside to operate the door. Rim lock/latches are still used for period properties in conjunction with doorknob sets. Cylinder night latches are mainly used on entrance doors to domestic properties. The cylinder fits through a hole bored through the door and the night latch is fixed to it via a back plate. The door is opened via a key from the outside and by turning a handle from the inside. The latch bolt engages with a staple keep that is fixed to the door jamb. Better-quality night latches have a double-locking facility that improves their security. They are available to fit narrow door stiles.
Panic bars 	Panic bars are a type of espagnolette bolt, sometimes referred to as crash bars or push bars. They are fitted to an exit door, typically a fire door, and are designed to be fitted to doors used by people not familiar with the door's operation. A panic bar allows for safe egress through the door by simply pushing on the horizontal bar. They are typically used in shops, schools, hospitals and other public building allowing emergency egress through the door way.

CLOSERS

These are used to ensure doors close on their own to prevent the spread of fire, draughts and sound, or to ensure privacy throughout a building. Overhead door closers are fitted to the top of the door and/or the door frame above it. They work by means of either a coiled spring mechanism or a hydraulic system enclosed within the casing, with an arm to either pull or push the door shut.

Different strengths of overhead door closers are available to suit the size and weight of the door. Door closers to be used with fire doors must also comply with relevant British Standards. The table on the next page gives guidance on the correct door closer to use.

A typical overhead door closer

Door closer power size	Recommended door width (mm)	Door weight (kg)
1	750	20
2	850	40
3	950	60
4	1,100	80
5	1,250	100
6	1,400	120
7	1,600	160

Fitting overhead door closers depends on where the door is situated, the size of the door frame and the amount of room available above the door frame or within the head of the door reveal. They can be fitted on either side of the door, to suit the situation.

Overhead hydraulic door closers can be adjusted to speed up the closing action, and have a hydraulic check called a dampening action which prevents the door slamming shut. They come complete with instructions and paper templets to make installation easier.

Where rebated double doors are used, they must be closed in the correct sequence. To aid this, a door selector is used. Door selectors prevent the incorrect **door leaf** from closing fully by means of an arm that drops down onto a strike plate fitted to the top of the leaf that should close last. The leaf intended to close first has the selector striker fitted to its top, lifting the door selector arm up as it closes, allowing the second leaf to close behind the first door leaf.

Door leaf

One half of a pair of double doors

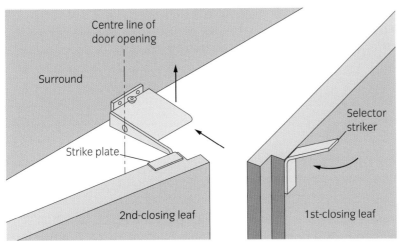

Double door selector

Concealed-spring door closer

Concealed-spring door closers work through a tensioned spring housed in a cylinder. This is morticed into the hanging side of the door and secured with screws. The anchor plate is recessed and screwed into a rebate on the door frame. Adjustment of the closing action is achieved by using a claw hammer to withdraw the chain and then inserting a metal plate to hold the chain in place. The anchor plate is unscrewed and turned clockwise to increase tension and speed up the closing action, or turned anticlockwise to slow down the closing action.

BOLTS AND SECURITY DEVICES

One door leaf will usually be held in place by bolts (the slave door), while the other door leaf (the master) will usually have a lock or latch of some sort. The type and number of bolts will depend on several factors but typically how the doors open and their construction method, location, appearance and type of finish. The table below shows some of the common bolts and security devices used with double doors.

Type of bolt/security device	Description
Barrel bolt  Monkey tail bolt  Cranked tower bolt  Spring-loaded bolt	Barrel bolts and tower bolts are used to secure external gates and garage doors from the inside. Two bolts are normally used, one at the top of the door and one at the bottom. They are available as cranked or necked for doors and gates that open out. For heavier-duty industrial applications and taller garage doors, there are tower bolts with bow and monkey tail handles that allow the bolt to be operated from within normal reach with minimal effort. The bolt is secured into a mortice in the head of the door frame, finished with a recessed striking plate. A bolt socket secures the bolt in the floor.
Padlock bolt 	Entrance gates are often fitted with padlock bolts, which are similar to barrel bolts but allow for a padlock to secure the handle to a heavy-duty plate when it is in the closed position.

Type of bolt/security device	Description
Brass straight bolt 	Slimmer and more detailed barrel bolts are available for dwellings. These are often finished in brass.
Flush bolt 	Flush bolts finish flush with the door and are used primarily for securing one half of a pair of doors within the rebated edge. One bolt is positioned at the top of the door and another at the bottom.
Mortice rack bolt 	Mortice rack bolts are deadbolts operated by a fluted key. They are morticed into the opening edge of external doors, positioned 150mm from the top and 225mm from the bottom. They are only operated from the inside.
Panic bolt 	Panic bolts and latches are used on the inside of emergency exit doors to bolt them closed. The bolts are espagnolette type, and have a double point locking system at the top and bottom of the door. Pushing the panic bar disengages the bolts from the keeps that are recessed in the frame.
Hinge bolt  	Hinge bolts are morticed into the hinge side of external-opening doors, just below the top hinge and just above the bottom hinge. They are made from hardened steel and improve the security of the door by preventing the door from being levered off its hinges. Security 100mm butts can also be used; these have interlocking pins to prevent hinge pin attack.

DOOR FURNITURE

Door furniture refers to items that are fixed to the door surface, typically to help open or protect the door.

Type of door furniture	Description
Lever furniture Euro lock brass handle Round bar handle Black japanned ornate latch/lock Internal door handle for a mortice latch 	Available in a wide variety of patterns and colours, they are used with mortice latches and lock/latches.
Knob furniture 	For use with rim locks and horizontal lock/latches. These should not be used with vertical lock/latches because the knob is very close to the edge of the door and may cause hand injuries when closing the door.
Thumb latch 	Usually known as Suffolk or Norfolk latches, thumb latches are used on ledged and braced doors. Traditionally they were made by a blacksmith from mild steel. There is a variety of designs available, usually black japanned.
Escutcheon 	Used to provide a neat finish to the keyhole on both sides of deadlocks and vertical mortice latch/locks.
Finger and kicking plates 	To prevent doors from damage, kick plates are positioned along the bottom rail and push plates on the opening or meeting stiles. In factories and hospitals, protection plates can also be fixed in other positions to protect the door from damage caused by trolleys or beds.

Type of door furniture	Description
Letter plates	Letter plates are usually positioned centrally in the middle rail but can be fitted in the bottom rail, and even smaller vertical letter plates can be fitted into the door stile. Traditionally, letter plates were fitted by drilling a series of holes and cutting out the shape with a pad saw. This is now often done with the aid of a jigsaw, or with a router and a jig to suit the size of the letter plate.
Information signs	Usually positioned centrally in the door's width and 1,500mm up from the bottom of the door.

Positioning door locks, latches, door furniture and signs

Positioning of door ironmongery will depend on several factors, typically including the construction method used in the doors, the material from which they are made, the specification and the ironmongery fitting instructions. However, the illustration shows the standard fixing positions for the main ironmongery items in use.

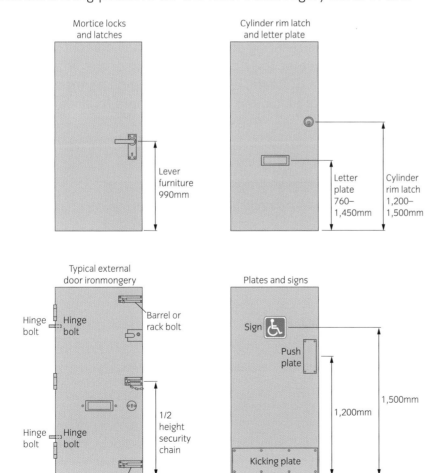

OUR HOUSE

You are to hang a set of French doors to the rear living room of 'Our House'. Calculate the size of the doors for the opening in the rear downstairs living room. Write out an order for the types and quantities of materials and ironmongery you would suggest. State the reasons for your choices.

HANGING DOUBLE DOORS

INDUSTRY TIP

To aid cutting out for rebate locks, it is good practice to fit a temporary batten into the rebate where the lock is to go. This will allow better marking out and cutting with both drill bits and chisels.

INDUSTRY TIP

Always ensure both doors are swinging correctly with the correct amount of clearance before fitting any locks.

Hanging side-hung double doors that have butt and tee hinges is very similar to the process of hanging single doors: the positioning and fitting procedures will be the same, as covered in Chapter 8 of *The City & Guilds Textbook: Level 2 Diploma in Site Carpentry and Bench Joinery*.

However, side-hung double doors require greater care during all stages of fitting the frames and doors in order to ensure the doors close correctly at the meeting stiles. Even a small amount of twist or misalignment to either the doors' positioning or the fitting of the frame will have a dramatic effect where the doors meet, and make fitting the rebate lock and getting the doors to close flush difficult.

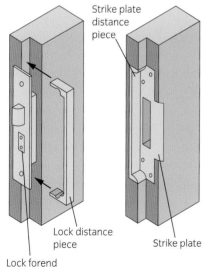

Rebated lock fitted

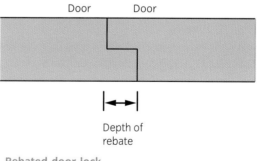

Rebated door lock

FITTING A REBATE MORTICE LATCH/LOCK

The following is one method of fitting a vertical mortice lock/latch (sash lock).

STEP 1 Determine the height of the spindle – refer to the specification or, where this is not available, the standard height would be 990mm to the spindle from the bottom of the door. With framed doors, the lock's mortice hole must not interfere with any of the tenons from the middle rail and/or glazing bars to the door stiles.

STEP 2 Wedge the door leaf open to a convenient position; gently wedge it from both sides to prevent it from swinging. Fit temporary packers to level out the rebate; this makes it easier to drill the holes for the lock in the centre of the door.

STEP 3 Holding the lock against the doors' meeting stile, lightly mark the upper and lower edges of the lock case. An additional allowance of 5mm will give the lock a little clearance and aid the lock's fitting and removal. With a combination square, mark the centre of the door leaf and the amount of backset.

STEP 4 Measure down the backset the distance for the keyhole. Using an auger bit or flat bit, bore a 16mm spindle hole and a 10mm and 6mm keyhole. Ensure you work from both sides of the door. Using a pad or keyhole saw, leave the removal of the keyhole waste until the lock case mortice hole has been cut out. For speed, often a larger hole is drilled for the keyhole, eliminating the need for shaping it. This process, though, causes increased problems with locating the key smoothly in the lock.

STEP 5 Using an auger bit or a flat bit of a diameter slightly larger than the width of the lock case thickness, bore a series of holes to a depth equal to the full depth of the lock body plus 3–5mm.

STEP 6 Set a mortice gauge to the edge of the drilled holes and mark down the sides of the mortice lock. This will give a neat position for the chisel to locate in and help produce a good clean edge to the edge of the lock mortice.

Forend face plate

The part of the lock that is seen when the lock is housed in the door

STEP 7 Chop and clean out the mortice hole using a mallet and chisel. Make sure you maintain clean accurate sides to the mortice, which will allow the lock to be fitted easily and allow for its correct function.

STEP 8 Position the lock into the hole. Mark around its **forend face plate**; using a marking gauge to carefully score along these lines helps prevent the edges from splitting when chopping out the forend housing. Chop out a recess until the forend fits flush. Remove the temporary batten, fit the rebate lock mortice kit and screw the lock into the door. Fit the spindle and fix the handles to the door, ensuring they are parallel to the edge of the door. Check that the latch operates and that the key activates the dead bolt.

STEP 9 Close the door and mark the position of the latch and lock onto the other meeting stile. Gauge the distance from the back of the door to the front edge of the lock and mark on the door stile, ensuring you allow for the additional thickness of the rebate kit. Form the keep by chopping out the latch and bolt mortice hole in the closing edge of the stile. Recess the striking plate into the door frame until it is flush with the front lip of the door; it may require a deeper bevelled housing to help the latch close smoothly when the door is pushed shut. Note the use of a cramped block to the edge of the door; this will give additional support to the door and help prevent thin weak edges from breaking away.

STEP 10 The finished doors should finish flush with each other, with a smooth closing operation. The finished doors should finish flush with each other, with a smooth closing operation. The handles should operate smoothly without any stiffness to their operation. The key should enter the lock smoothly from both sides allowing the key to turn and operating the lock smoothly.

DOUBLE-ACTION SWINGING HINGES

Double-action swinging hinges are designed to make the doors self-closing. They are typically used in corridors in public buildings such as schools and hospitals, and between kitchen and dining areas in restaurants and public houses.

To allow the door to close on its own, a helical spring is fitted in the large knuckle of the hinge. The spring's tension is adjusted by using a hardened steel bar (supplied with the hinge) that is placed in one of the holes in the top of the hinge and turned to apply the tension, moving the ring until the required tension is achieved. Pins are then inserted into the hole, maintaining the tension.

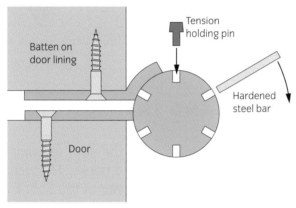

Adjustment for double-action swing door hinge

After tensioning, swing the door open and allow it to close by the hinges. Adjust the tension so that the door overswings by approximately 200–300mm – the door should have an easy and comfortable resistance when pushed open.

Double-action hinges are usually fitted onto a batten that is the same width as the door's thickness. This batten is fixed centrally to the door lining. One half of the hinge is fixed directly to the batten while the other half is attached directly to the edge of the door. Although the hinge can be recessed into the batten and door like traditional butt hinges, it is difficult to make a neat fitting around the hinge and knuckles of the hinge.

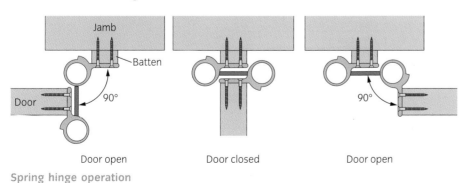

Spring hinge operation

The meeting stiles of the door will need to be fitted with a radius known as 'heeling'. This allows close fitting along the meeting stiles while still allowing either door to swing in either direction without one door leaf contacting and sticking on the other door leaf. Typically this type of door set involves the use of a brush seal on the meeting stiles.

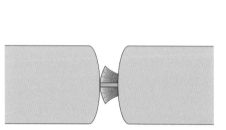

Meeting stiles for double doors using double-action swinging hinges

Hawgood double-action swinging hinge

PIVOTED FLOOR AND TRANSOM SPRINGS

Double-action floor spring hinges allow the door to swing in both directions and are common in shops and banks. The weight of the door is passed through to the floor instead of the door frame or lining. There are two main types of floor spring hinges:

1 a spring mechanism in the bottom of the door with a pivot point in the floor, used with timber doors

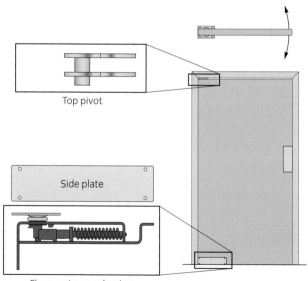

Top pivot

Side plate

Floor spring mechanism

Double-action floor spring

2 a spring mechanism in the floor with the pivot point fixed to the bottom of the door, typically used with glass doors.

Single-action floor spring   **Double-action floor spring**

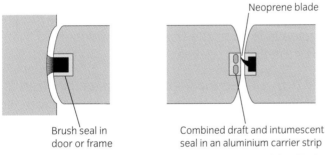

Single-action top centre

Single-action bottom centre

Cover plate

Spindle

Mechanism

Springbox

Double-action adjustable top centre

Double-action bottom centre

Cover plate

Spindle

Mechanism

Springbox

Single- and double-action floor spring concealed in the floor

Timber doors using this type of pivoting mechanism need both edges of the door shaped, as well as the door frame, with the bottom of the door chopped out to take the spring mechanism. The fitting instructions for floor spring hinges will come with the hinges but will follow similar stages as those shown below for installing double glass doors using double-action pivot floor springs.

Neoprene blade

Brush seal in door or frame

Combined draft and intumescent seal in an aluminium carrier strip

Shaped door edges and door frames (often termed heeling)

INSTALLING DOUBLE GLASS DOORS USING DOUBLE-ACTION PIVOT FLOOR SPRINGS

1 Mark out the position for the floor spring pivot box in the floor, ensuring the pivot is positioned centrally to the frame and the correct distance away from the frame.

2 Cut out the hole to take the floor spring pivot box.

3 Mark out and fit the top pivot in the frame, ensuring correct positioning with the bottom pivot by using a plumb line and measurements provided with the instructions.

4 Fit shoes to the doors.

5 Engage the shoes in bottom and top pivots.

6 Lock off when correctly positioned.

7 Adjust the tension to the springs to achieve the required operating sequencing.

8 Fit the spring box dust cover.

SLIDING DOOR MECHANISMS

Sliding a door across an opening provides a useful alternative to hinging when there is limited or restricted floor space. The disadvantage with this method of door opening is the restricted access to wall space. There is a considerable range of proprietary sliding door systems available that all come with the appropriate fitting and installation instructions.

Most systems work along the same principle, with an overhead track either fixed to the wall or ceiling from which the door is suspended. The door runs along the track on rollers of nylon (in the case of cheaper lightweight systems) or heavyweight roller bearings (for large industrial doorways). The whole of the top of the sliding gear can be covered with a **pelmet** to provide a suitable decorative finish.

The bottom of the door can be guided in three ways:

1 *A groove, allowing the bottom guide to run along the groove:* the bottom guide is fixed to the floor and positioned so that it runs centrally to the doors' thickness.

2 *A channel tray cut into the floor with a guide pin fixed to the bottom of the door:* this method is typically used with industrial doors but can be problematic with the tray filling with debris, etc.

3 *A guide block screwed to the floor on the outer edge of the door to prevent the bottom of the door from pulling away from the wall:* typically used in domestic situations.

Pelmet

A strip of wood or cloth that is fixed in front of the running gear to hide the track of a sliding door and provide a decorative finish

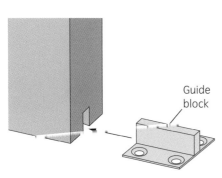

Grooved bottom door

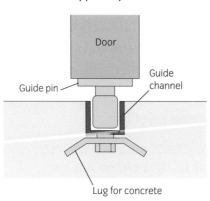

Channel tray cut into the floor

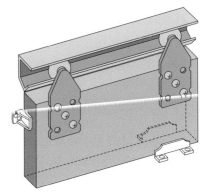

Guide block screwed to the floor

The same principles are applied to pairs of doors where they are allowed to pass over the face of the wall. If the doors are to sit within an opening then a double track would need to be used, allowing the doors to pass one another.

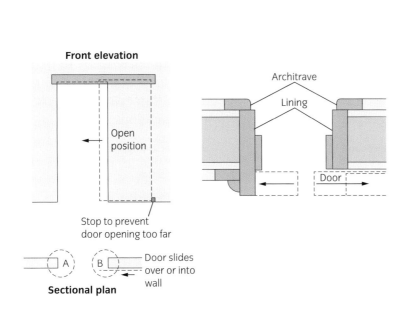

Front elevation

Open position

Stop to prevent door opening too far

A   B   Door slides over or into wall

Sectional plan

Architrave

Lining

Door

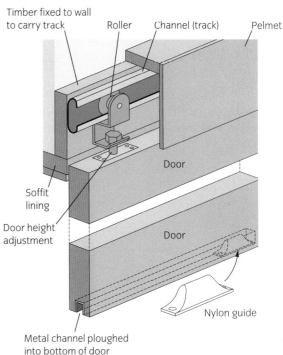

Timber fixed to wall to carry track

Roller   Channel (track)   Pelmet

Door

Soffit lining

Door height adjustment

Door

Nylon guide

Metal channel ploughed into bottom of door

Sliding door fitting up to a wall or stop

FOLDING DOORS

Like sliding doors, folding doors are a useful alternative to traditional side-hung doors. They are particularly useful for dividing up larger rooms using temporary partition screens, with fitted wardrobes, and are now also popularly used as fully glazed rear access doors to a garden.

These types of doors have the advantage of closing within the opening or up against the wall, and as a result take up little wall space compared with sliding doors. They also do not have the problem that large doors have of needing an area in which to swing.

Folding doors typically run along a suspended roller system that is fixed to the head of the door lining. The door can be further supported by a roller at the bottom of the door that runs in a track sunk into the floor, or by using guide rollers that have the bottom hinges attached and also run in a sunken track way.

Folding doors can be either end-folding or centre-folding.

Folding and sliding doors

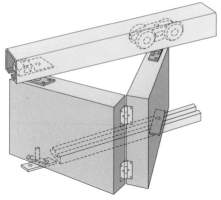

End-folding and sliding door

END-FOLDING DOORS

The pivoting points for these types of doors is at the beginning and end of each pair of doors. To achieve this type of sequence, each pair of doors has hinges that are attached to both door leaves and the supporting roller, while the other edge of each door is attached to the next using butt hinges.

CENTRE-FOLDING DOORS

The pivoting position for this type of folding sequence needs to be in the centre of the door. The pivoting sequence for these types of doors starts with a half leaf door hinged to the framework, typically with butt hinges, as shown in the following illustrations.

Vertical section

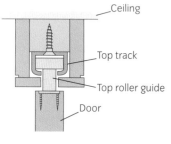

- Ceiling
- Top track
- Top roller guide
- Door

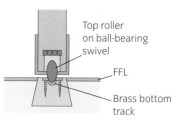

- Top roller on ball-bearing swivel
- FFL
- Brass bottom track

Centre-folding door sections

Horizontal section

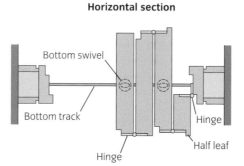

- Bottom swivel
- Bottom track
- Hinge
- Hinge
- Half leaf

Typical arrangement

- Top guide
- Hinges
- Bottom track
- Bottom roller
- Half leaf

OUR HOUSE

Suggest a suitable design of double door that could be used for the front entrance to 'Our House'. List suitable types and quantities of ironmongery that would be required. Source suitable suppliers for the ironmongery and produce a purchase order listing the required ironmongery, the cost per item, the total cost excluding VAT, the cost inclusive of VAT, and a 15% handling charge.

CURVED AND RAKING MOULDINGS

The profiles of mouldings used today are based on the classical profiles of Roman- and Grecian-period architraves.

- Roman-based profiles have curves and arcs based on circles or parts of circles, and tend to produce moulding with clearly defined profiles.

- Grecian-based profiles have curves based around ellipses, parabolic (U-shaped) curves and hyperbolic (mirror image) curves, which produce mouldings with smoother and softer flowing profiles.

Shown below are some of the standard profiles of mouldings.

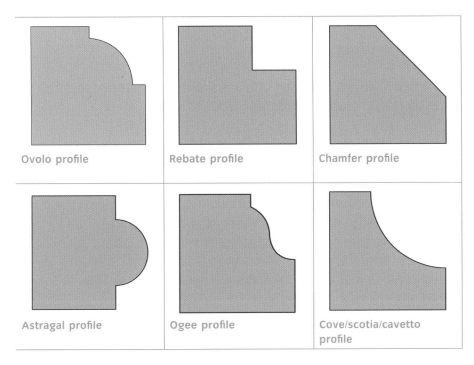

Ovolo profile	Rebate profile	Chamfer profile
Astragal profile	Ogee profile	Cove/scotia/cavetto profile

Production of curved mouldings is usually carried out in the woodworking machine shop using vertical spindle moulders or overhead routers, incorporating **jigs** and **saddles** to enable the correct curved profile to be achieved. The exception to this is when standard straight lengths of mouldings require fixing back to curved surfaces; a typical example of this is a walk-in bow window or segmental bay window commonly used in housing.

Curved moulding is commonly found in the following shapes:

- segmental

- semi-circular

- elliptical

- Gothic.

Setting out these types of moulding is covered in the following sections.

Jigs

A means of safely holding components while carrying out machining operations

Saddle

A means of holding and presenting materials for machining operations, allowing repetitive reproductions

CURVED MOULDINGS

The shape of curved mouldings usually follows the same architectural principles as the frames they surround. The Gothic design, which dates back to medieval times and is typically found in churches and cathedrals, has characteristics that are based on semi-circular pointed arches.

To enable the accurate setting out of shaped mouldings, a full-size drawing needs to be produced. To achieve this, three pieces of information are usually required:

- the required shape of the curvature
- the width of the moulding
- the rise and span of the curve.

In Chapter 6, pages 311–323, you will find step-by-step guides to show you how to set out different types of curves.

INTERSECTION OF MOULDINGS

Obtuse angle

An angle between 90° and 180°

Acute angle

An angle between 0° and 90°

INDUSTRY TIP

'Reflex angles' are angles between 180° and 360°.

Bisection

To divide an angle or line in two geometrically

The rule to follow when setting out mitres is that the line of the mitre is always straight and it always bisects the overall angle of the mouldings that are to be mitred (provided they are both equal widths and straight lengths). If the overall angle is 90° then the mitred angle would be 45°, or if the overall angle is 60° then the mitred angle would be 30° and so on.

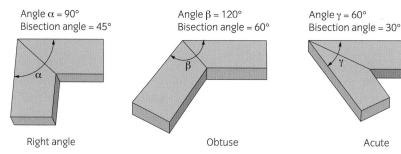

Angle α = 90°
Bisection angle = 45°

Angle β = 120°
Bisection angle = 60°

Angle γ = 60°
Bisection angle = 30°

Right angle   Obtuse   Acute

Bisection angles for equal-sized materials

Where two mouldings of equal size meet but one of the sections is curved, the bisection line will not be straight. Therefore a straight mitre does not correctly halve the overall angle. To achieve the correct **bisection** angle throughout the mitre, the mitre must have a curved cut.

SETTING OUT A CURVED MITRE

The following two examples show how to bisect curved and straight mouldings to produce the true curved shape of the mitre line.

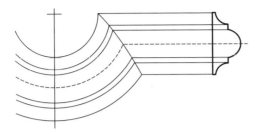

Example 1

1 Draw the section of the moulding.

2 Draw horizontal lines extending from to the profile section.

3 Draw the curved profile equal to the section, with point A as the centre of the inner radius and point B the outer edge of the curved profile. Continue the radius of the arc to meet the horizontal lines to find points C and D.

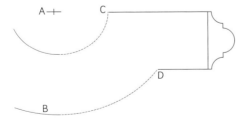

4 Mark in the centre lines on both profiles.

5 Draw in the profile on the curved section.

6 Mark in suitable horizontal lines to represent suitable points on the profile (1–6).

7   Transfer points 1–6 on line A–B. With A as the radius, draw radii A0–A6 until produced, they intersect with lines 0–6 of the horizontal moulding.

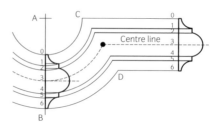

8   A curve drawn through each of the intersecting lines produces the shape of the curved mitre.

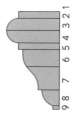

9   Cut a paper or card templet of the mitre line for marking onto both curved and straight mouldings.

Example 2

1   Draw the section of the moulding.
2   Divide up the profile at appropriate points (1 to 9 in this example).

3   Draw the inside and outside moulding outlines.

Section of the moulding used in Example 2

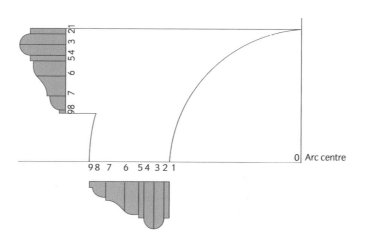

4   From the centre of the curve at 0, draw radii passing through these points to intersect with the straight lines. This produces the true shape of the curved mitre.

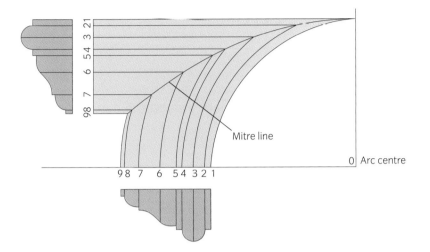

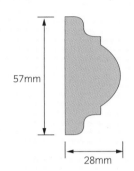

FUNCTIONAL SKILLS

Using the Example 2 steps and the profile in the image below, draw to full size the profile and a curved moulding with a 600mm radius with the same profile.

Work on this activity can support FM2 (L2.1.2).

5   Make a templet for both the concave and convex profiles.

6   Make saddles to fit the mouldings and with the templet mark and cut the mitre.

DETERMINING THE REQUIRED SHAPE OF A MOULDING USING A STRAIGHT MITRE

Although producing a curved mitre requires a considerable amount of time and skill, it is usually preferable to having straight mitres and adjusting the profile and size of the moulding to suit. Although at first glance both mouldings might appear the same size and profile, in fact they can be considerably different; this difference will depend on the original profile size, the shape and the radius of the curved mouldings.

Note: You will not see the true difference between the mouldings until drawn out at full size.

The following is a guide to setting out straight the correct curved profile for use with straight mitres.

1   Draw the section of the moulding. Draw horizontal lines extending from the profile section.

2   Draw the curved profile equal to the profile section at points C and D, with point A as the centre of the curved radius profile.

3   Continue the sweep of the arcs to meet the horizontal lines at points C and D. Then draw a straight line between them.

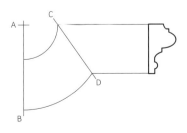

4 Draw horizontal lines representing suitable points on the profile.

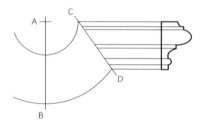

5 Using point A as the centre, mark around the curved profile lines from line C–D to line A–B.

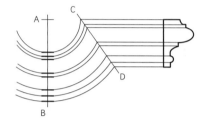

6 Where each line cuts line A–B, project horizontal lines.

7 Draw a corresponding set of lines to the right of line A–B; the intersections of the two sets of lines are the positions of the profile intersections.

8 Draw in the profile shape.

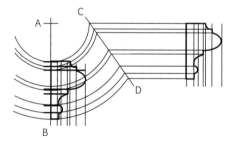

KERFING

Kerfing

The act of cutting a series of kerfs (the total width of saw cuts) into a piece of wood in close proximity to each other so that the wood can be bent

Short grain

This is where the general direction of the wood fibres lies across the direction of cut, increasing the likelihood of the timber breaking up during cutting operations

Where straight lengths of mouldings such as skirting boards need fitting to internal (concave) or external (convex) walls, the back of the moulding will usually require **kerfing**.

It is important when selecting material for curved mouldings, both those where kerfing is required and those that are used in circular profiles, that the quality of the material is high. In the case of material used in the production of curved mouldings, the avoidance of **short grained** and knotty timber is vital. Any short grained timber is more likely to break up, particularly during the machining process, risking serious injury to the operative.

Although the quality of the material used when kerfing is equally important, it is for different reasons. In the case of kerfing, if the material has any knots or other surface defects within the kerfing area there is an increased risk of the material snapping or not bending in a smooth manner and providing a neat continuous curve. The depth of cut must be consistent. The cut must be not so deep that the material snaps through and that instead of bending the material cracks at each kerfing point. But it must be deep enough to allow the remaining fibres of the material to freely bend.

Kerfing

As a rule of thumb, the depth of the kerfs (saw cuts) should be within approximately 2–3mm of the finished outer face of the material. This depth will depend on the type of material being cut, its moisture content and its quality. The approximate spacing of the kerfs will depend on the sharpness of the curvature that the moulding is to be fitted to, but will generally range from as close as 4mm with fine kerf saw blades, where the moulding needs to be fitted around small tight bends, to 50mm with thicker kerf saw blades and larger radius bends.

To calculate the number of saw cuts required to enable the timber to fold smoothly around a curve, you need to find the circle's radius of curvature to calculate the enclosing circle's circumference. This then must be divided by four to produce the length of quadrant required. This length must be divided by the kerf-producing saw blade width and spacing width added together. This will finally give you the number of cuts to make along the length of the curve (see example below).

To find the distance between kerfs, use the following steps.

1 Mark the required length of the curve in its correct position, either on the material to be fitted or on a test piece.

2 Make a saw cut to the correct depth at the start of the curve. Set in from the end as shown in the illustration on the following page. Always start the curve of the corner with a saw cut to better ease the material into the curve.

3 Hold the material down with the sawn kerf facing upwards and carefully lift the material up until the saw cut closes up, then measure the distance the timber has lifted at the end of the required curved section. This could be used as either an alternative to or check on calculations, such as those on the following page.

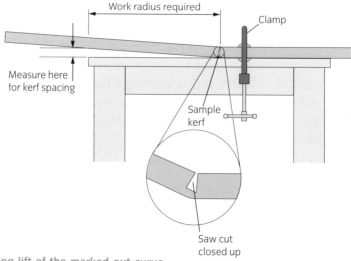

Measuring lift of the marked out curve

Formulae

- Circumference of a circle = 2πr

- Quadrant length (at corner) = circumference ÷ 4

- Number of cuts = quadrant length (corner) ÷ (kerf spacing + sawn kerf width).

Now that you know the distance between kerfs (kerf spacing), you need to work out the number of cuts. Let's look at an example.

> ### Example for finding the number of cuts
>
> This example calculates the number of cuts for a 300mm corner length using a saw blade having a kerf of 3mm and spacings of 20mm.
>
> *Step 1*
>
> Work out the number of cuts using the formulae above. Before you get there, remember that you always start the corner with a cut. So, reduce the corner length by the sawn kerf first:
>
> number of cuts = 300mm − 3mm = 297mm
>
> *Step 2*
>
> Now you can apply the formula:
>
> number of cuts = quadrant length (corner) ÷ (kerf spacing + sawn kerf width)
>
> 297 ÷ (20 + 3) = 12.91
>
> As you cannot have part cuts you need to round the number of cuts up. In this case it is now 13.
>
> Therefore **13 cuts** are required that are **20mm apart**.

FUNCTIONAL SKILLS

Using the formulae here, calculate the number of cuts required for a corner having a length of 450mm with a spacing of 22mm and a sawn kerf of 4mm.

Work on this activity can support FM2 (C2.5).

Answer: 18 cuts rounded up that are 22mm apart.

INDUSTRY TIP

Always experiment with a sample piece of material to determine the required spacing for kerfing cuts as the distance apart often requires reducing, typically with standard readily available lengths of skirting board.

DECORATIVE FINISHES

When finishing a moulding with a decorative face, the kerfing will usually need stopping before the saw blade cuts through the decorative part of the moulding. This is particularly important if the moulding is not to be filled and painted, with a clear surface finish being applied instead. To overcome this problem, two different methods can be applied:

- stopped kerfing

- built-up moulding.

STOPPED KERFING

This is where the sawn kerfs are stopped before the saw cuts into the decorative part of the mould or reaches the top portion of the moulding.

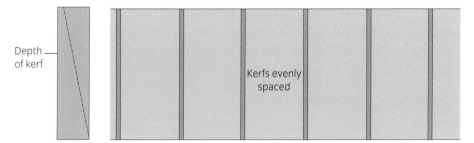

Back face of skirting

BUILT-UP MOULDING

With this method the decorative part of the mould is 'ripped' from the required length of moulding and is kerfed using the stopped kerfing technique described above, while the lower part of the moulding has its kerfs running fully across its width to achieve a constant depth of saw cut. This is then fixed in place with the decorative part of the mould planted on top to complete the finished moulding.

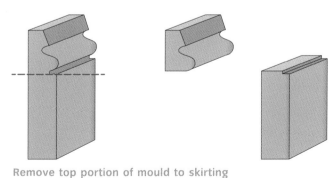

Remove top portion of mould to skirting

When fixing mouldings that have been kerfed, always add plenty of glue to the kerfs; this helps support the moulding and gives it strength.

If you are getting consistent failures, the chances are that you are not cutting deep enough or the cuts are too far apart, which results in the curve resisting the bend and fracturing.

LAMINATED MOULDINGS

Laminating is one option to enable mouldings to follow the curve of a wall without the need for kerfing. This applies particularly with delicate mouldings that could break easily or heavy bulky moulding that even with kerfing would be difficult to bend, or for mouldings that require tight turns. Laminated construction requires the gluing of multiple thin layers of timber, each about 3mm to 5mm thick, that are built up to the required thickness. These laminates are glued together and cramped around a jig to the required shape.

Try to gauge the thickness of each laminate so that some of the joints fall in line with the moulding profiles, if possible. In reality this will prove quite difficult as you will be working with more layers than the profile is likely to have.

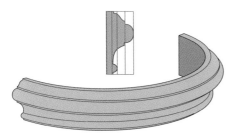

Laminated mouldings: the red lines show the glue joints in the laminated section

Advantages of laminating:

- strong
- the grain will always run along the curve
- no heading joints
- no short grain.

Disadvantages of laminating:

- requires the construction of a jig
- requires a substantial number of cramps
- is very time consuming
- requires a selection of good-quality timber without knots and holes along the required thickness
- thin laminated timber is more likely to split while bending unless it is good quality.

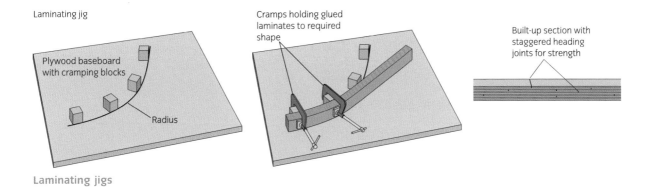

Laminating jig

Plywood baseboard with cramping blocks

Radius

Cramps holding glued laminates to required shape

Built-up section with staggered heading joints for strength

Laminating jigs

RAKING MOULDINGS

A moulding that is inclined at an angle to the horizontal is termed a raking moulding. They are used to bring continuity of mouldings and to join up more than one level, and typically to follow the pitch of staircases where panelling and dado rails are used. Where raking mouldings join up with horizontal mouldings at the mitred corner, the profile of the mouldings will be different, requiring one of the profiles to change. This is a complicated process usually requiring the purchase of expensive, specially produced mouldings or the operative to make the return moulding using hand tools. It has gone out of fashion as a result.

A more common approach today is to return the moulding at the mitre in the horizontal plane and then proceed with the incline up the stairs. This enables the moulding to be of the same profile. We will look at this in detail over the next few pages.

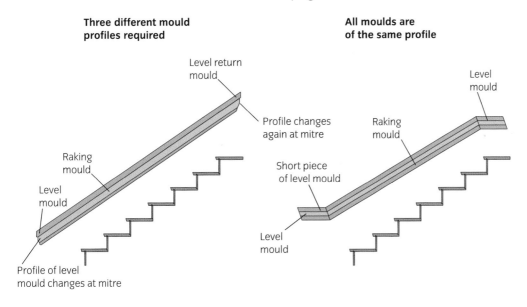

Three different mould profiles required

Level return mould

Raking mould

Level mould

Profile of level mould changes at mitre

All moulds are of the same profile

Level mould

Profile changes again at mitre

Raking mould

Short piece of level mould

Level mould

Raking moulds

The four options available for the transition of moulds to inclines are:

- rake to horizontal **transition**

- rake to vertical transition

- rake to custom hand-made profile

- true geometrical profile.

Transition

A line that connects two workflow elements

RAKE TO HORIZONTAL TRANSITION

The rake to horizontal transition method is where the moulding is returned around a corner on the horizontal plane, with the incline then mitred at the appropriate angle. This method is used where there is very little room before the start of the corner and enables use of the same size and profile of mouldings for both lengths.

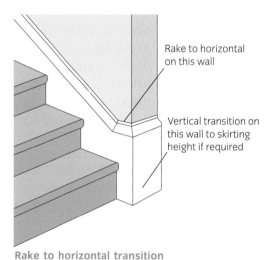

Rake to horizontal on this wall

Vertical transition on this wall to skirting height if required

Rake to horizontal transition

RAKE TO VERTICAL TRANSITION

With rake to vertical transitions, the moulding is again returned around the corner in the horizontal plane, then uses a vertical moulding that is mitred at the bisection angle. This is an option when there is slightly more room available beyond the raking mould.

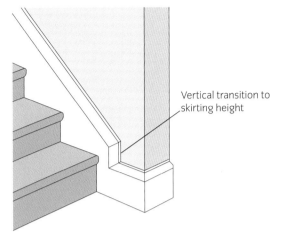

Vertical transition to skirting height

Rake to vertical transition

RAKE TO CUSTOM HAND-MADE PROFILE

The rake to custom hand-made profile is used when the return mould needs to be made to fit the raking moulding. It is really only practicable when there is a short piece of return moulding required with a relatively simple profile. Geometry will give the most accurate profile for reproducing the moulding but non-geometrical means can also be used.

This is a useful method when only a short length of material is required and could be made using simple hand tools. The following is a simple guide for obtaining the required profile.

1 Use the wall, or a block of timber, to represent the wall.

2 Mitre the horizontal material as normal.

3 Offer up the mitred raking moulding to the horizontal mitre.

4 Plot the profile from the raking moulding onto the horizontal material.

5 Use these points to mark parallel lines down the horizontal material (these will act as guide line when forming the required profile).

6 Using suitable hand tools form the required profile and finish with abrasive paper.

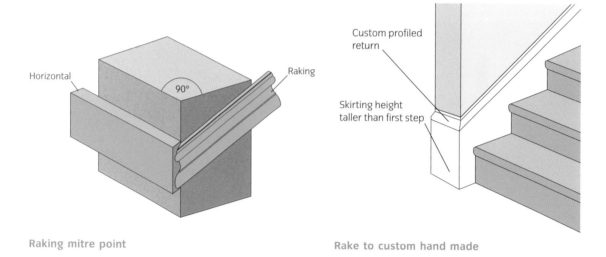

Horizontal

90°

Raking

Raking mitre point

Custom profiled return

Skirting height taller than first step

Rake to custom hand made

TRUE GEOMETRICAL PROFILE

In the true geometrical profile process we develop the true profile of the moulding we require. Either the horizontal moulding or the raking moulding can be achieved. The simple example below shows how the rectangular moulding at A changes shape as it returns horizontally at the bottom at B and at the top at C. The following example describes the process for the given moulding shown in Step 1.

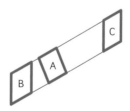

Process when given the raking profile.

1 Draw the profile at the required angle and divide it's thickness up in to any number of sections. Be aware that the more sections you use, the more confusing the drawing will become. Conversely, having only a few sections means you will have less detail to form the finished profile. In this case we have 6 sections numbered 1 to 7.

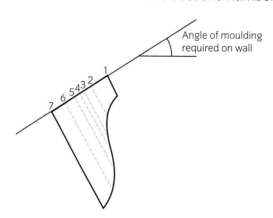

Angle of moulding required on wall

2 Project a horizontal line from point 7 to point X. Using point 7 as your centre, draw arcs from each point to touch the horizontal X–7 line.

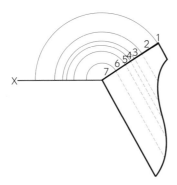

3 Divide the width of the moulding into an equal number of parts, and give them letters. In this case we have A–G. Draw lines square to A–G (shown here with solid grey lines).

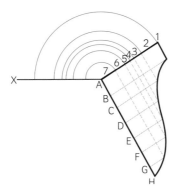

4 Draw perpendicular lines from line X–7.

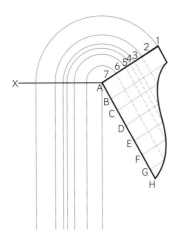

5 Extend the raking moulding lines A–G to points A–A1, B–B2, C–C3 and so on.

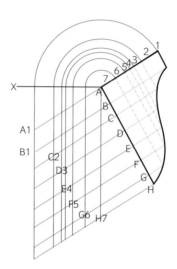

6 The new moulding profile can be drawn by connecting the intersection points.

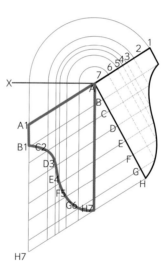

7 Producing the top horizontal moulding profile from the given
   raking profile will require the moulding to be developed above the
   raking profile. The same principles are used to develop the
   horizontal top profile. As can be seen when viewing the two
   profiles, the bottom projected profile becomes more compressed
   while the top profile becomes more stretched compared to the
   given raking profile.

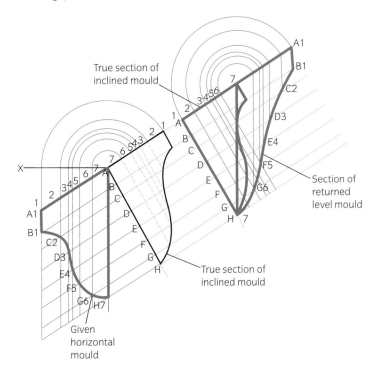

Case Study: Mark

Mark has been asked by Mr and Mrs Butler for advice on fitting decorative mouldings to the top of plain skirting boards to enhance the appearance of their rooms. They were wondering what the mouldings would look like where it turns the corner and joins the staircase. They also explained that the budget is limited so cheaper options would be preferred.

Mr and Mrs Butler have asked Mark for sketches of the options to help show the finished views. Along with the sketches, they have asked for an outline of how easy one method is to produce compared with the other, as well as any cost implications from their choice.

Looking at the requests from Mr and Mrs Butler, Mark has decided to put together two pieces of information for them:

- drawings of three options for finishing the moulding for the top of the skirting where it intersects with the stair string

- an outline of the savings to be made by not having the true development profiles made.

Putting yourself in Mark's place, answer the following three questions.

- What three drawings do you think Mark should produce?

- Which method would you recommend due to the cost implications?

- Do you think Mark should give Mr and Mrs Butler any other advice on the processes involved and, if so, what do you think should be covered?

Work through the following questions to check your learning.

1   What does the bracing used in framed, ledged and braced doors help to provide?

   a   Additional resistance to distortion.

   b   A means of fixing a lock.

   c   Additional fixing for hinges.

   d   A means of fixing a wicket door.

2   Rebate locks are normally fitted to

   a   hanging stiles

   b   top rails

   c   meeting stiles

   d   bottom rails.

3   What is an escutcheon used for?

   a   To cover the lock.

   b   To assist with door closing.

   c   As a type of closing hinge.

   d   Finish to a keyhole.

4   Which one of the following is a type of double door set often found in domestic dwellings?

   a   Spanish doors.

   b   English garden doors.

   c   French doors.

   d   American doors.

5   Which of the following is **not** classed as safety glass?

   a   Float.

   b   Laminated.

   c   Toughened.

   d   Georgian wired.

6   What is the purpose of an intumescent strip?

   a   To seal the floor, preventing water gaining access.

   b   To prevent water running down and under the door.

   c   To seal the door when the door is exposed to fire.

   d   To seal around the letter plate.

7   Which term is used to express heat loss?

   a   U-value.

   b   K-value.

   c   A-value.

   d   E-value.

8   Timber can be described as 'hygroscopic'. This means it can

   a   absorb moisture only

   b   lose moisture only

   c   both absorb and lose moisture

   d   neither absorb nor lose moisture.

9   What is the **minimum** gap required around external double doors?

   a   7mm.

   b   5mm.

   c   3mm.

   d   1mm.

10   An obtuse angle of 120° has a bisection angle of

   a   45°

   b   60°

   c   90°

   d   240°.

11 Where **two** mouldings of equal size meet the bisection line will **not** be straight when

   a   the bisection angle is an obtuse angle

   b   the bisection angle is an acute angle

   c   one of the mouldings is curved

   d   both of the mouldings are curved.

12 Moulding on a panelled wall that is fixed to follow the angle of a staircase is described as

   a   raking

   b   angled

   c   sloping

   d   geometric.

13 What is the **best** scale to use when drawing the setting out for curved moulded sections?

   a   1:10.

   b   1:20.

   c   1:1.

   d   1:2.

14 Which one of the following hinges encourages a door to self-close?

   a   Butt hinge.

   b   Expanding hinge.

   c   Parliament hinge.

   d   Hawgood hinge.

15 Which one of the following hinges allows a door to fold back against a wall and past a thick architrave?

   a   Butt.

   b   Expanding.

   c   Parliament.

   d   Hawgood.

16 Which one of the following Building Regulations Approved Documents provides information on the locations for the use of safety glass?

   a   Part K.

   b   Part N.

   c   Part L.

   d   Part A.

17 Which one of the following statements is true about raking mouldings?

   a   They are fitted horizontally to the staircase.

   b   They are fitted vertically to the staircase.

   c   They are fitted at an incline with the staircase.

   d   They are fitted as an arc around the staircase.

18 Mouldings that have curved profiles based on circles and parts of circles are described as

   a   Roman

   b   Grecian

   c   geometrical

   d   technical.

19 How thick is a fire door with the stamp mark FD60 on it?

   a   24mm.

   b   34mm.

   c   44mm.

   d   54mm.

20 The term given to the process of applying saw cuts to the back of mouldings that fit against curved walling is called

   a   kerbing

   b   knowing

   c   knotting

   d   kerfing.

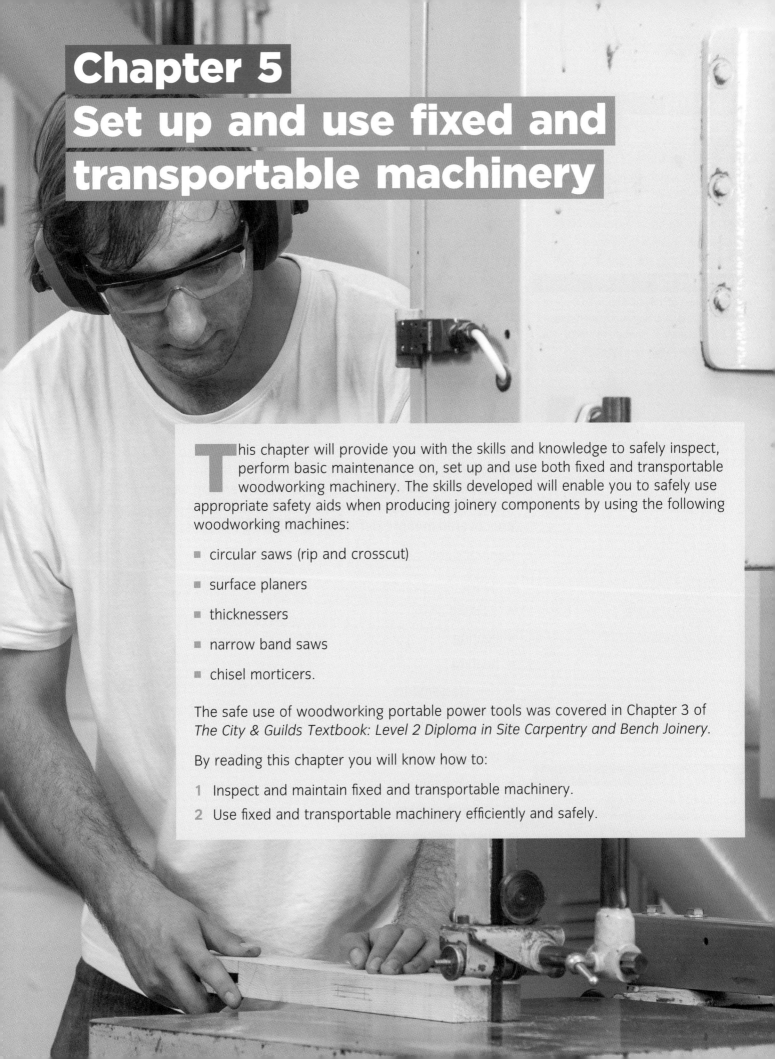

Chapter 5
Set up and use fixed and transportable machinery

This chapter will provide you with the skills and knowledge to safely inspect, perform basic maintenance on, set up and use both fixed and transportable woodworking machinery. The skills developed will enable you to safely use appropriate safety aids when producing joinery components by using the following woodworking machines:

- circular saws (rip and crosscut)
- surface planers
- thicknessers
- narrow band saws
- chisel morticers.

The safe use of woodworking portable power tools was covered in Chapter 3 of *The City & Guilds Textbook: Level 2 Diploma in Site Carpentry and Bench Joinery*.

By reading this chapter you will know how to:

1 Inspect and maintain fixed and transportable machinery.
2 Use fixed and transportable machinery efficiently and safely.

The primary piece of legislation that governs the use of work equipment, which includes woodworking machines, is the Provision and Use of Work Equipment Regulations 1998, commonly known as 'PUWER'. These regulations require the prevention or control of risks to people's health and safety posed by work equipment that they use while at work. Although PUWER deals with all work equipment in general, rather than specifically dealing with woodworking machines, the Approved Code of Practice (ACoP) for safe use of woodworking machinery gives advice on the precautions to take in order to ensure the safe use of woodworking machines. It is accepted that if you follow the ACoP you will also be complying with PUWER.

While PUWER specifically deals with work equipment, all other health and safety legislation must be adhered to as well, in particular the requirement for risk assessment. For further details, refer back to Chapter 1.

Other than machine-specific considerations that also need to be considered, there are more general considerations that must be taken into account when setting up and operating woodworking machines. Both these categories can be summarised as follows and are covered in more detail in the following sections.

General safety aspects:

- floors
- working areas
- lighting
- heating
- risk assessments.

Machine-specific aspects:

- training
- machine controls
- information plates
- dust collection/extraction
- braking
- maintenance
- noise
- tooling
- limitations.

GENERAL SAFETY ASPECTS

FLOORS

Floors need to be in good order, level and non-slip. The area around the machine needs to be kept clear of chippings and offcuts. Any cabling or ducting should either be at high level or below ground, otherwise they should be routed in such a way so as to not interfere in any way with the safe operation of the machine.

WORKING AREAS

A clear working area needs to be established around the machine to allow safe operation without the operator being knocked while operating it. Bins should be provided into which offcuts can be put. There should also be sufficient room to handle the material being machined and to safely store the machined components.

LIGHTING

All areas around the machine must be adequately lit. The light can be natural, artificial or a combination of both. Any artificial lighting that starts to flicker must be replaced straight away to reduce the risk of **stroboscopic effect**; this can give the effect that the machine tooling is stationary when in fact it is still moving. Double tubes and diffusers can help minimise the effects of strobing.

Stroboscopic effect

The illusion of making moving objects appear stationary, caused by a flickering light source

HEATING

A minimum temperature of 16°C is required for the safe operation of woodworking machines. If the temperature falls below this level then the operator can lose the ability to safely control the workpiece, as well as losing concentration.

RISK ASSESSMENTS

A risk assessment is required by Regulation 3 of the Management of Health and Safety at Work Regulations and should be carried out on all aspects of machine use. It should identify all significant hazards and put suitable control measures in place so as to reduce these hazards to an acceptable level. (See Chapter 1 for more information.)

ACTIVITY

Use the HSE website to find exemplar risk assessments. Compare these with the ones in your organisation. List any changes and recommendations that you think appropriate and discuss these differences with your tutor.

MACHINE-SPECIFIC ASPECTS

TRAINING

Training must be provided to everyone involved in setting up, operating and carrying out maintenance operations on woodwork machines. A list of authorised operators should be kept and displayed, stating which machine they have been authorised to use.

E. G. Redfern Ltd   List of authorised machine operators

The authorised trainer of _____ is _____
                                    (the company)   (name of trainer)

Date _____

I certify that:

    (a) I have carried out training, as indicated on the machines listed

    (b) I am satisfied that the people named below have demonstrated competence in the operation of the machines listed and have met all the training objectives for those machines, including:

        (i) correct selection of machine for type of work to be done;

        (ii) purpose and adjustment of guards and safeguards;

        (iii) correct selection and use of safety devices – push sticks, push spike, jigs and work-holders;

        (iv) practical understanding and application of legal requirements;

        (v) safe working practices to include feeding, setting, cleaning and taking off.

Signed _____ (Trainer)

Machine												
Operator's name	Circular saw	Cross-cut saw	Dimension saw	Surface planing machine	Thickness planing machine	Single-ended tenoner	Single-moulder	High-speed tenoner	Four-sided planer/moulder	Narrow band saw	Band re-saw	etc
J. Brown	✓	✓	✓	✓	✓	✓						
J. Bahri	✓	✓	✓	✓	✓					✓		
T. Yeung	✓	✓										

Example of an employer's machinery training record

(Adapted from HSE)

Nobody should use any woodworking machine unless they have received sufficient training and acknowledged that they have understood the training. The training should include instructions concerning general aspects of woodworking machinery, as well as machine-specific instructions.

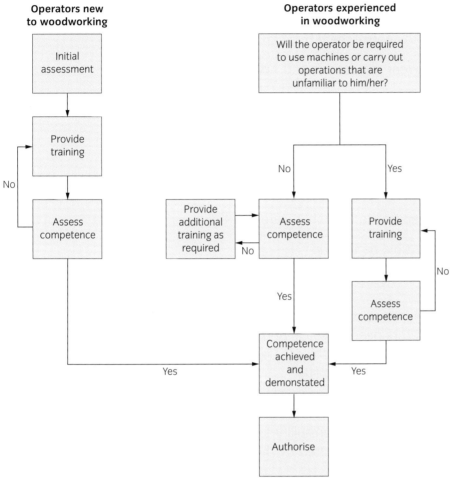

Competence assessment process for using woodwork machinery

Before using woodworking machinery, a risk assessment should be carried out to help identify the user's suitability, maturity, training requirements and supervision ratios. When training is in its early stages, it is expected that the supervision level will be high. As the user begins to demonstrate safe working practices and shows confidence in using the machines, a gradual reduction in supervision levels can be allowed.

All wood machining training schemes, including those as part of a joinery qualification, should include the following elements:

- general skills

- machine-specific skills

- machine familiarisation

■ demonstrating competence

■ competence checklists

■ record keeping.

General skills

General health and safety skills include an awareness of risks and how to control them through:

■ current regulations and Approved Codes of Practice

■ extraction

■ noise control

■ correct use of lifting aids

■ correct use of PPE for eyes, ears, hands, etc

■ keeping the workshop safe and tidy

■ sensible behaviour

■ awareness of other operators.

Machine-specific skills

Operators need practical and theoretical instruction in the safe operation of all machines. This should be covered within the training programme and include:

■ the main causes of accidents

■ operatives' responsibility for their own safety and that of others who may be affected by their working practices

■ the importance of reporting defects to responsible people

■ the dangers and limitations of working practices and ancillary equipment, for example:
  □ safety aids like push sticks and blocks
  □ the risks from in feeding and taking off material
  □ **dropping on**
  □ **kickback**
  □ **jigs**
  □ types and correct use of tooling
  □ timber selection
  □ curved working.

■ knowledge and demonstration of safe working practices for each stage of the process, for example:
  □ machine isolation, emergency stops, interlocks and speed controls
  □ purpose, use, limitation and adjustment of guards
  □ setting up, correct tool section and changing/replacing tooling

Dropping on

Starting or finishing a cut part way along a length of timber

Kickback

The material being forcefully pushed back towards the operator

Jig

A means of safely holding components while carrying out machining operations. Used to provide repeatability and accuracy in machining operations

- selecting and fitting correct guarding for machining process
- pre-start safety procedures including extraction, ear protection and safety glasses
- operation of the machine for different machining processes
- maintenance and fault reporting procedures.

Machine familiarisation

All operatives should be familiar with the machine, its ancillary equipment and machining processes, including through on-the-job training under close supervision.

Demonstrating competence

After the training has taken place, the operative's competence should be assessed to see if the training has been successful. The assessor must be someone who knows the machining process, its risks and the safe working practices that should be used.

Operatives can only be classed as competent when they can demonstrate that they use the required knowledge and safe working practices all the time.

Competence checklists

A competent operative should be able to demonstrate:

- that they can select the correct machine, tooling and protection devices

- the ability and confidence to say, 'This is the wrong machine for this job; it can be done more safely on …'

- what the guards do and how to use and adjust them properly, as well as any other protection devices, for example:
  - on a circular saw, why you need a riving knife and how to set it and adjust the top guard
  - on a planer, why you need to use the bridge guard, why and how it should be set

- knowledge of safe methods of working, including appropriate selection of jigs, holders, push sticks and similar protection devices

- understanding of the legal requirements for guards to be used correctly

- knowledge of the nature of the wood and the hazards that this can cause, such as kickback, **snatching**, short grain and ejection.

Record keeping

While undergoing training, it is good practice to keep written records for each operative detailing the types of training they have received to date. Once the operative has received the necessary training and has demonstrated their competence, it is good practice to authorise them in writing for the machines and operations that they can use.

Snatching

In a machining context, this is where the rotating cutter hooks into the split, causing it to break or the work to be thrown back at the operative. The operative could lose control or have an accident

An isolation switch

MACHINE CONTROLS

All machines should be fitted with a means of safe isolation from the electricity supply. This should be sited on or as close to the machine as possible. A safe means of stopping and starting the machine that is within easy reach of the operative's normal working position should also be included.

INFORMATION PLATES

All machines will carry specific information about the machine, for example the model number and operating speeds.

DUST COLLECTION/EXTRACTION

All woodworking machines should be fitted with a means of collecting dust and chippings. A ducted exhaust ventilation system is preferable but a local system is acceptable. In all cases the system needs to be regularly checked and inspected, with appropriate safety certificates issued.

BRAKING

All the woodworking machines covered in this chapter require braking of some form. All new machines are fitted with automatic braking systems but some older machines could have manually operated braking systems. Whichever system is incorporated into the design of the machine, it must be capable of bringing the machine tooling to a controlled safe stop in a **maximum time of 10 seconds**.

MAINTENANCE

Maintenance of woodworking machinery is a legal requirement under PUWER. A written record of the maintenance must also kept. Maintenance should be carried out only by those people trained and competent to do so. Maintenance is not only a legal requirement, it also makes good economic sense; like a car or any other machine, if you have failed to change the oil or replace the drive belts, after time they will break down or fail to perform correctly. Faults such as damage, DIY repairs, missing or damaged guards and poor wiring must be addressed immediately even if this means taking the machine out of use. Use only suitable replacement parts and tooling.

NOISE

All employees must be protected from the loud noise produced by woodworking machines and portable power tools. Exposure to loud noise will permanently damage hearing, resulting in deafness and **tinnitus**. Under the Control of Noise at Work Regulations 2005, duties are placed on manufacturers, employers and employees to reduce the risk of damage to hearing as far as reasonably practicable.

FUNCTIONAL SKILLS

Produce and complete a simple chart that will record the time it takes for each of your machines to stop.

Work on this activity can support FICT L2 (6.A6).

Wearing ear and eye protection when machining

Tinnitus

A constant and permanent ringing or buzzing in the ears which can be caused by exposure to load noise

Manufacturers and suppliers are required to provide low-noise machinery and information about the noise levels produced by the machinery. Employers must reduce noise levels to a reasonable level to protect their employees' hearing, with specific noise levels of 80 **decibels (dB(A))** and 85 dB(A) requiring action:

- At 80 dB(A) the Noise at Work Regulations require the employer to:
  - make a suitable assessment of the noise workers are exposed to
  - provide suitable ear protection to those employers who request it
  - provide information about the risks to hearing for their employees and information regarding the legislation.

- At 85 dB(A) the employer must:
  - reduce the amount of noise exposure as far as reasonably practicable by means other than ear protection, eg by erecting physical barriers like acoustic enclosures or changing to newer quieter tooling
  - designate the work area an 'ear protection zone' and establish a safe system of work to protect visitors and other workers, providing suitable signage stating, 'Ear protection must be worn in this area'
  - provide suitable ear protection and ensure that everyone who enters the designated area wears it.

Employees have a duty to wear and use all ear protection provided, as well as to comply with all other noise control procedures and to report any defects.

TOOLING

The choice of tooling is vital to the safe operation of the equipment. Cutters must be the correct type, kept sharp and in good condition, and suitable for the task in hand. Tooling should be the correct size and not exceed the correct safe operation speed as specified on the tooling or by the manufacturer. Only cylindrical blocks can be used on hand-fed machines with limited cutter projection.

Decibel (dB(A))

Decibels are used to measure sound levels, while the A weighting is applied to instrument-measured sound levels to try to account for the relative loudness perceived by the human ear

FUNCTIONAL SKILLS

Go to the HSE website and download wood information sheet number 37, 'Selection of tooling for use with hand-fed woodworking machines.' How does the tooling within your place of work or training establishment compare? Do you need to make changes? Produce a report for your tutor outlining your findings.

Work on this activity can support FICT L2 (9.A1) and FE 2.3.2.

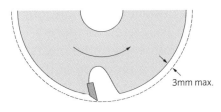

Solid profile block

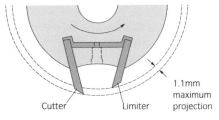

Chip limited cutter block

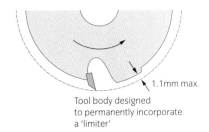

Cylindrical blocks

LIMITATIONS

Backer off

Someone who removes material from the machine after processing, also known as a 'taker off'

All operators and **backer offs** should be instructed in the limitations of both tooling and equipment. Because some types of woodworking machines are capable of carrying out several types of operations, ensuring that the most suitable type of woodworking machine is chosen for the specific task is vital. For example, a groove would be better machined on a vertical spindle moulder rather than on a circular rip saw, but if no vertical spindle moulder is available then a circular rip saw can be used provided the saw blade is fully enclosed in a tunnel guard by replacing the crown guard.

The following illustration shows a tunnel being formed using two shaw guards: one applying pressure downwards, forcing the material onto the machine bed, and the other applying pressure to the side, forcing the material up to the fence that has been pushed forwards.

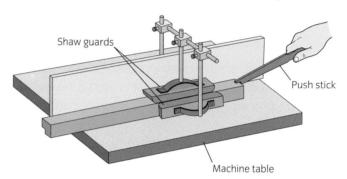

A tunnel being formed using two shaw guards

To help support the ACoP *The Safe Use of Woodworking Machinery*, the HSE has published a series of wood information sheets. These sheets contain practical advice and guidance on the safe use of woodworking machinery.

TIMBER SUPPORT SYSTEMS

Large and/or heavy materials not only require particular attention paid to supporting them through the machining process, but also to the techniques required in handling and supporting them when presenting them to the machine. Some machines, like crosscut saws, have large roller beds that are specifically designed to handle long and heavy materials, while other machines may need temporary support from rollers positioned at one or both ends of the machine. Examples of this would be long door jambs on a morticer, or the in-feed end of a rip saw and both sides of a narrow bandsaw.

Before any cutting of material is undertaken, the operator should ensure that an adequate dust extraction system is connected to the saw bench and is working correctly. Correctly designed and fitted dust extraction systems should remove the need for dust masks to

ACTIVITY

Go to the HSE website and download wood information sheet number 16 for circular rip saws and display this next to your circular saw.

Temporary support roller

be used. Circular sawing machines produce high noise levels, far in excess of the minimum exposure levels that require actions to be taken to reduce the risk of hearing damage (85 dB(A)). Your employer will have taken all reasonable steps to reduce this level, eg by using sharp tooling of the correct design and keeping the machine well maintained. But in practice high noise levels will remain and the operative should wear suitable ear protection, as should any support worker involved with the sawing operation.

It is recommended that eye protection is also worn while cutting operations are undertaken in order to reduce the risk from flying particles. Gloves can help prevent injury from splinters and are particularly useful if working with toxic timbers like Afrormosia and Douglas fir, whose splinters can easily cause the wound to become **septic**.

ACTIVITY

Locate the woodworking information sheet number 30, 'Toxic woods', on the HSE website and see whether you are likely to machine any of the timbers mentioned. Record the likely effects that could result from using these timbers if the correct precautions are not taken. Display your findings in your workshop.

Septic

When a wound becomes infected; can lead to septic shock which can lead to organ failure

CIRCULAR SAW BENCHES

Circular saw benches are among the most common in general use and account for around one-third of all accidents at woodworking machines. They can be divided into three broad types:

- rip saws

- dimension saws

- crosscut saws.

RIP SAWS

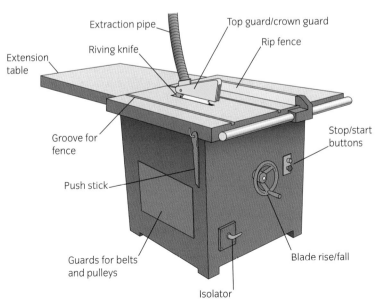

Extraction pipe

Top guard/crown guard

Riving knife

Rip fence

Extension table

Groove for fence

Stop/start buttons

Push stick

Guards for belts and pulleys

Blade rise/fall

Isolator

Parts of a hand-fed circular rip saw

Circular rip saws look very much like dimension saw benches but carry out three basic operations:

- *Flatting:* the timber is ripped down its length through its thinnest section (thickness) to the required width.

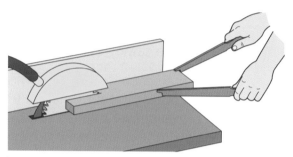

Flat cutting

- *Deeping:* the timber is ripped down its length through its thickest section (width) to the required thickness.

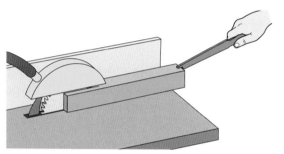

Deep cutting

- *Angled cutting:* the material is placed on a jig or saddle that holds the material stable at the required angle during the cutting operation. This ensures an accurate angle is produced on the material but more importantly also aids the safe delivery of the material and safe control of the 'offcut'.

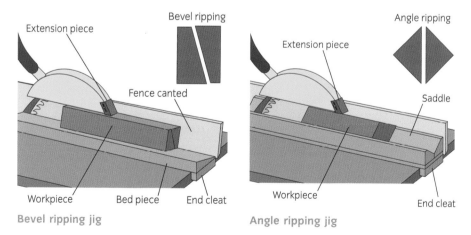

Bevel ripping jig

Angle ripping jig

Rip saws can only be used for ripping down the grain and are used mainly to rip timber to the required sizes. The processes of selecting tooling, setting up and operation are almost the same as those required for the dimension saw.

DIMENSION SAWS

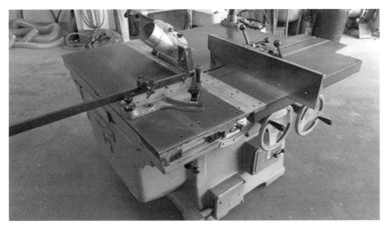

Parts of a dimension saw

The dimension saw is a more versatile machine. It incorporates a sliding table on which a crosscut fence can be fitted, allowing the dimension saw to crosscut as well as rip. The saw blade is able to **cant** over to an angle of 45°, allowing angle cuts, and by using the crosscut fence **compound cuts** can be performed without the aid of saddles or jigs. The dimension saw is usually used for fine, accurate work and for cutting man-made sheet materials like MDF.

Cant

Lean

Compound cut

A cut that consists of two angles: the bevelled angle from the canted saw blade and the mitre angle (or crosscut angle) from the fence

CROSSCUT SAWS

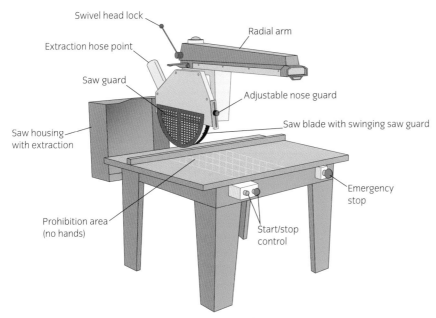

Components of a radial arm crosscut bench

Crosscut saws are used to cut across the grain of the timber, usually to a predetermined length from the adjustable end stops. Other operations carried out on the crosscut saw include:

- cutting tenons

- square cuts

- angled cuts

- compound angle cuts

- cutting birdsmouths

- cutting housings

- cutting halving joints

- notch cuts

- kerfing.

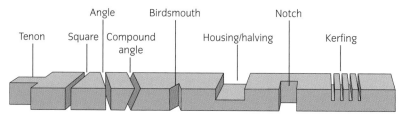

Operations carried out using a crosscut saw

SAW BLADE TERMS

The following are terms you need to know.

Term	Description
Pitch	The distance between tooth tip and tooth tip
Clearance angle	The distance between the heel (see page 250) and the cutting circle
Sharpening angle	The angle between the clearance angle and the angle of hook or rake
Hook or rake angle	The angle of the tooth in relation to a line taken from the tooth tip to the centre of the blade
Gullet	The area between the teeth that contains the saw dust during cutting
Point	The top of the tooth, which makes first contact with the material being cut

Term	Description
Heel	The back of the tooth
Face	The leading (front) edge of the tooth
Set	The clearance on either side of the tooth
Kerf	The total width of the saw cut, which comprises the thickness of the blade and the set on either side of the blade

TOOLING FOR CIRCULAR SAWING MACHINES

The choice of saw blade is critical for the safe operation of the saw bench. Saw blades are used for three main machining functions:

- ripping

- crosscutting

- combination or general purpose.

Most saw blades in general use today consist of a parallel plate steel blade onto which are brazed **tungsten carbide tips (TCTs)**. It is these TCT teeth that do the cutting, having almost totally replaced the use of parallel plate steel blades with sprung set teeth. TCT is a very hard material and will withstand the dulling effect of cutting hard **abrasive** timbers and man-made materials such as MDF, as well as softwoods, much better than the traditional plate steel blades. Parallel plate steel blades will cut softwoods very satisfactorily but will require regular maintenance to keep them working safely.

Tungsten carbide tips (TCT)

The cutting edges of a saw blade. They are very hard wearing but are brittle and can be easily damaged when changing the blade

Abrasive

A material that will quickly dull/blunt the cutting edge of tooling

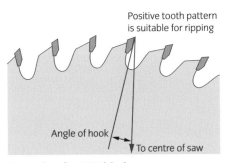

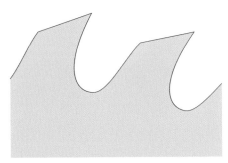

Example of a TCT blade   Example of a plate saw

The main difference between saw blades designed to rip down the grain and crosscut across the grain is the **angle of hook or rake**. This angle is referred to as:

Angle of hook or rake

The angle at which the face of a saw tooth slopes from the tooth tip, either down and forward from the tip, as in the case of negative tooth profiles for cross cutting, or down and backward from the tooth tip, as in the case of positive tooth profiles for ripping

- *positive angle of hook or rake:* to be used when ripping down the grain

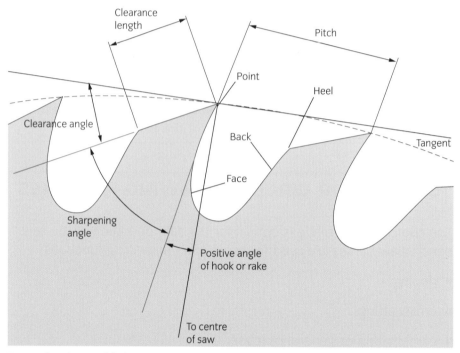

Parts of a rip saw blade (positive hook angle)

- *negative angle of hook or rake:* to be used for cutting across the grain.

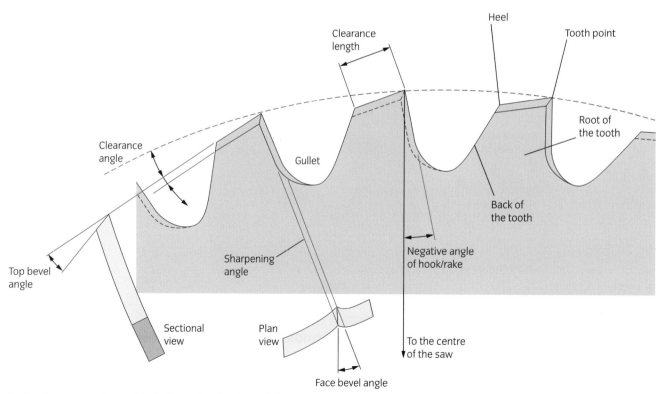

Parts of a crosscut saw blade (negative hook angle)

It is vitally important that you do not use a positive hook pattern blade for crosscutting operations. The tooth design has a tendency to grab at the material, causing the crosscut to snatch and try to run out into the timber, potentially risking injury and causing damage.

A typical rip saw blade has an angle of 20–25° positive hook for softwood, and 10–15° for cutting hardwood. In contrast, a crosscut blade will have a negative angle of hook of 5° for softwood and 10° for hardwood.

Combination or general purpose saw blades have a zero degree angle (also known as 'neutral angle') of hook or rake, and are used on dimension sawing machines where they are continually changing from ripping to crosscutting operations. A zero angle of hook only allows for a slow ripping operation.

The way the tooth face is ground also differs. The crosscut tooth design needs to sever the timber fibres across their length and therefore requires teeth with a more needle-point design. Teeth belonging to the rip saw blade have a chisel edge to them, allowing the blade to pull the fibres out as opposed to trying to sever them.

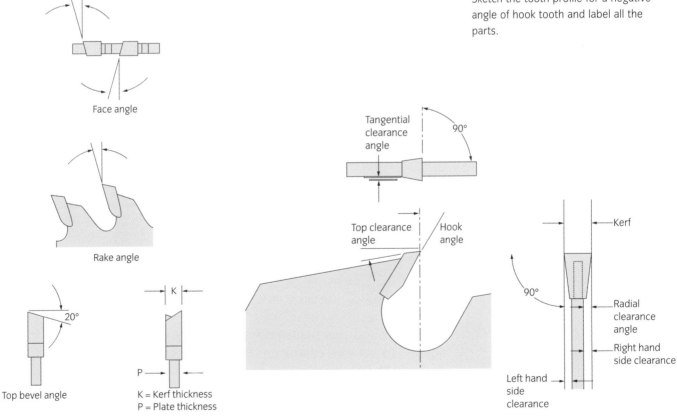

Crosscut teeth

Rip saw teeth

There are also other special types of blades of which the 'hollow ground blade' is the most common; this type of blade is typically used on crosscut and dimension saws. These types of blades give very fine finishes and are therefore typically used for finish cutting.

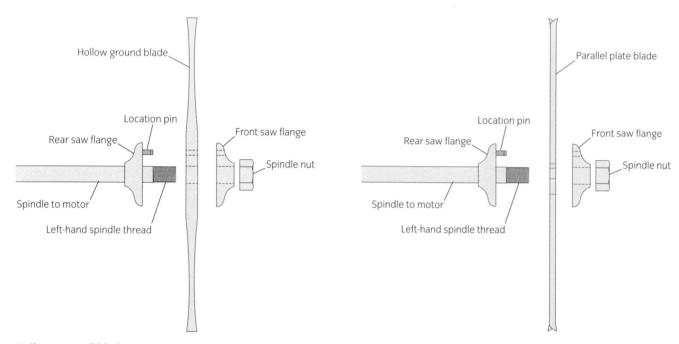

Hollow ground blade

Parallel plate blade

All saw blades require clearance during cutting. This is to prevent friction and binding and is provided by the set, the clearance cut made by the bent tooth either side of the saw.

Spring set

The bending of the top third of each tooth in alternate directions around a saw blade

Parallel plate saw blades achieve this clearance by applying **spring set**; this is the bending of alternating teeth in opposite directions around the saw blade.

Tungsten carbide tipped blades achieve set by the way the tungsten tip is ground, and the fact that the tip is wider than the thickness of the saw blade.

TCT blade showing built-in set from the teeths

Hollow ground saw blades do not have set applied as it is not needed to achieve clearance. This is achieved by the hollow ground shape of the blade.

Peripheral speed

The speed at the periphery of a blade; how fast the outer edge of the blade is travelling

Circular saws must show clearly on the machine the minimum diameter of saw blade that can be used on it. No blade that is less than 60% of the largest sized blade that the machine is designed to use can be used. This is because smaller diameter blades produce lower **peripheral speeds**, meaning the cutting speed of the blade will not be high enough to cut the timber efficiently. To efficiently and safely cut timber, it is recommended that a peripheral speed of 50 metres per second is reached.

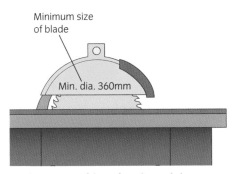

Notice on machine showing minimum size to be used

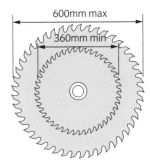

Smallest blade not less than 60% of the largest size

SAFE OPERATION OF CIRCULAR RIP SAWS AND DIMENSION SAWS

The principles for safe operation of circular rip saws and dimension saws are the same and can be dealt with together. Where differences do occur, they will also be described in this section. Refer back to pages 245 and 247 for illustrations of these types of machines.

INTRODUCTION TO HAND-FED CIRCULAR RIP AND DIMENSION SAWS

All hand-fed circular saws must be guarded. This is achieved in three ways:

- A guard must be fitted to the top and front of the saw blade above the table. This guard is known as the 'crown guard' and can have adjustable nose pieces fitted at the front.

- Guarding from the rear is achieved by the use of a riving knife and extension table.

- Guarding from below is provided by the frame of the machine bed and framework.

Saw guards are required to be:

- strong and rigid so they cannot be accidently moved or knocked out of position and into the saw blade

- easy to adjust and correctly set in position

- set to the current requirements of PUWER and the ACoP for woodworking machinery

- checked and maintained on a regular basis.

Crown guard

The upper part of the saw blade is guarded by the crown guard. Traditionally crown guards were made from steel and usually incorporated an extra adjustable nose section, while more contemporary designs are made from rigid transparent plastic and are easily adjusted via a spring-loaded mechanism.

The crown guard should cover both sides of the blade and be set as close to the top of material being machined as possible. The saw blade should always protrude through the material so that the lowest part of the gullet is always covered by the crown guard.

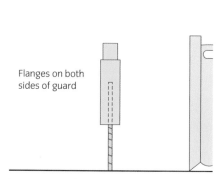

Flanges on both sides of guard

Correct guard position

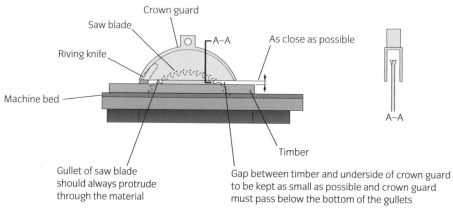

Crown guard

Saw blade

Riving knife

A–A

As close as possible

Machine bed

A–A

Timber

Gullet of saw blade should always protrude through the material

Gap between timber and underside of crown guard to be kept as small as possible and crown guard must pass below the bottom of the gullets

Positioning of the crown guard

Riving knife

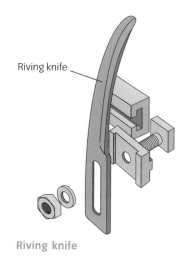

Riving knife

Riving knife

The rear of the blade is guarded by the riving knife. The riving knife has another function as well as acting as a guard, and that is to help stop timber from binding on the back of the saw blade. This would potentially forcefully throw the timber back towards the operative, risking serious operator injury and damage to the machine.

The riving knife must be correctly positioned, both to comply with the ACoP but also to aid the correct smooth operation of the machine. The correct positioning of the riving knife will depend on the size of the blade. For blades up to 600mm in diameter, the riving knife must be within 25mm of the top of the blade and 8mm from the back of the blade. For blades over 600mm in diameter the riving knife must extend above the machine bed by at least 225mm. In the following illustrations, the crown guard has been omitted for clarity.

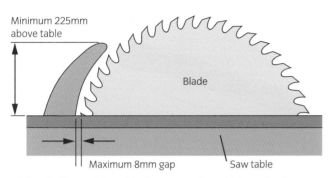

Riving knife position: blades more than 600mm in diameter

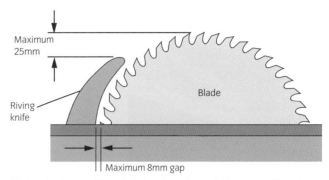

Riving knife position: blades less than 600mm in diameter

Riving knives should have a leading edge that has a slight taper or rounded edge, and should be thinner than the kerf but 10% thicker than the gauge of the blade.

Extension table

When an assistant is used to remove material at the out-feed end of the saw bench, an extension table must be fitted so that the distance between the saw spindle and the out-feed measures at least 1,200mm.

It is also good practice to have support rollers or a support table if long lengths are to be cut in order to provide support for the material during the cutting process.

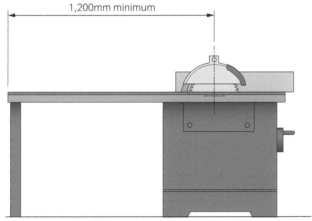

Positioning of extension table

> **INDUSTRY TIP**
>
> Always have a selection of wedges behind the saw blade. Wedges can then be easily and quickly inserted into the saw cut in the timber to prevent the timber from closing in any further.

Push sticks

At every circular sawing machine there should be at least one push stick to aid delivery and removal of components. Push sticks should be around 450mm long with a birdsmouth at one end to positively locate on the timber, and must be used to push the last 300mm of material through the saw.

Using push sticks while machining

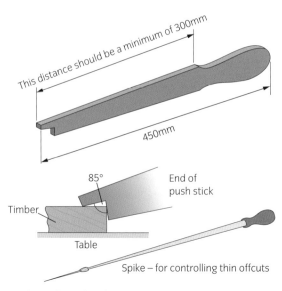

Push stick and spike

The operative's hands and those of any helpers who are assisting with either the delivery or removal of the components should not be in line with the saw blade or come within 300mm of the blade. A comfortable hand hold needs to be provided at least 300mm away from the birdsmouth to ensure the operative's hands are kept at least this distance away from the blade.

Mouthpiece and packers

Either side of large-diameter saw blades (typically those over 600mm in diameter), packers are inserted. These are usually made out of dense felt. They are there to help provide stability to the saw blade during cutting and can be used as a means of lubrication to help keep the blade cool and clean. (They are impregnated with thin oil or an oil/paraffin mix.)

On smaller capacity machines and dimension saw machines, timber is fixed to the machine bed either side of the blade. This is to keep the gap between the blade and the metal part of the machine as small as possible in order to aid safe cutting. By using timber inserts, the risk of damage to the blade if it was to develop a slight wobble (possibly due to loss of tension) is removed. Immediately in front of the saw teeth and wrapping around either side is a timber mouthpiece insert. This insert is also designed to reduce the gap around the teeth and the machine bed to the smallest amount possible, in order to give support to the material being cut and to reduce the chances of **spelching** or breakout on the underside.

INDUSTRY TIP

Close the dust extraction system gates to the machine before gaining access to the saw blade. This will reduce the likelihood of the packers being lost down the extraction system.

Spelching

Part of the material breaking away in an uncontrolled manner, producing damaged components

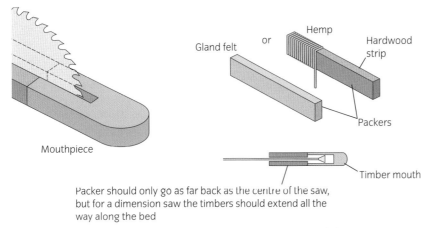

Timber mouthpiece and side packers for circular rip saw and dimension saw

SETTING UP AND OPERATION OF CIRCULAR RIP SAWS

Each piece of timber that is offered up to be cut poses its own problems depending on its length, section size, how bent or twisted it is and which operation is to be carried out. The following paragraphs give typical guidance on how best to safely process material through circular rip saw benches and typical problems that may be encountered.

Fence position and saw height settings

The fence position for ripping timber and cutting man-made sheet materials are different, as are the required saw heights. The normal correct position for the rip saw fence is with the curved front section of the fence in line with the radius of the saw blade. If the fence stops too much in front of the saw blade then the material will not remain stable at its end and could jump about, giving a poor finish as well as causing the push stick to slip. If the fence was pushed too far forward, the timber could become trapped and start rubbing on the saw blade, risking damage and kickbacks.

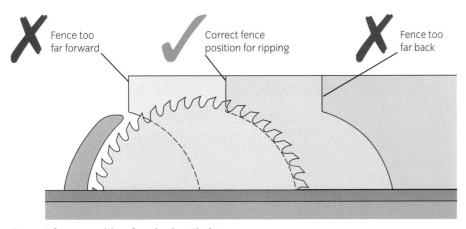

Correct fence position for ripping timber

When cutting man-made sheet materials, the fence can be pushed further forward, towards the centre of the blade. This will give extra guidance for the material during and after cutting. Because man-made materials are much more stable than timber, there is very little chance of it bending and twisting in the cut, unlike with timber.

Dimension saws usually have a smaller capacity than rip saws and so the blade is usually smaller in diameter than a circular rip saw blade. During the cutting of man-made materials, the saw blade can be lowered so that it only slightly protrudes above the material being cut; the saw teeth must still be covered by the crown guard, though. Keeping the saw blade lower when cutting sheet materials, particularly those that have face veneers, will help produce a cleaner cut with less breakout on the underside of the material. Because manufactured boards are less likely to distort than natural timber, there is less likelihood of kickback even though the material is being cut near the top of the saw blade; if you did this with natural timber the kickback risks would be greatly increased.

To help reduce breakout further, a 'running board' can be fitted to the machine bed. This can be a thin piece of plywood or similar that is part passed through the saw and fixed in position with a cleat at the in-feed end to prevent any further forward movement. The running board is then fixed at the out-feed end onto an extension table. This running board helps reduce breakout of the material by keeping the gap between the saw kerf and the machine bed (in this case the running board) as small as possible.

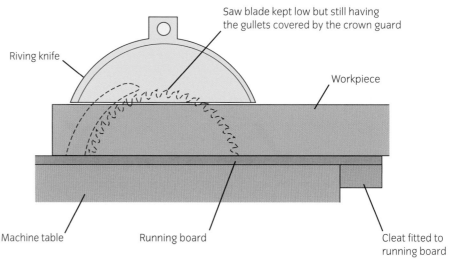

Running board position

Changing circular saw blades

The sequence for changing circular saw blades on a rip saw is similar for all types of machine design. However, you should always refer to the manufacturer's information booklet for specific information on any particular requirements for that machine.

STEP 1 Isolate the machine and close the dust extraction gates. This helps prevent any packers that may fall into the extraction ducting from being drawn away and requiring replacing.

STEP 2 Raise or remove the crown guard to expose the saw blade. Remove the finger plate and any timber packers to allow access to the saw blade locking nut. Remove the saw blade from the spindle after releasing the saw nut. For most standard machines, tighten in an anticlockwise direction and loosen in a clockwise direction.

STEP 3 Remove the riving knife. This step may not be needed if the riving knife is clean and the correct size and shape to suit the replacement blade.

STEP 4 Select the correct size and type of blade, eg TCT and tooth design. Fit the replacement saw blade, ensuring the teeth are facing in the correct direction. Make sure the blade is fully tightened using the correct spanners and collars for the machine; the tightening direction will be away from the saw blade's cutting rotation direction (ie for most standard machines tighten in an anticlockwise direction and loosen in a clockwise direction).

STEP 5 Reposition the riving knife, ensuring its height and the gap to the saw blade are correct.

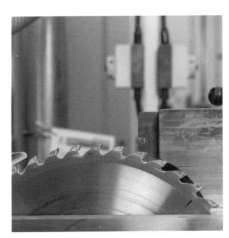

STEP 6 Re-fit the finger plate, ensuring it is flush with the machine bed. Re-fit timber packers, ensuring a good close fit to the size of the saw blade to help reduce spelching on the underside of the material. Reset the fence position to line up with the saw gullets at table height for cutting timber.

Incorrect ripping of timber (fence not pushed forward enough)

STEP 7 Set all guards including the extension table, ensuring it is at least 1,200mm long from the centre of the saw blade.

STEP 8 Ensure a suitable push stick is available that is at least 450mm long (see page 256). Conduct a trial run, listening for any unusual sounds or vibrations. Carry out a test cut.

Cutting lists

To enable accurate production of components, a suitable cutting list should be produced. Cutting lists provide the machine operator with vital information to enable them to produce components of the correct size and in the right quantities.

Cutting lists should specify:

- an overall job reference

- the type of component

- the number of components required

- the length of the component

- the type of material

- the sawn size of the component, both width and thickness (an allowance is added for planing to size)

- a planed or finished size for the component's width and thickness

- a section for additional information that may be relevant to the materials.

Cutting list for 'Our House' front door

Item Number	Component	Quantity	Length	Width sawn	Thickness sawn	Width planed	Thickness planed	Material	Comments
1	Stiles	2	2,020	100	50	95	44	Oak	Free from twist and bowing
2	Top rail	1	840	100	50	95	44	Oak	Free from splits
3	Bottom rail	1	840	175	50	170	44	Oak	Free from splits
4	Mid rail	1	840	175	50	170	44	Oak	Free from splits
5	Panels	2		400	665		12	Ext ply	
6	Panels	2		600	665		12	Ext ply	
7	Panels	2		600	665		12	Ext ply	

Example cutting list

Standard operating position

While cutting material, the operator should not stand directly in line with the saw cut but slightly off to the side. This is to prevent any timber that may be kicked back from being forcefully propelled at the operative and coming into contact with them.

Cutting material

The way timber is presented and fed through the saw bench is vital to its safe operation. Timber is subject to movement as a result of seasoning or atmospheric changes resulting from storage conditions. These changes in movement can take several forms, the most common ones being:

- *cupping:* where the timber 'cups' over the face of the board – wide boards that are cut tangentially will usually cup over time

- *bowing:* a curvature along the board's face (its widest section) from one end to the other

- *springing:* a curvature along the board's edge from one end to the other

- *twisting:* a curvature along both edges of its length, producing a propeller-shaped twist.

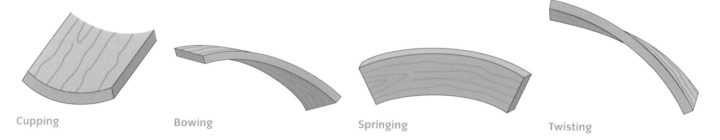

Cupping   Bowing   Springing   Twisting

Make sure you position your timber as shown in the following illustrations.

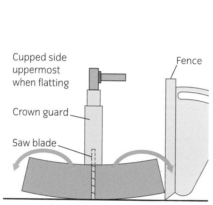

Cupped side uppermost when flatting

Fence

Crown guard

Saw blade

Cupped positioning when flatting

Correct timber position to deal with cupping when flatting

Position the timber with the cupped side facing upwards. This allows the timber to fall away from the saw blade after cutting.

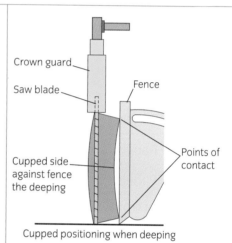

Crown guard

Saw blade

Fence

Cupped side against fence the deeping

Points of contact

Cupped positioning when deeping

Correct timber position to deal with cupping when deeping

Position the timber with its cupped side against the fence. This allows for safer feeding through the machine and allows the timber to fall away from the saw blade.

ACTIVITY

Research the moisture content for different types of timber and their intended uses, then present the information in an easily understood format. Discuss the issues that may occur if the moisture content is incorrect.

List the considerations that need to be borne in mind when selecting tooling for cutting down the grain of timbers like Afrormosia on circular saw benches. Present your findings to your tutor.

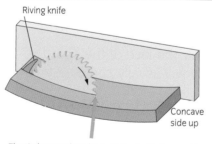

Riving knife

Concave side up

The timber needs to be in contact with the machine bed at the point of cutting with the large blue arrow

Correct timber position to deal with bowing

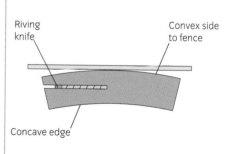

Riving knife

Convex side to fence

Concave edge

Correct timber position to deal with springing

Machining twisted timber can be problematic, so your best judgment needs to be used when following the above instructions. In the case of excessive twisting or doubt, do not use the material.

During the cutting operation timber can, from time to time, close up on the saw blade causing friction and for it to be forcefully pushed back at the operator. This is usually as a result of **case hardening** resulting from poor seasoning. To help prevent this effect, drive a wedge into the saw cut behind the riving knife, preventing the timber from closing in any further.

INTRODUCTION TO THE HAND-OPERATED CROSSCUT SAW BENCH

The primary function of the crosscut saw is to cut material to length; this is usually done before further machining operations take place. The timber is cut to minimise defects such as knots, shakes, bowing and twists in the material.

There are two main types of crosscut machines in use:

- *The travelling head crosscut:* in this design the head of the saw and the arm are at the back of the machine, requiring a larger amount of room. The arm and cutting head travel out towards the operator. This type of crosscut machine is sometimes called a 'pull-over' saw.

- *The radial arm crosscut:* this design has the arm on which the cutting head travels positioned over the table at the front (see image on the following page). As a result a lot of space is saved, as well as making it easier to carry out some operations.

All crosscut saws must be guarded in three ways:

1 Guards fitted around the back, top and front of the saw, often fitted with an adjustable frontal extension piece.

2 No access to the saw blade when it is in the rest (behind the fence) position; on older machines, this can be achieved by providing a saw housing into which the saw returns automatically.

3 Guards fitted below the machine table, usually formed by the main frame of the machine table.

All guards must be:

- strong and rigid so they cannot be accidently moved or knocked out of position and into the saw blade

- easy to adjust and correctly set in position

- set to the current requirements of PUWER and the ACoP for woodworking machinery

- checked and maintained on a regular basis.

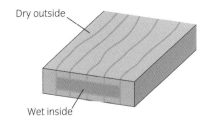

Dry outside

Wet inside

Case hardening

Case hardening

A defect caused by the timber being dried too rapidly, leaving the outside dry but the centre still wet. Typically causes the material to bend and twist during cutting, resulting in binding on the saw blade and kickback

ACTIVITY

Go to the HSE website and download wood information sheet number 36, 'Safe use of manually operated crosscut machines', and display it next to your machine.

Components of a crosscut saw

The major parts of the machines are outlined as follows.

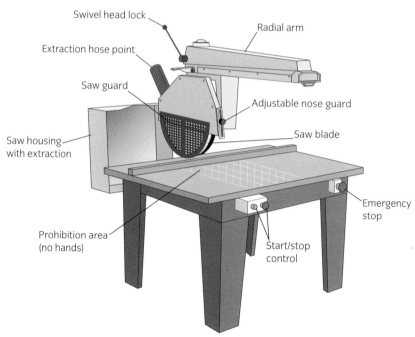

Major component parts to a radial arm crosscut bench

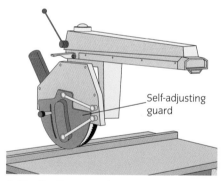

Self-adjusting guard

- *Swivel head lock:* used to lock off the radial arm at the required angle.

- *Extraction hose point:* used to collect and control saw dust.

- *Saw housing:* used as a safe resting place for the saw blade, acting as a guard while the saw is at rest.

- *Saw guard:* acting as a guard to the saw blade while in operation; must cover as much of the saw blade as possible and at least past the saw spindle. A modern version of the guard self-adjusts as the saw passes through the material being cut, thus eliminating the need to set the guard height.

- *Saw blade:* negative tooth patterned blade for cutting across the grain.

- *Start/stop switch:* used to start and stop the machine; must be within easy reach of the normal operating position.

- *Prohibition area:* an area of the machine bed 300mm each side of the saw's travel is a 'no hands' area. For short lengths of timber, a holding device is to be used.

- *Emergency stop:* used to stop the saw with your knee in an emergency, while your hands are holding the material and controlling the saw movement.

- *Adjustable nose guard:* used to reduce the gap between the guard and the timber to as small a gap as possible.

- *Radial arm:* enables the saw to rotate to the left or right so that a cut from 45–90° in either direction can be made.

- *Back fence:* a fence to prevent material moving into the blade and of sufficient height to support the material being cut. The gap for the saw cut should be kept as small as possible.

- *Machine bed:* a 'sacrificial' bed should be incorporated into the machine enabling the saw teeth to pass fully through the material being cut, reducing spelching on the underside.

- *Machine end stops:* used to determine the cut length of the material. These are either the standard lift (hinged) stops, or spring-loaded which become flush with the fence when the material pressed is pressed against the stop.

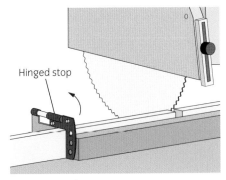

Hinged stop

Machine end stops

Distorted material

The correct way to deal with timber that is bowed, cupped, sprung or twisted is to ensure that the material is positioned with the **convex** side down and to the fence. This ensures stability during cutting and allows the rotational thrust of the saw blade to keep the material stable where it is being cut.

Convex

A face that bulges out in the middle

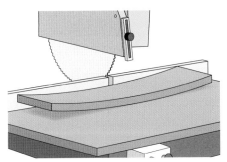

Bowed boards

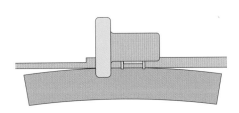

Sprung boards

Additional safety features

- *Self-return system for the cutting head:* it is good practice to have a spring-loaded return system fitted to the head to ensure constant return to the safe home position.

- *Cutting short lengths:* occasionally short lengths will need to be cut. It may be preferable to cut these in multiple lengths to aid further machining operations and to trim to the correct length later, either on a dimension saw or a crosscut saw. Safe holding of short lengths can be achieved by the use of a pivoted handle fixed to the back fence; alternatively a push stick can be used.

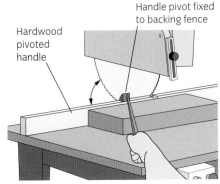

Handle pivot fixed to backing fence

Hardwood pivoted handle

Cutting short lengths with a radial crosscut saw

SAFE PROCEDURE FOR CROSSCUTTING MATERIAL

As well as correctly offering the material up to the crosscut, the operator needs to consider how they position themselves and how they operate the crosscut.

Fair ending

Trimming the end of a board to remove defects such as splits and to square it up

1 Bring the material up to trim the end of any defects and square up. This is called **fair ending**. The correct stance and positioning of the hands are vital to the safe operation of the crosscut. You should not cross your arms when using the machine. Instead, your arms should be kept parallel to the machine, and the hand not pulling the crosscut out should not be resting on the material within the marked danger zone 300mm each side of the saw cutting line. This greatly reduces the chances of accidental contact with the saw blade.

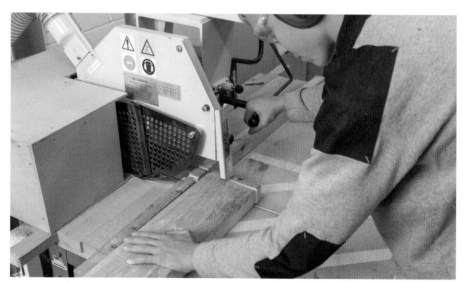

Fair ending timber

2 Pull the cutting head out at a steady, controlled rate, not allowing the cutting head to rush out towards you. If the blade begins to slow down during the cut, stop, do *not* pull it towards you but push slightly away, and allow the saw's speed to normalise. If the blade becomes jammed in the cut and stalls, stop the machine straight away and then carefully return the cutting head to its home position. Set any end stops to the required lengths as written on the cutting list.

Positioning the end stop

3   If the blade constantly becomes jammed or stalls, either the blade has become blunt or the incorrect tooth design is being used. In both cases the blade requires changing for a sharp blade of the correct design.

4   Push the trimmed length up to the required length position. If using an end stop, ensure you do not trap any debris between the stop and the material, and do not hit the end stop too hard or you risk the stop moving.

OUR HOUSE

You have been asked to help produce the information required to manufacture timber windows using fixed machinery. Produce a drawing for a design of window that would be suitable for the 'Our House' property. Along with this drawing, produce a cutting list containing all relevant information for the preparation of the timber.

THE NARROW BAND SAW

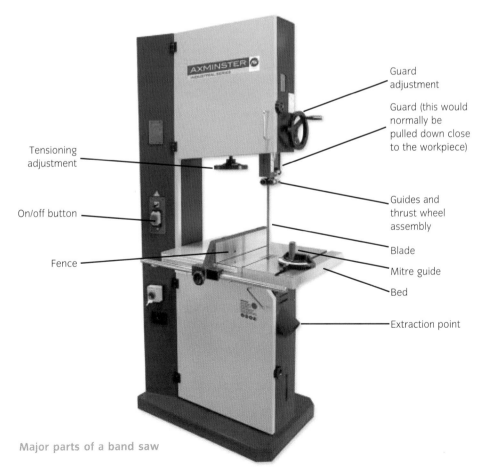

Major parts of a band saw

Throat

The horizontal distance from the cutting edge of a saw's blade to the machine casing

The narrow band saw is primarily designed for cutting a variety of curved shapes and joints. It can cut down the grain, across the grain, and also cut man-made materials. Narrow band saws are defined as 'narrow' if the size of blade is less than 50mm wide, not by the size of its **throat** or maximum cutting capacity. All narrow band saws cut with a continuous band of teeth running around two pulley wheels and cutting in a vertical straight line, rather than a circular band of teeth as with the rip saw and crosscut saw blades.

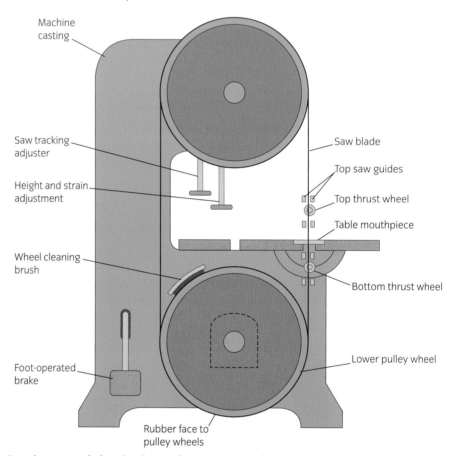

Machine casing

Saw tracking adjuster

Height and strain adjustment

Wheel cleaning brush

Foot-operated brake

Saw blade

Top saw guides

Top thrust wheel

Table mouthpiece

Bottom thrust wheel

Lower pulley wheel

Rubber face to pulley wheels

Guards removed showing internal components of a narrow band saw

The bottom of the narrow band saw's two pulley wheels is driven by the machine motor, while the top wheel is driven by the blade. Both wheels are rubber coated to help prevent the blade from slipping.

SETTING UP THE NARROW BAND SAW

1 The narrow band saw blade should run in the centre of the pulley when correctly tensioned. To tension or 'strain' the blade and keep it running straight, the top pulley wheel is raised. All machines have a tensioning gauge fitted which helps to apply the correct amount of tension.

2 To allow for different width blades and to help make the blade run in the centre of the pulley (known as **tracking**), the top wheel is

Tracking

Tilting the top pulley wheel to make the blade run in the centre of the pulley tilted or canted. When tracking the blade, the thrust wheels and guide assemblies should be moved clear of the saw blade to allow the blade to move freely on the wheel, both backward and forward, until the desired position is achieved. Always ensure the power to the machine has been isolated before attempting to track and tension any blade.

3 With the blade correctly tensioned and tracked, the saw guide assemblies can be positioned. The saw guides should be set as close as possible to the saw blade without rubbing or distorting the blade, and set with the front of the guides just back from the bottom of the gullets. The **thrust wheel** is set so that there is a gap between it and the back of the blade of 1–2mm.

Whenever the blade is changed, always check that the **timber mouthpiece** is in good condition, with as small a gap as possible around the blade. If the timber mouthpiece becomes damaged and worn, always replace it. This ensures bits of cut material cannot become trapped between the blade and the mouthpiece, which would risk breaking the blade, potentially injuring the operator and damaging the machine.

Thrust wheel

The small wheel behind the blade of a saw, used to prevent the blade from being pushed backwards during cutting operations

Timber mouthpiece

Used to prevent material becoming trapped between the machine bed and the blade of a saw, this should be replaced as often as required

ACTIVITY

Sketch the positions of the guides and thrust wheel in relation the blade.

SAW BLADES

Selecting the correct size blade is vital in the safe operation of the machine. The smaller the width of blade, the smaller the radii that can be cut. As a general rule, larger width blades are more suitable for cutting straight lines using a fence and power feed unit.

As well as the width of the blade, the pitch of the teeth is another vital consideration to correctly select the blade. The tooth pitch must be smaller than the thickness of the material being cut, with a general rule that coarse teeth are used for softwoods and fine teeth used for hardwoods.

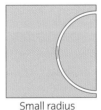

Small radius narrow blade | Large radius wide blade

Saw blades

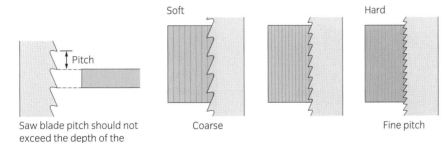

Saw blade pitch should not exceed the depth of the material being cut

Saw blade pitch

There are various design of blades available for cutting different types of material. Some have been hardened and are designed to be disposable; these are discarded when they become blunt and are

Saw doctor

A specialist who maintains machine tooling like wide band saw blades

particularly useful for cutting hard abrasive materials. Blades made from other materials, like carbon steel blades, require regular sharpening and setting, usually carried out at a specialist **saw doctor**.

GUARDING NARROW BAND SAWS

The narrow band saw is guarded in two ways:

In the first way, the top and bottom pulley wheels are guarded by **interlocked doors** that prevent the machine from starting if any of the doors are left open. The top door also locks into place, covering access to the side of the framework, which allows removal and fitting of the blade.

Interlocked doors

Doors that are impossible to open while a machine is in motion and prevent the machine from starting if the doors are open

Secondly, the blade is also guarded by an adjustable guard that is fitted to the guide assembly. This guard should fully enclose the front and sides of the blade. It is this guard that the operator has to adjust to suit the type of operation being performed. This adjustable guard needs to be set as close to the top of the material and the guide assembly as practicable. As with adjustment of the guide assembly, it should not be adjusted while the blade is moving.

CHANGING THE NARROW BAND SAW BLADE

The sequence for changing narrow bandsaw blades is very similar regardless of machine design. However, you should always refer to the manufacturer's information booklet for specific information on any particular requirements for that machine.

Before beginning this process, isolate the machine and close the dust extraction gates. This helps prevent any packers that may fall into the extraction ducting from being drawn away requiring replacing.

STEP 1 Open up the guard doors, exposing the saw blade.

STEP 2 Move the saw guide assemblies back away from the saw blade.

STEP 3 Loosen tension on the saw blade by lowering the top pulley wheel.

STEP 4 Fold the saw blade and put it safely to one side.

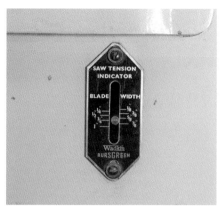

STEP 5 Clean down the saw guides and ensure all parts move smoothly. Fit the new saw blade and ensure it is tensioned correctly by moving the top pulley upwards and checking tension against the tension gauge.

STEP 6 'Track' the saw blade by tilting the top pulley. The blade should be running in the centre of the top pulley. This can be checked by slowly turning the top pulley and adjusting the tracking to ensure correct positioning.

STEP 7 Adjust the guide assemblies so the guides are just back from the bottom of the gullets and set as close to the blade as possible without making contact.

STEP 8 Adjust the thrust wheel so it is within 1–2mm of the back of the saw blade. Repeat for the lower guide assembly.

STEP 9 Re-fit the timber mouthpiece, ensuring it is as close as possible to the side of the blade and the back of the blade so as to reduce the risks of thin pieces of cut material becoming trapped against the saw blade and timber mouthpiece.

STEP 10 Set all guards and close both doors, ensuring the interlocks are located correctly or the machine will fail to start. Conduct a trial run, listening for any unusual sounds or vibrations. Carry out a test cut.

In Step 6, make sure the blade is neither too far forward or too far back. The correct position is shown in Step 9.

Blade too far forward

Blade too far back

There are several ways to fold a narrow band saw blade (see Step 4 on the previous page). The following steps are one simple and safe way to do it. The method shown not only keeps the blade away from your face but keeps the blade under a greater degree of control than other methods.

STEP 1 Remove the blade using suitable protective gloves.

STEP 2 Hold the blade approximately a third of the way down, with the teeth facing away from you. Your thumbs will need to be on the outside of the blade with one of your feet holding the blade down on the floor; this prevents unwanted movement from the lower part of the blade.

STEP 3 Keeping your elbows still, rotate your wrists so your thumb now faces you and your hands have moved from the vertical position to the horizontal position; the top part of the blade will loop over away from you. Next, keeping your elbows and arms still, rotate your wrists outwards: this will try to bring the top part of the blade back towards you. While the blade is trying to come back towards you, push down with your arm and the blade will form into a circle on the floor.

STEP 4 Push the blade down onto the floor after folding, forming a small band, enabling safe transportation. Practise the stages first without a saw blade until you are familiar with the required stages.

USING THE NARROW BAND SAW SAFELY

The narrow band saw is an extremely versatile machine. The type of work to be done will determine the choice of blade.

CUTTING USING THE FENCE

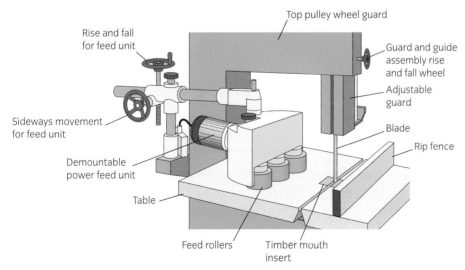

Narrow band saw machine set up for ripping timber

If you are required to cut large quantities of timber using the fence, it is more important to use a wider blade. This not only helps keep

Demountable power feed unit

Automatic feed systems that can be moved out of the way. Mainly used for continuous feeding of material during cutting or profiling operations

the cutting line parallel to the edge but also helps stop the blade from wandering. The extra strength in the blade will also help prevent breakages. When using the narrow band saw to rip material, a **demountable power feed unit** should be used; this helps feed the material in a safe, controlled manner, keeping the operator's hands away from the blade.

If there is no demountable power feed unit available, a guide block should be used. This helps keep the material up against the fence at the point of cutting without putting your hand close to the blade.

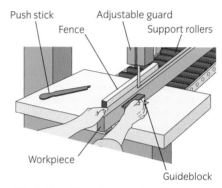

Straight cutting by hand with a narrow band saw

FREEHAND CUTTING

Often it is not practicable to use a fence, for example when cutting irregular shapes or curves. In these situations the material should be fed in with a steady and even forward motion, without applying any undue pressure to the blade. Hands should be kept out of the line of the cutting path and as far away from the blade as practicable.

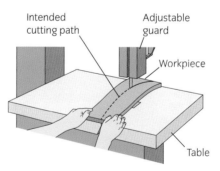

Freehand cutting with a narrow band saw

INDUSTRY TIP

You can safely cut wedges using a jig and fence. This helps with increasing the production and continuity of wedges.

Templet

Also known as a 'template', this is a thin pieces of hardboard, MDF or ply that is cut to the shape required and is then used to help reproduce the shape

Where possible when carrying out repetitive work, use a jig and guide pin. The guide pin is fixed in front of the blade and runs against the jig or **templet**; this will improve safety as well as productivity.

Further examples of using the fence and jigs which increase safety and accuracy are shown in the following illustrations.

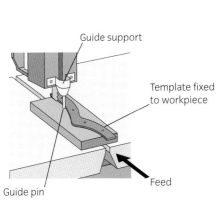

Using a jig and guide pin with a narrow band saw

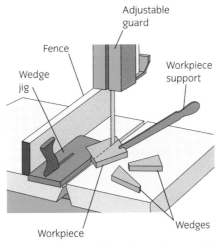

Using a fence and jigs (guard and guide moved up for clarity only)

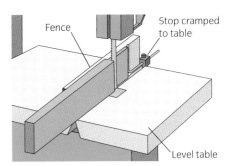

Cutting tenons using the fence and back stop

When cutting complex shapes it is important to choose your starting and stopping points with care. You generally do not want to be pulling the material out of the cut. Sometimes it will be unavoidable and extra care should be taken while doing so: there is a risk that if you do not follow your original saw path you will pull the blade from the pulley wheels or cause the blade to break. Always cut the shortest length first; this will ensure the risks are reduced to the minimum.

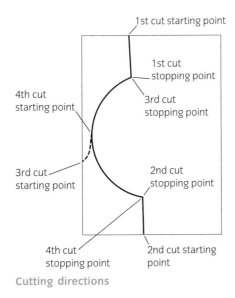

Cutting directions

Mobile support rollers

Always ensure that the material being cut is sufficiently supported both at the in-feed end but also at the back of the bandsaw. If no fixed extension table is available then use mobile support rollers.

ACTIVITY

Create a templet for a push stick on suitable material like 18mm MDF. Using a suitably sized blade, cut out the push stick for your own use.

PLANING MACHINES

There are three main types of planning machine:

- the surface planer
- the thicknesser
- the combination planer.

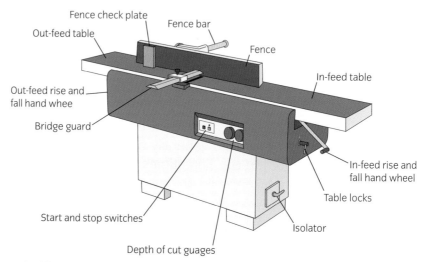

Fence check plate · **Fence bar** · **Out-feed table** · **Fence** · **In-feed table** · **Out-feed rise and fall hand whee** · **Bridge guard** · **Start and stop switches** · **Depth of cut guages** · **Isolator** · **Table locks** · **In-feed rise and fall hand wheel**

Typical hand-fed surface planer

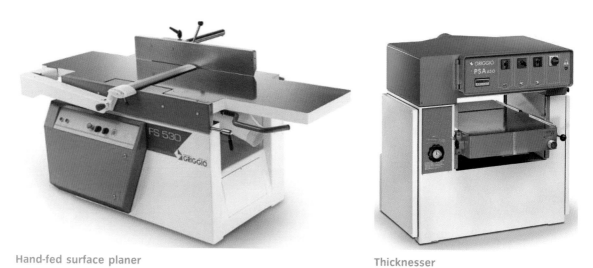

Hand-fed surface planer

Thicknesser

All planers put a smooth, flat, planed surface on the timber, but the three different machines do it in different ways.

SURFACE PLANER

The hand-fed surface planer carries out two main functions:

- *Facing:* this is where the planer produces a smooth, straight and flat surface along the widest face of the material, known as the 'face side'.

- *Edging:* this is where the planer produces a smooth, straight and flat surface along the narrowest face of the material, known as the 'face edge'.

Apart from being used to produce standard facing and edging work, the planer can be used to produce bevels, chamfers and, in older versions, rebates.

Surface planers (like thicknessers, which are covered later) have circular blocks that normally contain two cutting **knives**, although they can have more. The cutters revolve at speeds up to 6,000 **revolutions per minute (rpm)**, which is 12,000 cuts per minute. These circular cuts leave ripples on the finished surface called **pitch marks**; it is the sizes of these pitch marks that determine the quality of finish produced by these machines. The faster the timber is fed across the cutters, the larger these pitch marks will be and the poorer the quality of finish will be. A slower feed speed produces pitch marks that are closer together, giving the feel and appearance of the surface being totally smooth and flat. The better the quality of finish at this stage, producing small pitch marks, the more time is saved later in sanding and finishing.

Knives

The name given to the part of the block that does the cutting. Can also be known as 'blades' or 'cutters'

Revolutions per minute (rpm)

How many times the block rotates in one minute

Pitch mark

The name given to the small circular cuts produced by all rotary planing machines

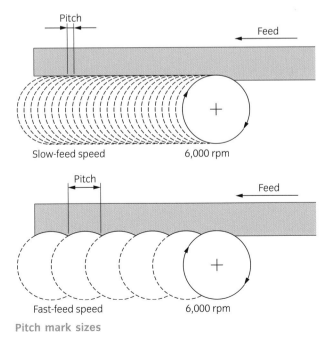

Pitch mark sizes

Cutter block

The machine block that holds the cutting knives or cutters

For general joinery, a pitch mark of 2–2.5mm is acceptable, while for better classes of work and cabinet work a pitch mark of 1mm will be required. To enable the pitch of cutter mark to be worked out, three pieces of information are required:

- *feed speed:* how fast the material is passing over the cutters

- *number of knives cutting:* the number of knives in the **cutter block**, typically two, three or four

- *how fast the cutter block is travelling:* usually 4,500 or 6,000 rpm.

Using the following formula, the pitch of cutter marks can be determined:

$$P = \frac{1,000 \times F}{N \times R}$$

where

F = feed speed

N = number of cutters cutting

R = speed of block (rpm)

P = pitch of cutter marks.

ACTIVITY

Calculate the pitch of cutter marks for a surface planer that has four cutters cutting a feed speed of 20 metres per minute and a block revolving at 6,000 rpm.

Answer: 0.83mm.

Example

Assuming a two-knife planer block runs at 4,500 rpm with a feed speed of 15 metres per minute, the pitch of cutter marks would be:

$$\frac{1,000 \times 15}{2 \times 4,500} = 1.66\dot{}mm$$

Therefore the pitch of cutter marks would be 1.7mm (rounded), suitable for general joinery products.

CORRECT POSITIONING OF THE PLANER TABLES

To ensure smooth, safe and accurate planing of the material, correct positioning of the in-feed and out-feed tables is vital.

The in-feed table controls the amount of material that is removed in one pass. As a general safety rule and to enable safe control of the material, a *maximum depth of cut of 3mm per pass* is recommended. If this amount fails to achieve the desired flat surface, for example due to the timber being twisted, cupped or just to remove some other defect, then send the material over the planer several times to achieve a smooth flat finish.

To ensure smooth transition from the in-feed table, over the cutter block and on to the out-feed table, the out-feed table needs to be set in line with the cutting circle at its maximum height. Failure to correctly set this position will result in a poor finish and in extreme cases could be dangerous.

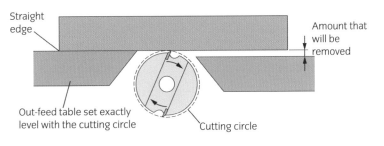

Correct setting of out-feed table

Only cylindrical cutter blocks can be used on hand-fed surface planning machines with a maximum cutter projection of 3mm, while for machines manufactured after 1995 this should not exceed 1.1mm.

GUARDING THE PLANER

Every planing machine must be correctly guarded. This is primarily achieved by the use of the 'bridge guard' situated on the operator's side of the machine. It should be:

- strong and rigid (to support heavy timbers)

- made from a material such as aluminium, so that in the event of contact with the cutter block neither guard nor cutter will disintegrate

- constructed so that it is not easily deflected, which would expose the cutter block

- long enough to cover the table gap with the fence at maximum adjustment (telescopic guards are available for large machines)

- sufficiently wide (at least equal to the cutter block diameter)

- easily adjustable, both horizontally and vertically, without the use of a tool.

Attached to the back side of the fence there is another guard that is fixed, meeting all the above criteria.

Guard positioning

Facing and edging timber requires the timber to be placed on the in-feed table; it is then passed over the cutters and on to the out-feed table. To ensure this is carried out safely, the in-feed and out-feed tables should be set as close as possible to the cutting circle of the block.

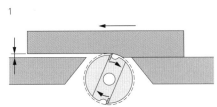

1

Note the gap between the planed surface and the top of the out-feed table.

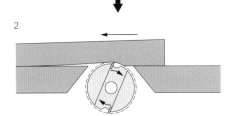

2

The cut deepens as the unsupported workpiece travels onto the out-feed table.

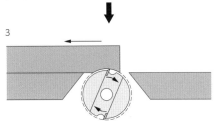

3

The trailing end drops onto the planer knife as it leaves the in-feed table.

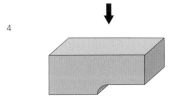

4

Drop in the trailing end of the timber.

Incorrect position because the out-feed table is set too low

ACTIVITY

Carry out research on the HSE website to find up-to-date information on the safe use of hand-fed planing machines. Consider the information you discover and state how you would put it into practice for the machining of softwood timber that is 3.6m long, 225mm × 75mm and cupped.

ACTIVITY

Sketch the correct relationship between the cutting circle and the out-feed table.

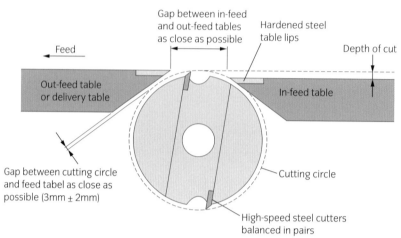

Table positioning for hand-fed planing machine

Bridge guard positioning

While feeding the material across the cutter block, the bridge guard needs to be correctly positioned. The guard should be set as close as possible to the top and side of the timber being planed. For combined planing of the face followed by the edge, the guard should be set as shown in the illustrations, but when facing ensure that the timber is fed through on the part of the machine bed that is fully covered by the bridge guard.

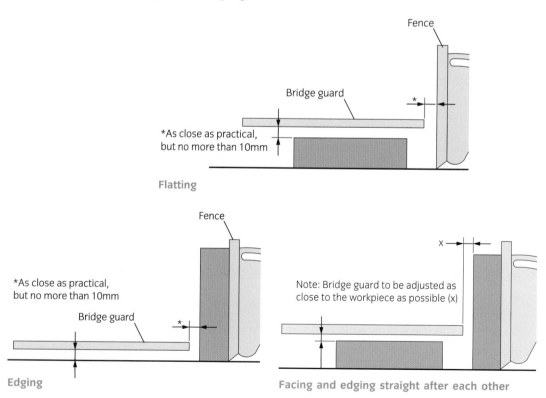

When facing and edging small sectioned square material, position the bridge guard as close as possible to the fence and the top of the timber. When facing and edging large sectioned square material, position the bridge guard as close as possible to the bed and the timber side.

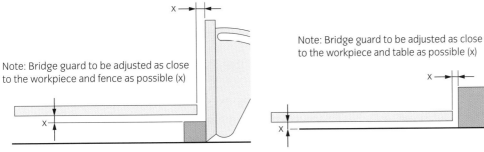

Note: Bridge guard to be adjusted as close to the workpiece and fence as possible (x)

Note: Bridge guard to be adjusted as close to the workpiece and table as possible (x)

Facing and edging small square sectioned material

Facing and edging large square sectioned material

When planing short lengths of timber, a well-designed push block should be used. It should incorporate handles that are suitable, fixed and give a firm grip on the push block, with a back stop preventing the push block from slipping along the timber.

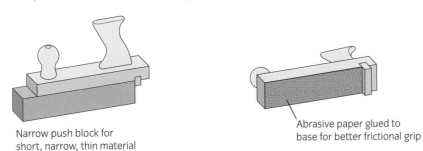

Narrow push block for short, narrow, thin material

Abrasive paper glued to base for better frictional grip

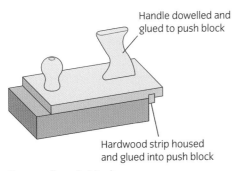

Handle dowelled and glued to push block

Hardwood strip housed and glued into push block

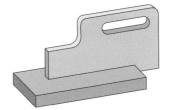

Alternative push block design made from MDF

Types of push blocks

When operating a hand-fed planer, it is important to correctly position your hands while passing the timber across the cutter block.

Facing

The workpiece should be fed by pressure with the right hand, with the left hand initially holding it down on the in-feed table. As soon as there is enough timber on the out-feed table, the left hand can pass safely over the bridge guard to apply pressure on the out-feed table, to be followed by the right hand to complete the feeding operation.

Face side and face edge

The two sides that have been planed squared to each other on the surface planer

Face side and face edge marked on timber

It is not necessary to exert feeding pressure directly over the cutter block and your hands should not pass over the cutter block while they are in contact with the timber. This surface is usually known as the **face side**.

Edging

Follow the procedure shown in the illustration. Again, your hands should not pass over the cutter while they are in contact with the timber – their main function is to exert side pressure to the workpiece to maintain it square to the fence. This surface is usually known as the **face edge**. If the workpiece is short, use a push block to feed it through.

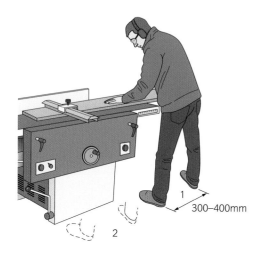

Facing timber

Feet should provide a firm base and allow forward movement to position 2

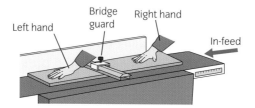

Left hand   Bridge guard   Right hand   In-feed

Left hand to apply pressure onto timber as it passes under bridge guard

Right hand to apply firm but not heavy pressure

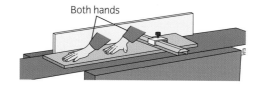

Both hands

Both hands on timber towards the end of the cut

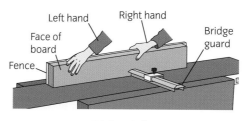

Left hand   Right hand   Bridge guard

Face of board

Fence

Note: at no time should your hands pass over the cutter block

Edging timber

Stance and hand movements for facing and edging timber

ACTIVITY

Sketch the correct position of the bridge guard when facing and edging timber in consecutive stages.

BEVELLING AND REBATING

PUWER 1998 requires that the most suitable machine available (the one posing the smallest risk) is used for every machining operation. For example, cutting a rebate on a correctly guarded vertical spindle moulder is safer than cutting it on a surface planer. New hand-fed planing machines by design do not allow rebating, but on pre-1995 machines rebating can be carried out provided:

- a tunnel guard is used: this is usually achieved by using two shaw guards

- the material is properly supported.

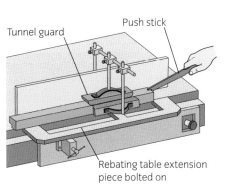

Tunnel guard

Push stick

Rebating table extension piece bolted on

Rebating on a hand-fed planing machine using a tunnel guard and push stick

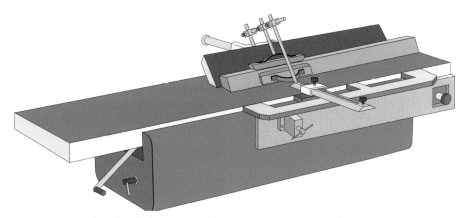

Bevelling on a hand-fed planing machine using a tunnel guard

PRESENTING THE TIMBER TO THE MACHINE

The timber to be planed should be offered up to the hand-fed surface planer in such a way as to make the timber as stable as possible while it is passing over the cutter block. This is achieved by placing the timber with its concave side sitting down on the machine bed.

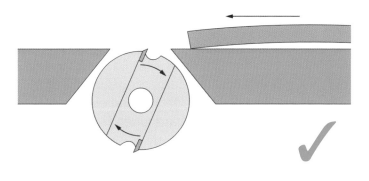

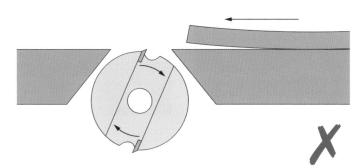

Correct positioning of timber when planing

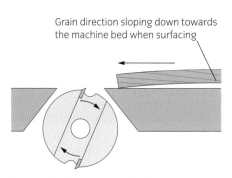

Grain direction sloping down towards the machine bed when surfacing

Grain direction when planing

Wherever possible you should feed the timber through the machine with the grain running with the cutter block.

The photographs below show the correct positioning of the bridge guard for machining material for facing only. Note that the bridge guard is pushed all the way over to the fence and kept as close to the top of the timber as practicable.

The photographs below show the correct positioning of the bridge guard for machining material on its edge. Note that the bridge guard is dropped down to the bed and kept as close to the timber as practicable when the timber is against the fence.

The photographs below show the correct positioning of the bridge guard for machining material while facing and edging. Note the correct positioning of the bridge guard when facing and edging operations are carried out consecutively.

THICKNESSER

The thicknesser planes the timber that has been previously faced and edged to the required finished size. Always bring the materials to the finished width first – this will enable greater stability while the timber passes through the machine. Some machines allow for the thicknessing of more than one piece of timber at a time; the thicknesser should have a label stating whether or not this can be done. Alternatively, the machine can be fitted with split sectional feed rollers to convert it to take more than one piece of timber at a time.

If you feed more than one piece at a time into machines that have one fixed serrated feed roller, it can result in the timber being forcefully thrown back at the operator.

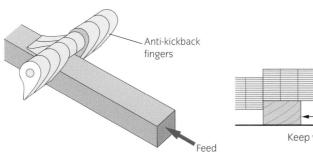

Anti-kickback fingers

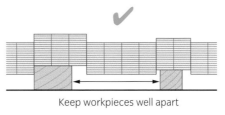

Keep workpieces well apart

Small section has no roller contact and so there is a real danger of it being ejected

Example of split-feed roller and multi-feeding

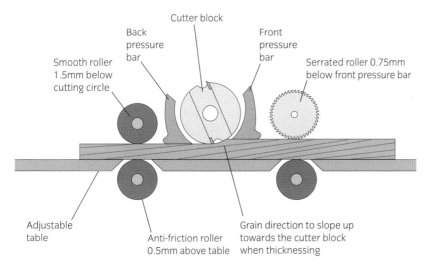

Section through thicknesser feed system

The machine bed of the thicknesser can have extra rollers incorporated into it. These **anti-friction rollers** aid the smooth delivery of the timber through the thicknesser.

GUARDING WHILE THICKNESSING

The cutter block is fully enclosed within the thicknesser, presenting little danger if the machine is used correctly. Combination planers need changing from their surfacing position to their thicknessing position; this is covered later in this chapter.

Anti-friction rollers

Adjustable free-running rollers set into the machine bed of a thicknesser, enabling the material to move through the machine without rubbing, which would cause excess friction

OPERATING THE THICKNESSER

The timber is fed into the machine with the side requiring thicknessing uppermost; these are the sides opposite the face edge and face side, in that order. The serrated feed roller grabs the timber, feeding it past the pressure bar. The pressure bar forces the timber onto the machine bed and helps reduce the size of the chippings produced by the cutter block. The material is then fed on past the cutter block, where another smooth feed roller continues to feed the timber through the machine.

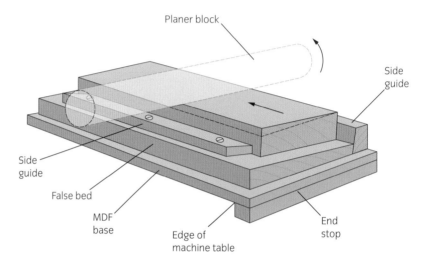

Jig for bevelling (false bed)

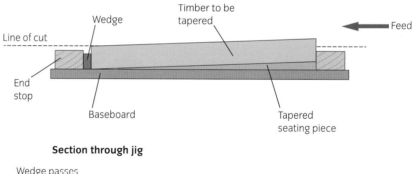

Section through jig

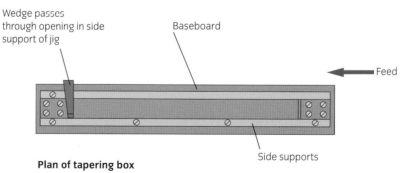

Plan of tapering box

Jig for tapered work

Tapered cuts

It is important not to try to remove too much material in one pass; a maximum depth cut of 4mm for each pass is good practice. The depth of cut is adjusted either by a hand wheel or electronically on modern machines, while the feed speed can be adjusted to cope with different species of timber and the required pitch mark finish.

The timber is fed into the machine so that the revolving cutters are rotating down, round and towards you.

Tapered cuts can be applied on the thicknesser provided you make use of a well-designed jig.

COMBINATION PLANER/THICKNESSER

The combination planer/thicknesser, as its name implies, combines all the functions of a surface planer and thicknesser in one machine. All the regulations, safety principles, setting up and operating techniques described for the individual machines on the previous pages apply to this machine.

The combination planer/thicknesser is particularly useful in workshops that have limited room because the machine has a similar footprint to that of a surface planer, thereby saving the room needed for a thicknesser. Other advantages include only having to pay for one machine (both purchase and installation costs), not needing additional ducting, and the fact that there is no need for separate tooling.

Although there are clearly advantages in only having a combination planer/thicknesser, there are also some disadvantages to be considered. The major disadvantage is if the machine breaks down or becomes unusable for some other reason, both planing and thicknessing operations would have to stop. Other disadvantages include:

- possibility of a backlog of other jobs with performing just one operation

- time taken to change guards, etc from planing to thicknessing and back

- increased risk of operatives not correctly changing guarding, etc from one operation to another

- small chips, etc to cutters affecting both planning and thicknessing operations.

Consider carefully all the advantages and disadvantages for your particular workshop and working practices before deciding on which type of machines to purchase.

As suggested above, with combination planers the machine will require changing from its surfacing position to its thicknessing

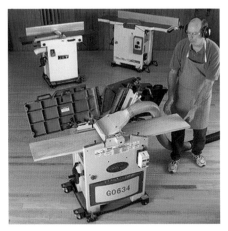

Example of combination planer set up for thicknessing with tables lifted out of the way

position. This requires removal of the bridge guard and the fitting of an additional guard that incorporates the chippings extraction point, to which the ducting needs fitting in order to take away the chippings produced. The reverse process will need undertaking once the thicknessing has been completed.

On some smaller models the machine beds are folded out of the way and then the additional guard incorporating the extraction point is fitted before thicknessing operations can be carried out. This additional guard must still completely cover the cutter block, restricting access and accidental contact with the cutter block and other guarding criteria, as stated previously.

REPLACING THE TOOLING IN SURFACE PLANERS AND THICKNESSERS

The type of machine will affect the type of cutters that the machine uses and how they will be changed. There are two basic methods involved.

POSITIVE LOCATION

This type of tooling is the most simple to change. It involves the use of disposable tooling with each cutter in the set (usually two) able to be turned over as one edge becomes dull. This method requires little in the way of specialist tooling or skill to replace the cutters. It is used a lot on newer machinery and is limited to a cutter projection of 1.1mm maximum. The cutters are held in place by either a wedge bar that tightens up onto the cutters, thereby holding them in place, or by **centrifugal force**; when released the cutters are either lifted out or slid out of the side of the block.

Centrifugal force

A force that is generated by the rotating cutter block; it forces objects away from the centre of rotation, thereby wedging the cutter tight in the block

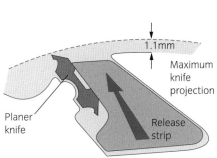

Centrifugal force tightens the tapered release strip into the slot and wedges the planer knife in position

Centrifugal force cutter holding system

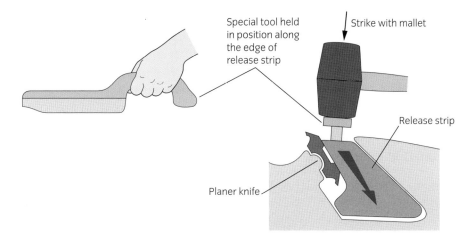

ADJUSTABLE POSITION TOOLING

This type of tooling allows for the cutter knives to be re-ground after they have become blunt. Because of the reduction in size resulting from re-grinding, the cutter position changes each time the cutters are changed. There is a great deal more skill and time required to correctly set this type of tooling.

The cutters are normally held in place by a wedge bar that tightens up against the cutters, preventing movement. The cutter projection for this type of tooling is a maximum of 3mm. The machine usually comes with a specific tooling setting device that enables correct positioning of the tooling.

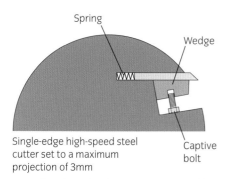

Spring

Wedge

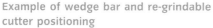

Single-edge high-speed steel cutter set to a maximum projection of 3mm

Captive bolt

Example of wedge bar and re-grindable cutter positioning

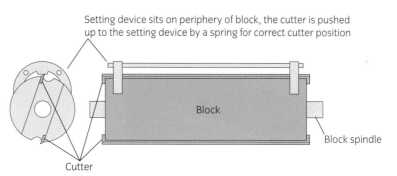

Setting device sits on periphery of block, the cutter is pushed up to the setting device by a spring for correct cutter position

Block

Block spindle

Cutter

Setting device located on cutter block with cutter pushed up to setting stand

TIGHTENING AND LOOSENING CUTTERS

Always start by isolating the machine before starting any adjustment to the machines tooling. Also remember to study the risk assessment to ensure you follow all other relevant health and safety requirements.

Wherever a wedge bar is used to secure the cutters, the correct tightening and releasing sequence must be used. This helps to eliminate the risk of distortion to the cutter knife. The correct sequence for releasing the cutters, regardless of the number of fixing bolts, is to start undoing the bolts at the edge of the block, working in from each end alternatively towards the centre.

Component parts of the cutter block

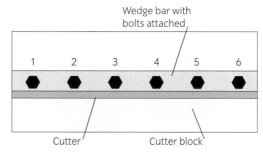

Wedge bar with bolts attached

1　2　3　4　5　6

Cutter　　Cutter block

A plan view of the cutter block with the cutter and wedge bar with the bolts in

1 Starting at the outer edge of the block, undo bolt 1 to release the pressure.

2 Move to bolt 6 and release the pressure.

3 Release the pressure from bolt 2.

4 Release the pressure from bolt 5.

5 Release the pressure from bolt 3.

6 Finally, release the pressure from bolt 4.

7 Remove the cutter and wedge bar.

After cleaning the block and wedge bar with the appropriate cleaning fluids, the tightening sequence is the reverse of the loosening sequence.

Remember to start in the centre when tightening the cutters and work outwards

Checking the level of the out-feed table to make sure it is in line with the cutting circle

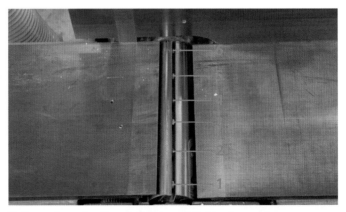

Bolt numbers for the undoing and tightening sequence

Undoing the cutter

MORTICE MACHINES

There are two main types of mortice machine:

- *Hollow square mortice chisel:* the most common machine used to produce a mortice, consisting of a square hollow mortice chisel in which revolves an auger bit. Using this method to produce a mortice can take longer than the chain morticer (see next paragraph) as several strokes could be required from each side of the material to produce the finished size mortice.

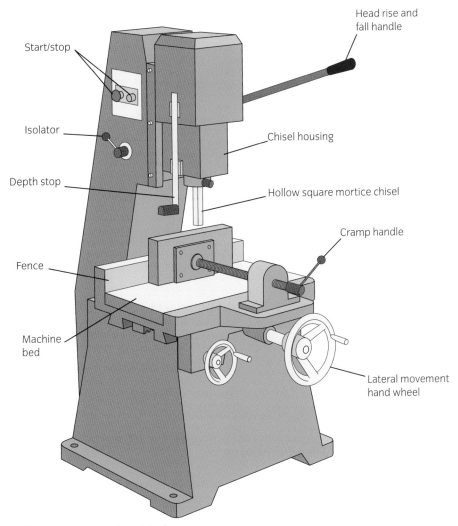

Start/stop

Head rise and fall handle

Isolator

Chisel housing

Depth stop

Hollow square mortice chisel

Cramp handle

Fence

Machine bed

Lateral movement hand wheel

Hollow square mortice chisel

- *Chain morticer:* a chain not unlike one used on a chainsaw runs around a guide bar. It is capable of producing in one go a mortice that penetrates all the way through the material, making it quicker than a hollow chisel morticer. As a result, it is often used for mass production. The quality of the finished mortice is not usually as neat as that produced by the hollow mortice chisel.

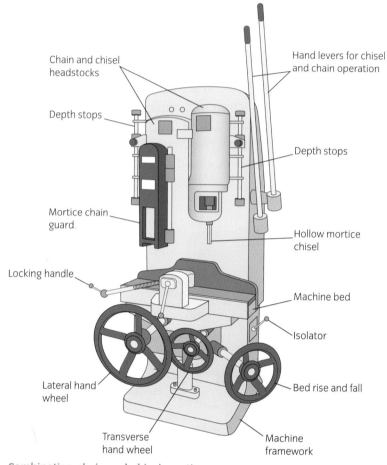

Chain and chisel headstocks

Hand levers for chisel and chain operation

Depth stops

Depth stops

Mortice chain guard.

Hollow mortice chisel

Locking handle

Machine bed

Isolator

Lateral hand wheel

Bed rise and fall

Transverse hand wheel

Machine framework

Combination chain and chisel morticer

In both cases the depth of cut can be controlled by an adjustable depth stop. In some cases the depth stop can be swung out of the way, allowing for more than one depth of cut to be set. This is particularly useful when mortising and haunching on the same component.

Advantages and disadvantages of the two different types of mortice machine are shown in the following table.

Mortice type	Advantages	Disadvantages
Hollow mortice chisel	Easy to use Safer to use Easy to maintain Easy to sharpen Produces neat mortice holes Bigger range of mortice hole sizes	Takes longer to produce a mortice Small sized chisels set can split easily if forced to cut too fast More marking out required unless end stops are available
Chain morticer	Quick to produce a mortice Able to penetrate all the way through the material	It is more dangerous to use It is more difficult to sharpen It is more difficult to maintain in good working order Limited in the size of mortice it can produce Small pitch chains can break if care is not taken

The hollow mortice chisel is generally the preferred method for cutting a mortice for the reasons on the opposite page. In production work, the need for higher production output can dictate that a chain morticer is used.

HOLLOW SQUARE MORTICE CHISEL

Hollow square mortice chisel sets are available from 6 to 25mm. For sizes below 16mm the set usually consists of four parts, while for sizes of 16mm and over only three parts are required. The component parts that make up the hollow chisel set are:

- *Hollow mortice chisel:* this part cuts the outer shape of the mortice.

- *Auger:* this part bores out the bulk of the mortice.

- *Chisel bush/**collet***: this part has a variable internal diameter to suit the size of the chisel but a constant external diameter to suit the collet size of the machine.

Collet

A means of centralising the auger in a chisel headstock

- *Auger bush/collet:* this part has a variable internal diameter to suit the size of the chisel set and a fixed external diameter to fit the machine. For chisel sets of 16mm and above an auger bush is not required because the shaft size of the auger is the correct size to fit into the machine housing.

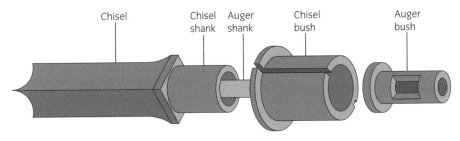

Chisel   Chisel   Auger   Chisel   Auger
         shank   shank   bush   bush

Hollow square mortice chisel components

When correctly set up, the bottom of the auger will protrude out of the bottom of the hollow mortice chisel about 1.5mm for sizes up to 12mm and about 2.5mm for sizes larger than this.

The side window (waste port) is to allow the chippings to escape and should be positioned on either the right or left side as you look at the machine. This is because you work away from the window so the chippings flow into the previously cut hole, reducing the chance of blockages and overheating and masking of the line you want to cut to.

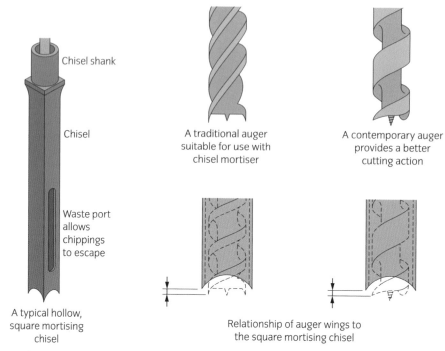

Chisel shank

Chisel

Waste port allows chippings to escape

A typical hollow, square mortising chisel

A traditional auger suitable for use with chisel mortiser

A contemporary auger provides a better cutting action

Relationship of auger wings to the square mortising chisel

Positioning of the auger in relation to the hollow chisel

CORRECT FITTING OF A HOLLOW MORTICE CHISEL

Before changing the tooling in any machine always refer to the risk assessment. This will ensure you are adhering to the correct safety procedures as required by the particular machine, its location and any relevant health and safety issues. The following steps assume you have referred to the risk assessment and are following the requirements as specified.

The following steps are a guide to the correct procedure for the installation of a 10mm hollow mortice chisel set (local machine variations may need to be applied).

1 Isolate the machine from the power supply.

2 Select the required size chisel set which will include the hollow mortice chisel with its matching auger, and auger and chisel bushes. Ensure the mortice set is in good condition with the chisel having sharp edges and corners. The chisel should not have any splits or other defects. The auger should have wings that are sharp and intact. Both the chisel and auger bushes should be the correct sizes.

3 Insert the auger bush into the mortice head ensuring the window of the bush lines up with the Allen key. Tighten the Allen key sufficiently only to stop the auger bush from falling out but slack on the auger.

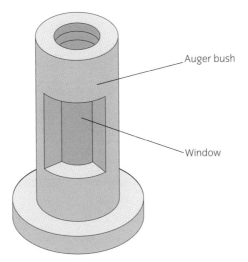

Auger bush

Window

4 Insert the chisel bush into the headstock with its split lining up with the split jaws of the headstock. This will allow for easier tightening around the chisel without putting undue force onto the headstock. Tighten only sufficiently to stop it dropping out.

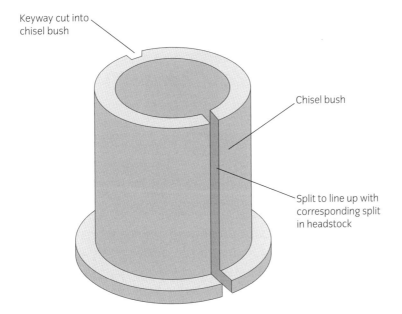

Keyway cut into chisel bush

Chisel bush

Split to line up with corresponding split in headstock

5 Insert the auger into the hollow mortice chisel and offer it up to the machine. The chisel should be pushed home and at this stage just tighten the chisel in place; there is no need to square it up.

6 The end of the auger will have a flat ground on it. The flat is to ensure that when tightened the auger will not slip during the cutting operation. It is this flat that should face the window in the auger bush, which allows the Allen key to tighten a grub screw firmly onto the flat of the auger. The auger should be set so the wings of the auger protrude out of the bottom of the chisel. The correct amount for its size in this case is about 0.5 to 1mm. Tighten the grub screw with the Allen key.

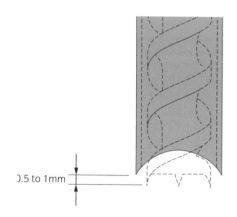

0.5 to 1mm

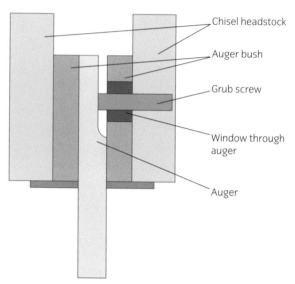

Chisel headstock

Auger bush

Grub screw

Window through auger

Auger

7 Slacken the chisel cramp and square it up with a set square against the fence. Make sure the window in the side of the chisel is either facing to the right or left. Tighten the chisel cramp keeping the chisel fully pushed home.

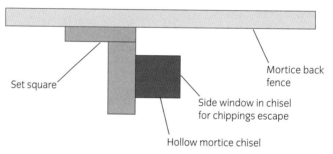

Set square

Mortice back fence

Side window in chisel for chippings escape

Hollow mortice chisel

8 Make a final check of the auger position in relation to the bottom of the chisel and remove all tools.

9 Turn the power back on and do a quick start–stop of the machine. The auger should run quite quietly even though it is steel running inside steel. If there is a loud squeaking sound from the auger it will probably be due to it being set too high within the chisel. If it is set too high within the chisel, then this could be due to the lack of an auger bush or that you have used the incorrect size auger bush.

10 A trial cut will allow you to check that the chisel is cutting correctly, that it is square to and the correct distance away from the fence.

Accurate positioning of material in the machine is vital to ensure mortice holes are cut accurately. Always ensure no chippings are trapped between the material and the machine fence and bed.

SEQUENCE FOR CUTTING THE MORTICE

The following step-by-steps assume that the mortice being cut is 50mm deep and 50mm long, and is being cut using a 10mm mortice chisel.

STEP 1 Place the timber in the machine with its face towards the fence.

STEP 2 Without turning the machine on, bring the chisel down to the timber. Adjust the fence so that the chisel is on the gauge lines.

STEP 3 If making haunches, the depth stop can be set by drawing a line on the end of the timber to be morticed, and bringing the chisel down to that point and setting the depth stop.

STEP 4 Tighten the stop.

STEP 5 After checking that everything is set up correctly and safely, turn on the machine. Bring down the chisel carefully at the end of the mortice and push it into the timber, taking care to go just over halfway through.

STEP 6 Cut the other end of the mortice. This ensures that the ends of the mortice go straight.

STEP 7 Remove the middle of the mortice and cut the haunch.

STEP 8 Turn over and repeat the process, remembering to start at the end.

STEP 9 Finish the cuts by moving to the end and finally the middle.

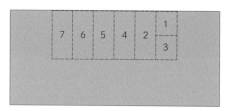

Order of morticing

If the chisel is pulled down to the intended depth (just over halfway for a through mortice), it can become trapped or jammed. This makes it difficult to pull the chisel back up without lifting the material out of the cramp. To prevent this, create clearance and minimise the effects of friction, particularly when morticing hardwoods. The order of chiselling shown will reduce the chance of this occurring.

Regular sharpening is required to ensure the hollow mortice chisel produces a clean mortice, without risking splitting the chisel or the breaking away of one of the small 'wing spurs' at the bottom of the auger. Sharpening can be divided into hollow mortice chisel sharpening and auger sharpening.

PROCEDURE FOR SHARPENING THE HOLLOW MORTICE CHISEL

Reamer

A type of rotary conical cutter used for cutting metal

Sharpening the hollow mortice chisel is carried out using either a **reamer** and pilot set or a conical grindstone.

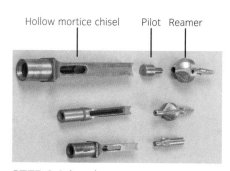

STEP 1 Select the correct size reamer and pilot and fit into a power drill.

STEP 2 Tighten the chisel in a vice, being careful not to over tighten and risk breaking the hollow chisel. Insert the pilot into the opening, ensuring you keep the drill parallel to the chisel, and re-cut the bottom of the chisel, cutting sufficient material away to develop a sharp edge with corners of equal length.

STEP 3 File the relief angles in all four corners.

STEP 4 File the relief angle of 25° to the four sides.

STEP 5 Remove the burr from the outside of the hollow chisel by placing it flat on a fine oilstone and rubbing it up and down the stone. Apply a small amount of grease to the inner part of the chisel, ready for use.

PROCEDURE FOR SHARPENING THE AUGER

STEP 1 Using a fine triangular file to suit the auger, file the end face of the auger, maintaining the original grinding angle.

STEP 2 File the wing spurs just lightly enough to ensure they are sharp.

STEP 3 The wing spurs must always project beyond the bottom of the auger so that they cut in advance of the chisel.

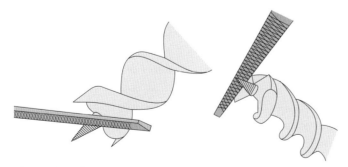

Sharpening the auger

FAULT DIAGNOSIS

Typical examples of faults found when using and setting up hollow mortice chisels are shown in the table.

Faults with hollow mortice chisel tooling	Possible reason	Corrective action
Mortice is not at right angle to the fence.	The back face of the chisel is not set parallel to the fence.	Adjust the chisel so the back edge is parallel to the fence.
The bottom of the chisel set leaves the corners of the chisel, producing an uneven bottom to the mortice.	The auger is protruding too far from the bottom of the chisel.	Reposition the auger at the correct distance below the chisel to suit the size of the chisel set.
The chisel shows signs of burning (blueing around the bottom cutting edge).	The auger is set too deep within the chisel, causing the auger to rub against the chisel.	Reposition the auger at the correct distance below the chisel to suit the size of the chisel set.
The chisel develops a crack.	The chisel has developed a crack that starts from the cutting edge, working its way upwards.	The chisel has become dangerous and should not be used. Always ensure that the chisel is regularly and correctly sharpened and the auger is set correctly before using, particularly with small chisel sets.
The bottom of the auger has a broken wing.	Poor or lack of sharpening of the auger.	Dispose of the auger and ensure that regular and correct sharpening of the replacement auger takes place.

MAINTAINING MACHINES

Maintenance is not an optional activity but a legal requirement. Maintenance carried out with the correct parts and at the correct time intervals will also in the long-term benefit the user.

Apart from the legal reasons, other advantages to carrying out maintenance are:

- prolonged machine life

- fewer breakdowns and so the machine becomes more productive

- less risk of injury to operatives

- better quality work produced through correct functioning of all parts of the machine.

All the information required to carry out maintenance on your machines should be available in the manufacturer's handbook for that type of machine and model number. The following is an example of the type of information available from the maintenance section:

- lubricants required, eg oil, grease, water

- how much to use each time, eg 500ml, 'three depressions of the grease gun'

- where they should be used, eg 'point A and B'

- suitable replacement parts, eg belts, bearings

- how often parts should be replaced, eg weekly, monthly, yearly

- how to check conditions and tensions of drive belts, eg amount of free movement, signs of cracking or fraying, stopping times.

Remember that while carrying out maintenance operations attention should be paid to the PPE that you will need. This could include:

- *barrier creams:* to protect the skin from oils, etc

- *clothing:* to keep contaminants away from the skin

- *safety glasses or goggles:* to keep dust and chippings, etc away from the eyes

- *gloves:* to keep contaminants away from the hands and for when handling tooling

- *a face mask:* to prevent dust and fumes from being breathed in

- *ear defenders:* to protect against loud noise.

ACTIVITY

Produce a maintenance plan using the manufacturer's information for a machine that has been covered in this chapter. Include in your plan all equipment, materials and PPE required.

The typical contents of a maintenance schedule are shown in the table.

Step	Notes
1 Inspect and follow the risk assessment.	Always locate the risk assessment and identify all significant risks along with relevant control measures. Ensure you are authorised to carry out the maintenance.
2 Isolate the machine.	Always ensure every machine is correctly isolated before undertaking any type of maintenance operations.
3 Clean down the machine using extraction or a vacuum cleaner.	Do not blow down the machine as this puts harmful dust into the atmosphere.
4 Follow the manufacturer's instructions on maintenance required.	Always dispose of waste oils, liquids and parts in the approved way. *Do not* dispose of waste oils, etc down drains or into skips – use only authorised methods of disposal.
5 Replace tooling with suitable and legally compliant replacements.	Replacements should conform with PUWER 1998 and the ACoP for woodworking machines.
6 Check all safety devices, eg guards, push sticks.	These should conform with all requirements discussed earlier in the chapter.
7 Check extraction hosing for splits, etc.	As a temporary fix, heavy-duty tape could be used to repair holes or splits but this will soon fail, so suitable replacements need to be arranged straight away.
8 Replace any damaged timber lippings or mouthpieces, eg band saw or table saw.	Tight-grained hardwoods are best for this purpose due to their ability to withstand wear.
9 Set guards and carry out test run.	Always carry out a test run after any maintenance. Listen carefully for any unusual noises or vibrations; these are usually signs of a fault developing.

Case Study: Joe

Joe has been using a circular rip saw for several months. He has demonstrated his competence in using and setting up the machine.

He has now been tasked with preparing the circular saw in readiness for ripping 200mm × 50mm English oak planks to form 25mm × 50mm laths that are then to be moulded as part of a high-specification panelling job. As part of the preparation he is to carry out routine maintenance on the machine, as well as select and fit suitable tooling in readiness for the task. He has been told to consider specifically:

■ the type of saw blade that should be selected and why he has chosen it

■ any particular safety aids that should be provided and their requirements for use.

While carrying out the maintenance on the machine, Joe notices that the drive belt to the saw spindle has started to fray and reports this problem to his supervisor. He is told to ignore it for now, as a new belt would have to be ordered direct from the manufacturer and an immediate replacement cannot be guaranteed, but to consider the consequences of not replacing the drive belt and make a note about this.

Looking at the tasks Joe has been set, answer the following.

■ What type of saw blade would you select and why?

■ What type of safety devices would you recommend and why?

■ What are the short- and long-term consequences of poor machine maintenance, and what procedures could be implemented to improve this?

TEST YOUR KNOWLEDGE

Work through the following questions to check your learning.

1 PUWER is short for

   a Provision and Use of Work Equipment Requirements

   b Provision and Use of Work Equipment Regulations

   c Provision and Understanding of Work Equipment Requirements

   d Provision and Understanding of Work Equipment Regulations.

2 Before you change tooling on a circular saw bench, the **first** action should be to

   a isolate the power supply

   b remove the guard

   c check the riving knife is set correctly

   d turn on the extraction.

3 What is the guard covering the top of a rip saw blade called?

   a Bridge guard.

   b Landing guard.

   c Shaw guard.

   d Crown guard.

4 What is the noise level that requires the mandatory wearing of ear protection?

   a 80 dB(A).

   b 85 dB(A).

   c 90 dB(A).

   d 95 dB(A).

5 What **three** pieces of information should be obtained when carrying out maintenance on fixed and transportable machinery?

   a Manufacturer's maintenance schedule, risk assessment, authorisation.

   b Manufacturer's maintenance schedule, authorisation, time sheet.

   c Risk assessment, authorisation, HSE permission.

   d Manufacturer's maintenance schedule, risk assessment, HSE permission.

6 What is the **maximum** stopping time for the cutter blocks on a surface planer?

   a 12 seconds.

   b 10 seconds.

   c 6 seconds.

   d 3 seconds.

7 A saddle used on a circular rip saw bench will assist with

   a deep cuts

   b angled cuts

   c using negative tooth pattern saw blades

   d using positive tooth pattern saw blades.

8 Guards on woodworking machines should be

   a strong, easily adjusted and always used

   b flexible, fixed position and used when required

   c adjustable and used if required

   d set to the largest capacity of the machine.

9 What is the **maximum** permitted distance from the leading edge of the riving knife to the uprising teeth of the saw blade at table height?

   a   12mm.

   b   10mm.

   c   8mm.

   d   6mm.

10 What is the **maximum** projection of the knives on a circular cutter block fitted to a surface planer pre-dating 1995?

   a   1mm.

   b   2mm.

   c   3mm.

   d   4mm.

11 During pre-start up inspection, faulty equipment has been identified. What is the correct course of action?

   a   Isolate machine, put a warning sign on the machine, inform your supervisor.

   b   Put a warning sign on the machine, inform you supervisor, start the repair.

   c   Isolate machine, put a warning sign on the machine, start the repair.

   d   Carry on with the job, inform you supervisor, start next job.

12 What is the **maximum** cutter projection for planer blocks manufactured after 1995?

   a   5.5mm.

   b   2.1mm.

   c   1.5mm.

   d   1.1mm.

13 Where would you find a thrust wheel?

   a   Behind the circular saw blade.

   b   Fitted above the planer block.

   c   Fitted at the back of the band saw blade.

   d   At the top of the hollow mortice chisel.

14 Which one of the following lists the four component parts to a 10mm standard hollow square mortice chisel?

   a   Chisel, chisel bush, gouge, gouge bush.

   b   Chisel, chisel spacer, gouge, gouge spacer.

   c   Chisel, chisel bush, auger, auger bush.

   d   Chisel, borer bush, auger, auger bush.

15 What **must** the distance from the back of the out-feed table to the centre of the rip saw spindle be?

   a   1,200mm minimum.

   b   1,200mm maximum.

   c   12,000mm minimum.

   d   12,000mm maximum.

16 The purpose of a riving knife is to help

   a   stop the out-feed operative from touching the saw blade and to stop the timber from closing up

   b   stop the out-feed operative from touching the saw blade and help the timber to close up

   c   speed up the feeding of timber into the saw and to enable the timber to close up

   d   enable the out-feed operative to touch the saw blade and to help stop the timber from closing up.

17 Which one of the following is a legal requirement to complete when carrying out maintenance operations?

   a   Manufacturer's instruction book.

   b   Maintenance log book.

   c   HSE permission booklet.

   d   Time sheets.

18 Which type of tooth hook should a crosscut saw have?

a   Single.

b   Negative.

c   Sharpened.

d   Positive.

19 What is the purpose of the guide assembly on a narrow band saw?

a   To allow uncontrolled movement of the blade.

b   To stop sideways movement while allowing the blade unrestricted backward movement.

c   To restrict backward movement but allow uncontrolled sideways movement of the blade.

d   To restrict uncontrolled backward and sideways movement of the blade.

20 Too little tension applied by the top pulley on a narrow band saw is likely to result in

a   the blade being under too much strain

b   the blade being unable to move

c   increased risk of the blade coming off the pulley

d   the saw guides being in the way during cutting operations.

Chapter 6
Manufacture shaped doors and frames

Doors and frames vary in style, but the majority that are constructed are flat and rectangular in shape. This is largely due to their ease of manufacture and therefore the comparatively low final cost of the item. Shaped joinery is generally only produced for bespoke architect-designed residential, commercial or public buildings. Shaped work of this nature will be considerably more expensive but will provide very pleasing architectural features.

The principles that you have already learned form the basis of all joinery construction. This chapter will provide the additional knowledge required to take on shaped work successfully. For most operatives, the issue is very often not lack of ability to be able to produce this type of work but lack of opportunity to do so. While shaped work is very enjoyable to produce, it does inevitably provide its own challenges.

By reading this chapter you will know how to:

1 Set out shaped doors and frames.

2 Manufacture shaped doors and frames.

3 Assemble and finish shaped doors and frames.

SETTING OUT SHAPED DOORS AND FRAMES

SETTING OUT SHAPED DOORS AND FRAMES

INDUSTRY TIP

Where ironmongery is specified, it is recommended that the items are researched using manufacturers' catalogues or websites in order to see whether any special considerations need to be taken into account when manufacturing the product, as some ironmongery products require these.

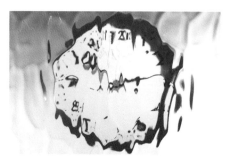

Example of patterned glass with a low level of obscurity

Example of patterned glass with a high level of obscurity

Obscured

'Frosted' glass that provides a level of privacy by obscuring the view through the glass. Different patterns are available depending on the level of privacy required, graded from 1 to 5, where 1 is the least and 5 is the greatest.
This type of glass is commonly used in bathroom windows and front entrances to residential properties

ACTIVITY

Go to www.pilkington.com and look at the range of glass types available. Which types would be suitable for a front entrance door?

INTERPRETING INFORMATION

As with all setting out, you will have to interpret a range of information to produce a product that meets the client's needs and wishes but also meets the requirements of the building regulations. Generally, you will be provided with architect-produced scaled detail drawings, showing the required elevations and sections of the product, and a specification providing information not usually shown on these drawings. The specification includes information such as the required species of timber, the finish required (painted, varnished, colour tint, etc) and any specified ironmongery to be used.

On other occasions, the details of the work will be agreed with the client and the joinery manufacturer. Where this is the case it is important that the scope and detail of the work are signed off by the client before starting work. This will help manage their expectations of the finished work.

The shaped frame to be constructed (regardless of whether it is for a door or a window) may also be defined within a door or window schedule. You will remember from Chapter 2 (see page 46) that a schedule provides information about the product in table form. Commonly it gives the item a reference, eg D1 ('door 1') or W1 ('window 1'). This will be referenced to the site plans and shows the final position of the product. It will also state the type of glass required (whether single/double glazed), any pattern type if the glass is to be **obscured**, and whether the glass is float, laminated or toughened, together with the thickness required.

It will take a little time to interpret this information and take on board what is required, before having a sound understanding of what has to be produced. It is quite common at this stage to have a couple of queries, or even a whole list of them, to discuss with the architect. The architect will have a great deal of knowledge about many aspects of building but will not be a specialist in any of them. With this in mind, they may have drawn something that it is not practical to make. A discussion at this early stage will manage expectations of the completed product. Very often they are only concerned about the final look of the product and will be happy to leave the construction detail to the joinery manufacturer, but it must be agreed at this early stage.

SITE SURVEY

Before any setting out takes place, a site survey is generally required to obtain accurate information. This could include:

- taking detailed measurements

- creating **templets** (the traditional name for 'templates') of any door/window opening shapes

- noting details of any mouldings or profiles that have to be matched

- making notes on any access problems that will affect delivery, such as parking restrictions, etc.

Generally the survey details for the arch shape can be quite easily ascertained, particularly for most new arches, but sometimes the dimensions of the rise and span are not what was expected. For example, the drawing might state that it should be a 2,080mm span with a 1,040m rise, yet the site measurement for the rise is 1,000mm. This should raise concerns and will require you to double check the dimensions. In this example, these dimensions would indicate either that the arch centre used was not accurate, or that the centre for the arch was eased too early by the bricklayer and has sagged. Alternatively, if it is an old arch then settlement has probably taken place. In this case a templet will be required. It should be taken back to the joiners' shop and a frame made to fit it. It is more likely for a low-rise arch to drop or settle, such as ones with a segmental or elliptical shape.

Often the survey site is a long way from where the product will be manufactured. It is always best to take as many details as possible and take plenty of photographs to refer back to. The last thing you want is to have to travel back to the site again because you have missed a vital detail. Too much information is better than too little!

After carrying out the site survey, if there are discrepancies between the information supplied by the architect and the information recorded during the survey, all appropriate parties must be informed. Initially you would inform your supervisor; depending on the size of the joinery manufacturer this could be the:

- joinery shop supervisor

- joinery works manager

- senior setter out

- owner.

Whoever you report to should then liaise with the architect to agree how any discrepancies found are to be overcome. This must be confirmed in writing by the architect to prevent any disagreements at a later date. This confirmatory letter, sometimes known as a 'variation order' or 'architect's instruction', will then ensure that the agreed changes form part of the contract and the expectations of the finished contract have been managed effectively.

Templets

Also known as a 'template', this is a are thin pieces of hardboard, MDF or ply that is cut to the shape required and is then used to help reproduce the shape

INDUSTRY TIP

To ensure an accurate brick arch shape is produced, the centring of the arch is best manufactured in the joinery shop rather than being made on site. This way there is a greater chance of a good fit to the finished product and a templet of the arch shape will not be required.

A templet being drawn from an arch shape

Very often some form of access equipment is required to measure arch openings. A risk assessment that is being produced for the setting out operation should also cover the survey. (More information about risk assessments can be found in Chapter 1, see pages 5–6).

ARCH TERMS

Most shaped frames will be fitted into a brick arch, although some will be fitted to a concrete or timber-framed structure. For this reason we should know the standard arch terms used; see the illustration pointing out the parts of an arch. While the shape of the arch may alter, the terms remain the same.

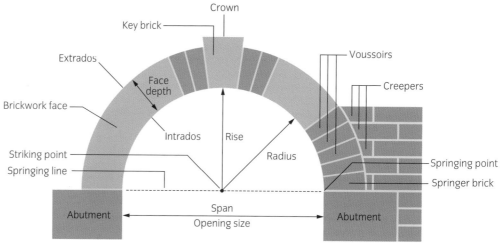

The parts of an arch

We also need to know the standard circle terms given in the following table, also shown in the diagram.

Term	Description
Circumference	The line bounding the circle, or around its edge
Diameter	A straight line passing through the centre point of the circle and touching the circumference on both sides of the circle
Radius	A straight line drawn from the centre of the circle to its circumference
Arc	A portion of the circumference
Chord	A straight line (shorter than the diameter) that touches two points on the circumference of a circle
Tangent	A straight line touching the circumference of a circle
Normal	A straight line drawn through the circumference from the centre of a circle (which will always be at 90° to the tangent at that point)
Quadrant	A quarter of a circle
Sector	A part of a circle contained between two **radii** and the circumference of the circle
Segment	A portion of a circle contained between an arc and a chord

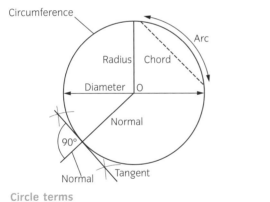

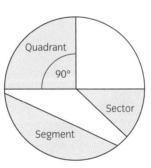

Circle terms

Radii

The plural of 'radius'

ARCH SHAPES

There are many different arch shapes that shaped joinery can be created to fit to. This chapter will limit itself to those listed in this unit in the qualification. They are:

- segmental

- Gothic

- semi-circular

Pseudo

Not genuine

- elliptical (true and **pseudo**).

The unit also mentions shaped joinery work, segmentally shaped in plan.

The following sections cover each of these types of arch, showing the step-by-step stages required to set them out full size using given rise and span dimensions. The examples are drawn using one, two or three centres, with the exception of the elliptical arch which is drawn without fixed centres and is therefore much more complex to draw.

All arch **geometry** requires the understanding and use of **bisection**. Study the instructions on the following pages to ensure you can follow and reproduce the arch shapes listed on the previous page.

Bisecting a level line to produce a perpendicular line along the centre of its length

Geometry

A branch of mathematics that deals with points, lines, angles, surfaces and solids

Bisection

The division of an angle or line in two geometrically

STEP 1 Draw the required span A–B.

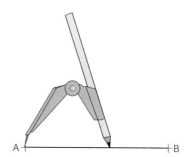

STEP 2 Set the compasses to a distance that is more than half the length of the line A–B.

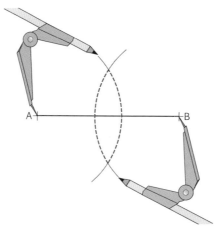

STEP 3 Keep the compasses locked and draw arcs through the centre of the line from points A and B.

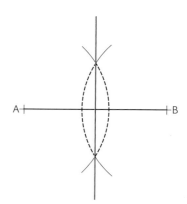

STEP 4 Draw a straight line through the arc intersections.

SEGMENTAL ARCHED JOINERY IN ELEVATION OR PLAN

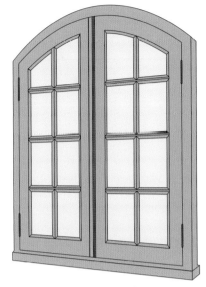

Segmental headed window

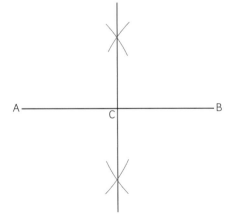

Segmental headed door frame

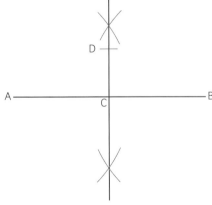

Segmental headed panelled door

This type of arch is still in common use today. It gives some architectural shape to the **facade** of a building but is not as expensive to produce compared with semi-circular work because the curved work is limited. As stated previously in this chapter, a segment is a portion of a circle. To be able to draw a segment, we must be able to find the centre of the circle that contains the segment.

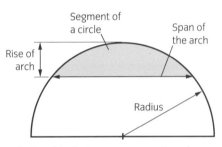

Relationship between segment and semi-circle

Drawing a segmental shape

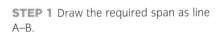

STEP 1 Draw the required span as line A–B.

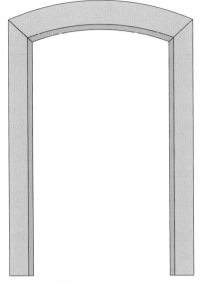

STEP 2 Bisect line A–B to find the centre, point C. Extend this line above and below the centre point.

STEP 3 Measure the rise D above the centre point on the bisection line.

Facade

The exterior face of a building, generally referred to as the 'front face' of the building

ACTIVITY

Following the method outlined here, draw a segmental arch shape with a span of 150mm and a rise of 50mm.

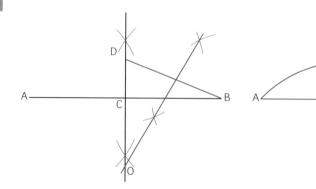

STEP 4 Draw chord B–D and then bisect it. Extend the bisection line to cross the extended line D–C at point O.

STEP 5 With point O as the centre and the distance from O–D as the radius, strike the arc A–D–B for the arch.

Method of calculating the radius from a given span and rise

The radius of an arch can be calculated if you know the span and rise. The mathematical formula, where R = rise, is:

$$radius = \frac{(\frac{1}{2} \, span^2 \div R) + R}{2}$$

ACTIVITY

Calculate the radius of a segmental arch where the span is 1,800mm and the rise is 300mm.

Answer: 1,500mm.

Example

Let's work out the radius where the span is 900mm and the rise is 75mm, working out one step at a time.

Step 1

$$radius = \frac{(450^2 \div 75) + 75}{2}$$

Step 2

$$radius = \frac{(202,500 \div 75) + 75}{2}$$

Step 3

$$radius = \frac{2,700 + 75}{2}$$

Step 4

$$radius = \frac{2,775}{2}$$

$$radius = \mathbf{1,387.5mm}$$

Where segmental arches form part of an architectural feature, a problem can occur where openings of different widths are required. Rather than the same radius being used, to give a uniform appearance the arch rise is maintained. This is achieved as shown in the following illustration.

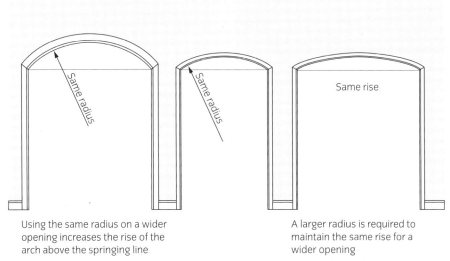

Using the same radius on a wider opening increases the rise of the arch above the springing line

A larger radius is required to maintain the same rise for a wider opening

Problematic effect of having differing spans on segmental arches adjacent to one another

There is one last way of producing a segmental curve without a compass or trammel rod. This method uses a segment frame. This is made up using three laths of timber or ply. The following shows this method step by step.

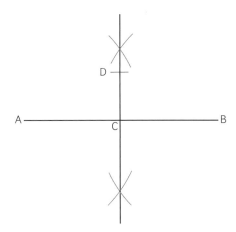

STEP 1 Follow Steps 1–3 as shown on page 313.

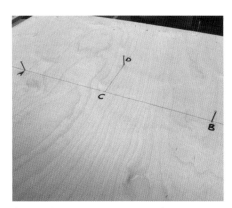

STEP 2 Drive temporary nails at points A, B and D.

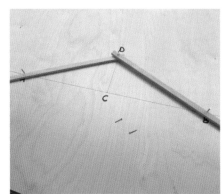

STEP 3 Rest ply or MDF laths against the nails at lines A–D and B–D, and secure the two laths together at point D.

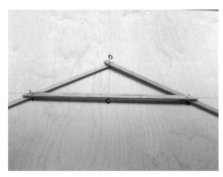

STEP 4 Tack a third lath between the first two to create a triangular frame.

STEP 5 Remove the nail at point D and replace it with a pencil or pen. Slide the triangular frame to describe the segmental curve, being careful to keep the laths against points A and B.

This method is only practical where the span is so great that a trammel is difficult or impossible to use. It was commonly used to set out a segmental **centre** or turning piece for arches to be formed on site. A turning piece is made from a solid piece of stout timber and is used as a former for the bricklayer to turn and lay the bricks around its shape. It can be made on site but can be cut more easily in the joinery shop. A turning piece is only used for low-rise segmental arches where the shape can be cut from solid timber.

Centre

A temporary structure made by a carpenter or joiner to allow a bricklayer to turn a brickwork arch around them. When the mortar has set, the centre may be 'struck' (removed)

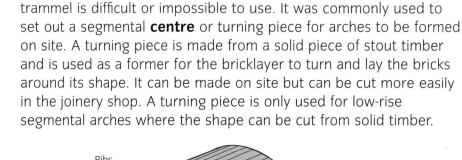

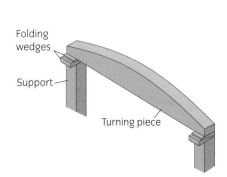

A turning piece used for low-rise segmental arches

A semi-circular arch centre and its parts

OUR HOUSE

Draw full size a templet for a segmental arch with a rise of 200mm to suit a door opening used in 'Our House'.

GOTHIC ARCH

This arch shape is most recognised from **ecclesiastical** buildings. Gothic architecture was a style of architecture used in Western Europe in the Middle Ages. It began in France in the 12th century and was commonly used until the 16th century. It is also the architecture of many old castles, palaces, town halls, universities and some houses. In the 19th century, the Gothic style became popular again during the Victorian era, particularly for building churches and universities. This style is called 'Gothic Revival' architecture.

There are three commonly used Gothic shapes:

- *drop Gothic:* where the radius is less than the span

- *equilateral Gothic:* where the radius of the arch is equal to the span

- *lancet Gothic:* where the radius is greater than the span.

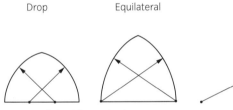

Drop   Equilateral   Lancet

Types of Gothic arch shapes

Ecclesiastical

Of or relating to a church

A pair of glazed drop Gothic doors

Equilateral Gothic

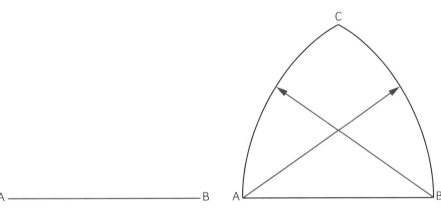

A ———————————— B   A ———————————— B

STEP 1 Draw the span, line A–B.

STEP 2 Using the span as the radius, draw arcs A–B and B–A to intersect at point C.

The procedure required to draw the drop and lancet arches is very similar in that both radii will be struck from the base line A–B. The radius will generally be taken from the architect's drawings. If the arch has been built, we will have to take the rise and span dimensions from the survey. You should never take manufacturing sizes from a drawing as they may not accurately represent what has been built. There are no specific proportions used for either; it will be decided by the individual opinion of the architect.

Drop Gothic arch

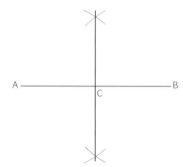

STEP 1 Bisect span A–B to produce point C.

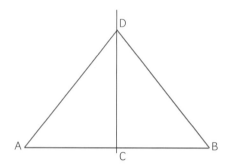

STEP 2 From point C, measure the rise to produce point D. Draw lines A–D and D–B.

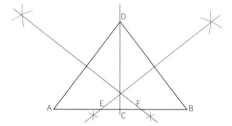

STEP 3 Bisect lines A–D and D–B to produce points E and F on the span A–B.

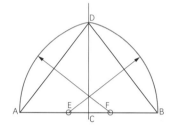

STEP 4 Draw arcs from points E and F to intersect at point D.

ACTIVITY

Draw a lancet Gothic arch with a span of 600mm and a rise of 800mm. Research a pointed segmental arch shape to see how this differs from those shown.

Lancet Gothic

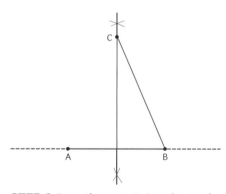

STEP 1 Draw the span A–B and extend along the springing line beyond the span. Bisect line A–B. Mark the rise as C, using the measurement taken from the drawing. Draw chord B–C.

STEP 2 Bisect chord B–C, where the bisecting line falls on the springing line past point A: mark this as point D (the striking point). To obtain the opposite striking point, measure equal distances from point O–D to point O–E.

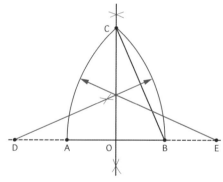

STEP 3 Using a radius E–A and D–B, draw arcs to intersect at point C.

SEMI-CIRCULAR ARCH

Semi-circular arches are sometimes known as Roman arches, as they were used in that era. They are the easiest to draw as the rise of the arch is half the span. In other words, the rise is the same as the radius.

Semi-circular window

Drawing a semi-circular arch

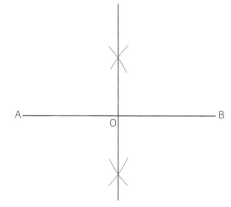

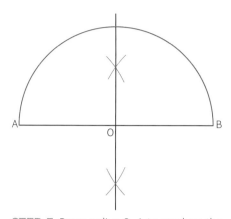

A———————————B

STEP 1 Draw span A–B.

STEP 2 Bisect line A–B to produce the radius centre, point O.

STEP 3 Draw radius O–A to produce the semi-circular shape.

ELLIPTICAL ARCH

An elliptical arch is a flattened version of a semi-circular arch. They have a significant advantage compared with a semi-circular arch as they allow much more headroom over a greater width. Therefore, an elliptical arch can be lower.

True ellipses are rarely used in construction unless the shape is stand alone (nothing has to fit it), such as a railway arch or the arched opening shown in the photograph. The problem lies mainly with the fact that no two ellipses can be drawn exactly parallel to one another so in joinery, where you have to draw many parallel lines, it becomes a considerable challenge. To overcome this, a pseudo (false) ellipse is drawn with the aid of a compass or trammel. This overcomes the problem of being able to draw a number of parallel lines.

An ellipse has additional terminology that we need to know prior to drawing one:

An elliptical arched opening

- *major axis:* the widest part of the ellipse

- *minor axis:* the narrowest part of the ellipse

- *focal points:* the ellipse is a curve surrounding two focal points where a straight line drawn from either of the focal points to any point on the curve and then back to the other focal point will have the same combined length, regardless of which point on the curve it touches (see the illustration of ellipse terms for a visual example).

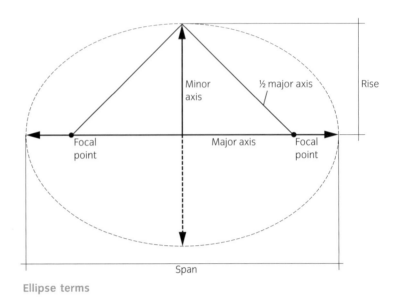

Ellipse terms

Drawing a true ellipse

There are several ways of drawing true ellipses; unfortunately they are not practical to use for setting out purposes, as you will see with the first two methods shown.

Pin and string method

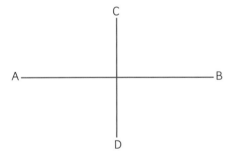

STEP 1 Draw the major and minor axis, lines A–B and C–D.

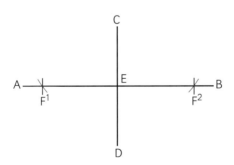

STEP 2 With a centre at C and a radius of A–E, mark focal points F¹ and F² on the major axis.

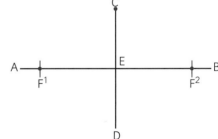

STEP 3 Place pins at points F¹, F² and C.

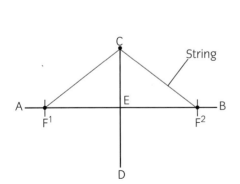

STEP 4 Place taut string around the pins.

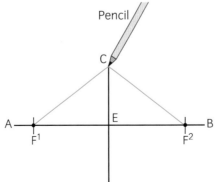

STEP 5 Remove the pin at point C.

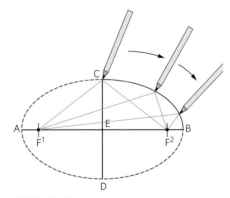

STEP 6 Place a pencil tight up against the string and use it as a guide to draw the ellipse.

Concentric circle method

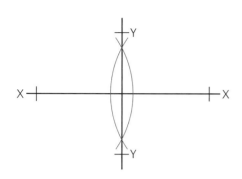

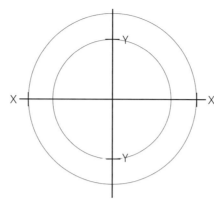

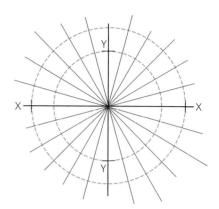

STEP 1 Draw the major axis (line X–X), then bisect this to produce the minor axis (line Y–Y).

STEP 2 From the centre, draw radii that are half the minor axis and half the major axis.

STEP 3 Using a set square, draw a number of radiating lines.

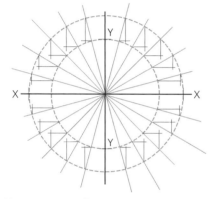

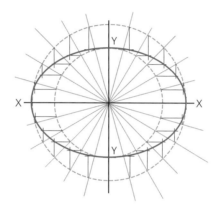

STEP 4 From the points where the radiating lines cross the smaller circle, draw horizontal lines towards the larger circle, and vertical lines to intersect with the horizontal lines from the larger circle.

STEP 5 Carefully draw a freehand curve that joins all these intersection points to produce the elliptical shape.

Trammel frame method

This is by far the most practical workshop method. Once an accurate frame has been made it can be kept for any future requirements as it is not size specific. Furthermore, MDF board versions do not need to be braced. The frame is used in conjunction with a set of trammel heads and a trammel beam.

1 Fix the frame down to the setting out board.

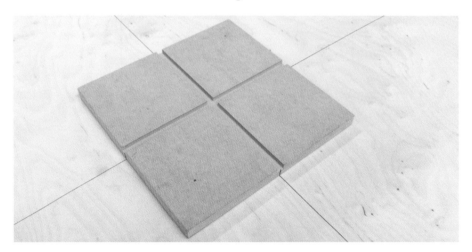

2 Set trammels up as shown below, with the distance between the pencil point and the first trammel pin half the minor axis and the distance to the second pin half the major axis. Insert the trammel points into short wooden glides that fit snugly to the groove on the trammel frame. Gently slide the trammel frame to allow the glides to slide within the grooves and draw the outline of the elliptical curve.

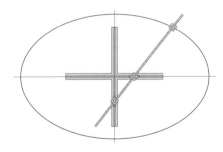

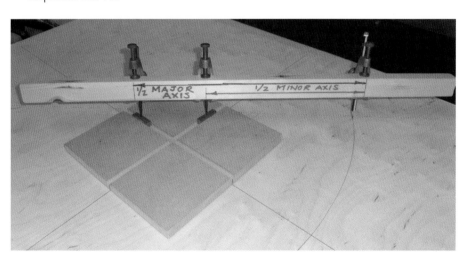

THREE-CENTRED ARCH

Also known as a pseudo ellipse, as previously mentioned this approximate ellipse shape is used more often than not when constructing joinery that requires an elliptical shape. Being drawn from three fixed centres, the drawing and construction are made simpler.

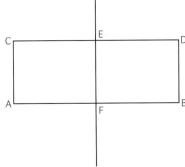

STEP 1 Draw a rectangle A–B–C–D, where line A–B is the major axis and line A–C is half the minor axis. Next bisect line A–B to produce points E and F.

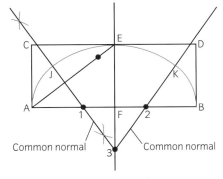

STEP 2 Draw a line between points A–E. Using point C as the centre and line C–A as the radius, draw an arc to cut line C–D at point G. Using point E as the centre and line E–G as radius, draw an arc to cut line A–E at point H.

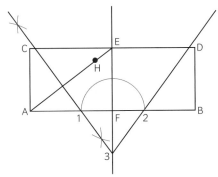

STEP 3 Bisect line A–H to produce point 1 cutting line A–B, and point 3 crossing extended line E–F. Using point F as the centre, draw radius F–1 to produce point 2 on line A–B.

STEP 4 Using point 1 as the centre, draw radius 1–A to cut the bisector at point J. Using point 2 as the centre, draw radius 2–B to cut the bisector at point K. Using point 3 as the centre, draw radius 3–J through point E to point K.

Lines 3–J and 3–K are called 'common normals' and provide the ideal jointing place when producing the curved component.

ACTIVITY

Draw a three-centred false ellipse with a span of 600mm and a rise of 200mm.

SETTING OUT

With CAD drawings becoming more popular, joinery shops are making good use of its benefits in setting out joinery work. Unfortunately, while large sheets and rolls of drawing paper can be used and printed on, paper by its very nature will shrink, expand and distort, and will become damaged in use. For these reasons all shaped work should be set out full size on a flat and a lightly coloured board.

ACTIVITY

Find the listed equipment on an appropriate website and cost them.

Birch ply is the best material to set out on. MDF can be used but it should be lightly sanded with a very fine abrasive, as the surface as supplied is 'waxy' and cannot be drawn on easily.

Keeping the material flat is very important as the circular part of the construction is marked out, fitted up and assembled over the full-size drawing to ensure that it will be the correct size and shape.

The following equipment will be required when setting out curved work:

- trammel points and beam
- 2m steel rule
- accurate parallel straight edge
- large perspex set squares
- line runner.

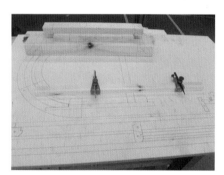

Trammel points and beam

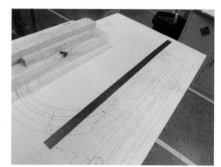

2m steel rule

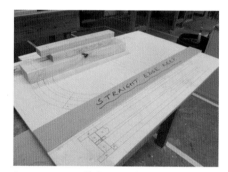

Accurate parallel straight edge

Large perspex set squares

Line runner

Before starting, take a moment to think about where the setting out will be positioned on the board. There is nothing worse than starting the setting out only to discover that you have run out of board because you have started in the wrong place.

Generally we only need to draw the elevation of the shaped part of the frame, with just enough straight to align the work while marking out, fitting up and assembling. The straight part of the door or frame can be drawn using standard height and width sections.

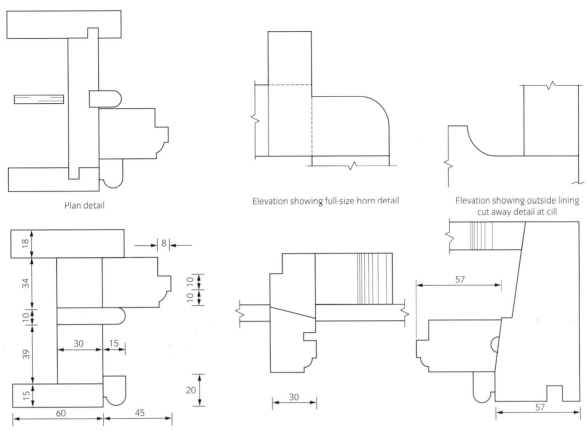

Plan detail

Elevation showing full-size horn detail

Elevation showing outside lining cut away detail at cill

Height and width sections of semi-circular headed box frame

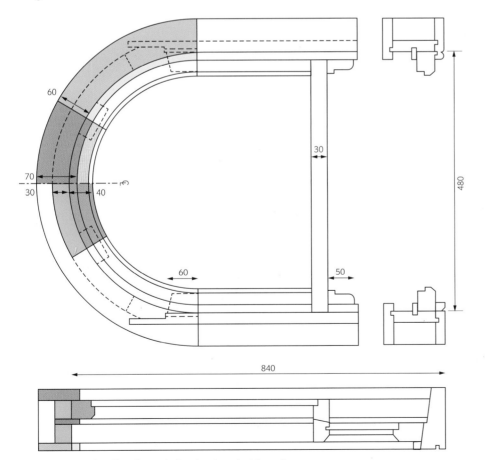

Construction details of semi-circular headed box frame

Box frame

Name given to vertical sliding windows

Meeting rail

Name given to both the top rail of a bottom sash and the bottom rail of a top sash as they meet when closed

Single curvature

Where work is shaped in one view only, either elevation or plan

The photograph shows a completed rod of a semi-circular headed **box frame**. You will see that only the top half of the frame is drawn to just below the **meeting rail**, as this is all that is required to be drawn of the elevation in order to manufacture the window. Note the colour coding of the vertical section and the corresponding components in the elevation. This makes the drawing easier to interpret, and is very useful where there are a number of components lying on top of each other. In this example, there are seven components to the shaped top sash and head construction. A semi-circular headed box frame is probably one of the most complex examples of **single curvature** work.

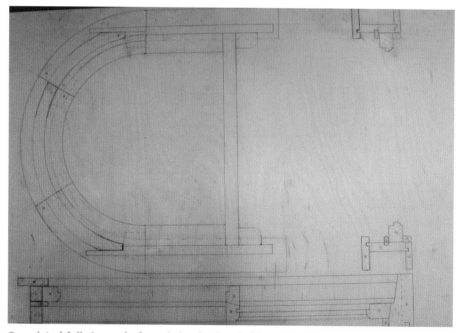

Completed full-size rod of semi-circular headed box frame

CUTTING LISTS

Once the drawing is complete, a cutting list can be produced. A cutting list gives the timber sizes needed for all the components that are required to manufacture the product. The following method can be used to accurately **take off** the material requirements and record them on a cutting list:

Take off

Taking the length, width and thickness of components and recording them on the cutting list

1 Using a circle templet, draw a circle in every component shown on the rod.

2 Starting with the longest component, take off the dimensions and record them on the cutting list as 'Item 1'.

3 Once taken off and recorded, place a '1' in the circle showing that component.

4 Take off any other components of that section size.

5 Repeat Steps 3 and 4 until every circle has a number within in it. This is the 'double check' that all items have been 'booked up' (recorded).

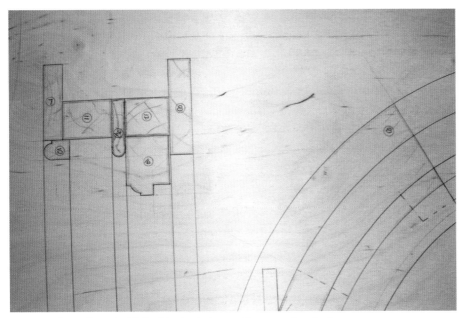

Item numbers inserted into blank circles as they are booked up

The curved items need to be ordered as blanks (timber in its rectangular form that the curved shapes then are cut from) that are large enough for the curved component to be cut from.

Below is an example cutting list for the semi-circular headed box frame window.

FUNCTIONAL SKILLS

Research the current cost of unsorted European redwood and calculate the cost of the timber required in the example cutting list.

Work on this activity can support FM2 (C2.2).

Item	Qty	Description	Mat.	Length	Sawn size		Planed size		Instructions
---	---	---	---	---	W	Th	W	Th	
1	2	Pulley stiles	S/W	730			95	20	
2	1	Cill	H/W	620			116	57	
3	3	Staff bead	S/W	600			20	15	
4	2	Parting bead	S/W	600			21	10	
5	2	Inside linings	S/W	650			60	15	
6	2	Outside linings	S/W	650			75	18	
7	1	Curved in' lining	S/W	1100	105	20		15	Cuts 3 in length
8	1	Curved out' lining	S/W	1100	115	23		18	Cuts 3 in length
9	1	Curved parting bead	S/W	1100	90	15		10	Cuts 3 in length
10	1	Curved staff bead	S/W	1100	60	25		20	Cuts 3 in length
11	1	Curved inner head	S/W	1100	75	45		39	Cuts 3 in length
12	1	Curved outer head	S/W	1100	75	38		34	Cuts 3 in length
13	1	Curved top rail	S/W	850	85			33	Cuts 3 in length
14	1	Stiles (top/btm sash)	S/W	1600			45	33	Cuts 4 in length
15	1	Bottom rail	S/W	500			43	33	
16	2	Meeting rails	S/W	500			57	30	

Preparing timber from a cutting list

Always select the timber carefully before cutting to length. Pull out a number of boards to examine them, rather than cutting as you go. This allows you to select for consistent colour and grain characterisation. This is particularly important for polished hardwood jobs. Try to cut all the blanks for the curved work out of one piece; this ensures consistency of colour and grain and helps produce a harmonious-looking frame.

The general rule is that the longest lengths on the cutting list should be cut first, leaving the offcuts for the shorter lengths. (If you cut the short pieces first you may not be left with any timber in the rack long enough to cut the longer lengths required.) Try to cut between defects (knots and shakes, etc) if possible.

The order of machining is as follows and the page numbers afterwards show where these were covered in detail in Chapter 5:

1 crosscut to length (pages 264–267)

2 rip to width (page 246)

3 surface plane face and edge (pages 277–284)

4 plane to width the thickness (pages 284–287).

INDUSTRY TIP

When cutting timber, cut the longest lengths required first, using the straightest timber. Any bent, sprung or twisted timber can be used for shorter lengths where these defects will be less noticeable.

CONSTRUCTING CURVED COMPONENTS

TEMPLETS

When the cutting list is complete, the curved component templets can be produced. Templets are made to allow the blanks to be marked out, cut and spindled to shape and profile. Traditionally they were produced by marking the shape with the trammel heads and beam, bandsawn slightly oversize, and brought to shape with a compass plane or **spokeshaves**.

The best way nowadays is to produce the templets using a portable power router and trammel bar. As the router cutter takes the place of a pencil in a compass, it can produce very accurate templets. This method also produces the templets very quickly. A templet will need to be made for each shaped component.

Templets are usually made from 12mm birch ply or MDF. They need to be cut to the exact shape required to make the curved component. They are initially used to mark out the shape of the

A templet being trimmed to the exact shape required using a compass plane

Spokeshaves

Used to shape curved surfaces, consists of a blade fastened between two handles. The sole can be flat or round bottomed for planing concave or convex shapes

component onto the wide boards prior to being bandsawn to shape. They need to be between 75 and 100mm longer than required to allow for 'lead on' when spindling later.

A templet being produced using a trammel bar and router

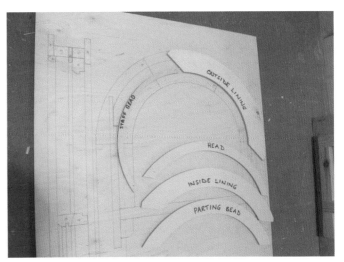

The templets required to produce a semi-circular headed box frame

METHODS OF CONSTRUCTING CURVED COMPONENTS

The setting out would show the method to be used for constructing the curved components. There are three methods that could be used:

- solid
- built up
- laminated.

Which of the three construction methods is chosen will depend on a number of factors including:

- the strength required
- the component type
- whether the finish is painted or polished
- the radius of the component.

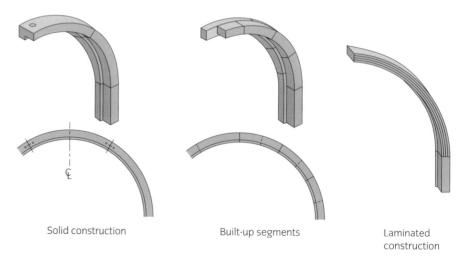

Solid construction

Built-up segments

Laminated construction

Methods of forming curved components

SOLID CONSTRUCTION

With the solid (and the built-up) method of construction, we need to be aware of the problems of 'short grain'. Short grain occurs where the shape is made in too few segments or where the templet is not applied 'with the grain' of the timber.

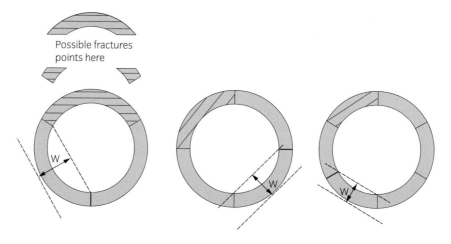

Possible fractures points here

W = width of timber required for blank

Frames made in three, four and six parts. The short grain is reduced as more parts are used

As you will see in the illustration, using fewer parts to make the circular frame will lead to more short grain being encountered. This will increase the chance of the grain splitting and of breakages due to **pick-up** on the spindle or router where it is weakest. The other advantage of using more parts is that the timber section is made from narrower material, minimising any shrinkage and therefore distortion of the shape.

Four parts to a circle should be the minimum number used, with six being good practice and eight being the best. Obviously the more parts that are used the more costly it will be to construct due to the extra labour required to joint and manufacture the sections of the

Pick-up

Where timber is fed into a cutter against the grain

frame. However, the main advantage of this method of construction is that it is the most simple to produce and therefore the least costly.

Traditionally, the templet would be applied to a full-width board that has been planed to finished thickness to minimise waste and short grain. It can still be done this way if the components are to be shaped by hand using a compass plane or spokeshaves. This would be the case if it was to be a 'one-off' or a job with no heavy rebates required to be produced on the spindle. In this case, the mouldings, grooves, etc can quite safely be **stuck** using a portable power router.

Stuck

A traditional term for the working of the moulding, grooves and rebates onto solid timber using hand tools and machines

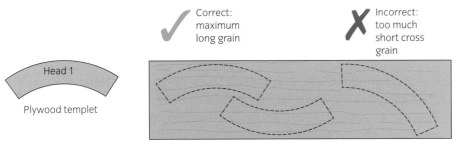

Templet applied to board when planing components by hand

It is a requirement of the wood machining Approved Code of Practice (ACoP) that all circular work be held in secure holding jigs while being machined. This being the case, the blanks must be cut to an exact size in order for them to fit into the jig and be safely machined on the spindle moulding machine.

BUILT-UP CONSTRUCTION

The difference between solid and built-up construction is that the thickness of the curved component is made up of more than one piece. It is quite common for two or three layers of segments to be assembled with the joints staggered to create a strong curved component.

This method suits frame construction where the rebate forms a natural break. It is not particularly suitable for shaped top rails to doors or sashes, as when you plane the edges to fit to the frame you will get considerable pick-up as the grain in each layer is running in a different direction. Another disadvantage is that the timber sections will move (shrink or expand) in different directions, distorting the finish to the face. This can show up particularly when the work is gloss finished.

The semi-circular head of a box frame is an ideal example of where this type of construction works well due to the construction method required.

ACTIVITY

Next time you see shaped doors and frames at a local building of interest, have a look to see how they have been constructed and what construction and jointing methods have been used.

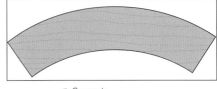

Correct: blanks cut to an exact size to fit spindle moulding jig

Rectangular blanks required when using a spindle moulding machine

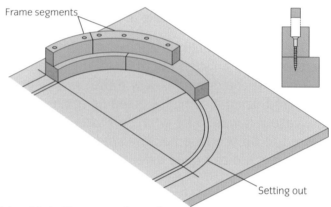

Frame segments

Setting out

The curved head is built up over the rod

LAMINATED CONSTRUCTION

Laminated construction (sometimes referred to as **glulam** construction) is the most labour intensive but can produce very stable components with no short grain. The section required is made up of a number of thin laminates (thin strips of timber) that are glued and bent around a former. The former would usually be made of two parts, one male and one female. The glued laminates would be placed between the two with cramps applied until the adhesive has thoroughly cured.

A synthetic resin adhesive should be used rather than a PVA to prevent **spring-back**. As there is considerable tension involved in bending the laminates around the former, it needs to be very robust in construction. The thinner the laminates, the easier they are to bend around the former.

There needs to be a balance between the number of laminates used and the extra material required to produce the laminates. Each laminate has to be planed up so will require being sawn with a planing-up allowance, adding to the cost of the material used. The thickness of laminate used will depend on the species of timber. Generally hardwood laminates have to be thinner than softwood laminates.

Working out the thickness of laminate required

As a rule of thumb for softwood (European redwood), divide the radius by 200. For example, if the radius was 1,000mm the thickness of laminates would be 1,000mm ÷ 200 = 5mm.

For hardwood, we would divide by 250. For example, for a radius of 1,000mm in mahogany the thickness of the laminates would be 1,000mm ÷ 250 = 4mm.

It is not practical to use laminates less than 3mm thick as this is generally the thinnest that a thicknessing machine will plane to. This determines the minimum radius that laminated construction can be used for. Laminates can be sawn thinner but they will not be of a consistent thickness and the quality of the **joint line** will be poor.

Glulam

A commonly used abbreviation for 'glue laminated'

The laminates are cramped between a male and female former

Spring-back

Where the tension in the timber wants to pull the curve flat and it loses its intended shape

Joint line

The fit of a joint where two pieces are bonded to each other

The laminates will need to start about 5mm wider than required as the finished edges will need to be surfaced and brought to thickness using planing machines.

Once the former is made, it is advisable to do a trial run for the thickness of the laminates, particularly when using some dense hardwoods, to make them sufficiently pliable to bend around the former. When preparing softwood laminates it is advisable to prepare about 20% more than required. This is because many of the laminates will have shakes or knots that will either break or split during bending.

A vacuum bag can be used for laminating certain components, as long as they are small enough to fit in the bag. This method is best when laminating plywood structures. A male former is constructed and placed in the bag. The laminates have adhesive applied to each bonded face and are then put on top of the former, held with two light tacks (punched so as not to puncture the bag) that hold them to the top of the former to prevent slipping. The compressor is then started, which will slowly suck the air out of the bag. The polythene will gradually pull the laminates tight to the former. The compressor remains switched on until the adhesive has dried.

A laminated component being formed in a vacuum bag

OUR HOUSE

Look at the project drawings for 'Our House'. The owners have decided to convert the loft and want a bullseye window inserting into the gable wall. Draw up a full size rod for a fixed glazed window that is 675mm in diameter. Which of the three construction methods do you think is most suitable?

JOINTING METHODS

The method of joint chosen will depend on the nature of the work, whether the joints will be seen, and how the components are arranged within the structure.

HEADING JOINTS

Heading joint

The name given to a lengthening joint between two components

The **heading joints** used between parts forming a continuous curved component could be formed using any of the following jointing methods:

- hammer headed keys
- dovetail keys
- handrail bolts
- loose tenons
- kitchen worktop connectors.

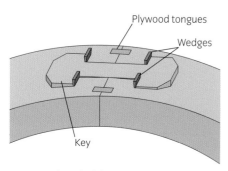

Hammer headed key

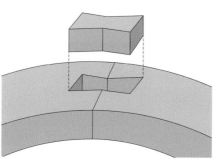

Dovetail key

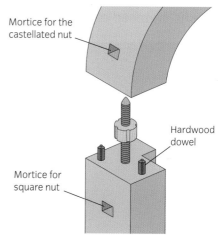

Handrail bolt

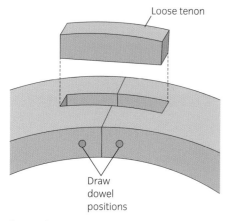

Loose tenon

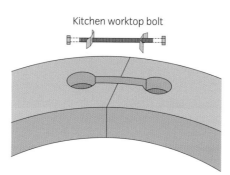

Kitchen worktop connector

Hammer headed and dovetail keys are both labour-intensive joints to produce. The hammer headed key is the best of the two as it uses wedges to pull the joint up tight.

Handrail bolts are very efficient as they also pull the joint up, but can be quite fiddly to fit. A pair of dowels or loose tongues should be incorporated into the joint to prevent it twisting. (These bolts are now difficult to obtain.) Below is a traditional handrail bolt with its modern replacement, a Zipbolt™. The handrail bolt has two nuts, a captured nut (square) and a castellated (grooved) nut. The bolt is tightened on the castellated nut side. The threaded side of the Zipbolt™ is turned into one side of the joint and is tightened with an Allen key on the other side.

Handrail bolt with a captured (square) and castellated nut

Loose tenons are now probably the most commonly used joint as they are easy to produce. They require pinning through the face as, unlike the other jointing methods listed, they are not **mechanical joints**.

Kitchen worktop bolts can be used as a substitute on large frames but do require large holes to be bored, weakening the component, and should only be used on the back of a frame.

JOINTING ON SPRINGING LINES

Usually a variation of a mortice and tenon will be used when jointing on **springing lines**, with the tenon extending from the jamb of the frame or stile of the door. This joint is very easy to produce and can be draw pinned to cramp the joint together. If there is more than one component coming together at the springing line, the joints of the different components will have to be offset to avoid clashing. A twin tenon is commonly used between the transom and the jamb or the rail and the stile, with a single tenon/bridle on the jamb to the head or stile to the top rail.

A more complex and time-consuming joint could be used if specified, such as a hammer headed tenon joint or handrail bolt. Alternatively, the joint position could be moved further towards the **crown** of the frame or door; this avoids seeing the grain running in three directions at one point, which can look ugly on polished work, and also avoids weakness at the same point.

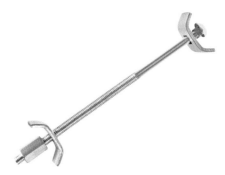

Worktop connecting bolt

Mechanical joints

Joints that, due to their design, hold or pull themselves together

Springing line

Where a curved section starts to 'spring away' from a straight line

Springing joint and transom intersection using a twin tenon and hammer headed tenon joint

Crown

The uppermost part of a shaped headed door or frame

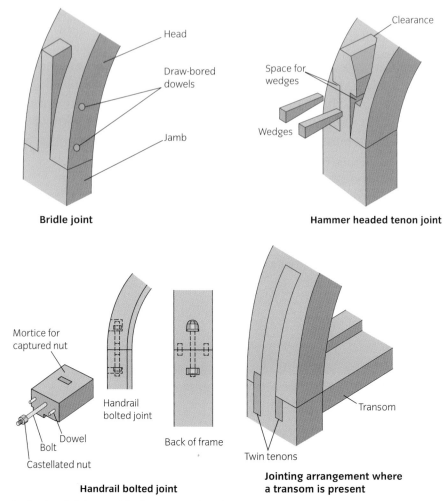

Bridle joint

Hammer headed tenon joint

Handrail bolted joint

Jointing arrangement where a transom is present

Springing joints used in frame construction

Springing joints used in door construction will generally be limited to tenons, as bolts are unsightly even when plugged.

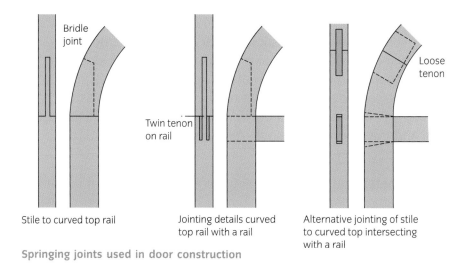

Stile to curved top rail

Jointing details curved top rail with a rail

Alternative jointing of stile to curved top intersecting with a rail

Springing joints used in door construction

The springing joints of segmental headed doors and frames are different, as the jamb/stile does not run at a tangent to the head/rail. The most common joints used in this situation are shown below.

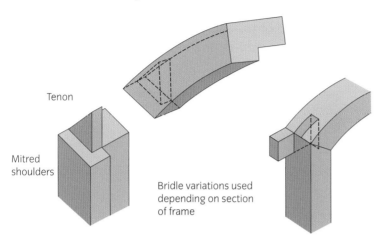

Tenon

Mitred shoulders

Bridle variations used depending on section of frame

Springing joints used for segmental headed frames

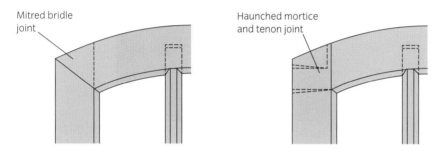

Mitred bridle joint

Haunched mortice and tenon joint

Joints between curved top rail, stile and bar

CROWN JOINTS

For the frame, any of the heading joints already mentioned can be used. Joints used for single width doors can be similar.

ACTIVITY

Draw an isometric view of the crown joint for a semi-circular headed door.

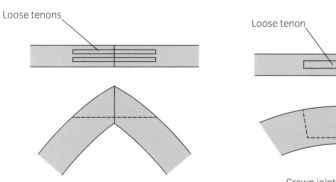

Loose tenons

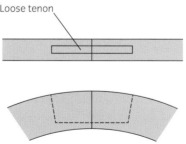

Loose tenon

Crown joint for a Gothic headed door

Crown joint for any curved headed door or sash

Crown joints suitable for single doors

Where the frame is wide enough to require a pair of doors, a different jointing arrangement is required where the curved rail meets the meeting stiles on the door. Inserting false tenons removes the problem of the short grain that would occur if a tenon was machined on the crown end of the rail. The false tenon should be inserted and glued to the curved rail, then allowed to dry. The joint can then be fitted up in the usual way.

Pair of glazed Gothic doors

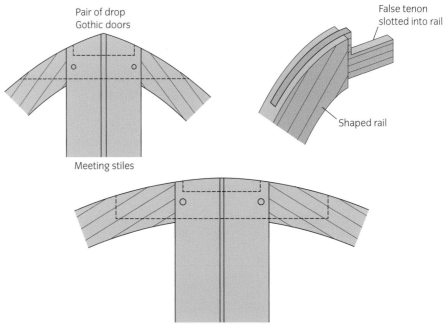

Pair of drop Gothic doors

False tenon slotted into rail

Shaped rail

Meeting stiles

Pair of semi-circular headed doors

Using false tenons at pairs of curved headed doors

SEGMENTAL BAY/BOW WINDOWS

Work on segmental bar/bow windows is fairly rare due to the expense of building curved brickwork and the fact that most replacement windows are now high-performance uPVC windows. Occasionally, however, this type of window is still specified for new work or as a replacement in a conservation area where an identical window must be installed.

Technically, a true bow window has curved glass but, again due to the expense of making the formers to enable the glass company to bend it, flat glass is usually used instead. The glass will sit in straight rebates and form **facets** around the face of the window. This being the case, the head, cill and any horizontal bars will have the inside and outside faces following the curve, and the rebate produced as a chord to the shape to allow for the flat glass.

Facets

Flat faces

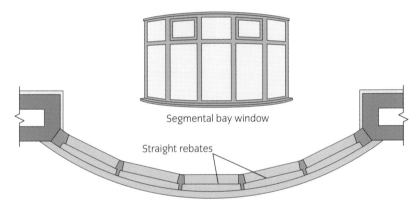

Segmental bay window

Straight rebates

Segmental bay window

Segmental bay/bow windows can be constructed in one of two ways.

LAMINATED

The window can be made with a continuous laminated head and cill. In this case, a former is made to allow the head and cill to be bent and glued around. The finished components then need to be brought to thickness and marked off the plan in the usual manner. The jointing arrangement will be standard through mortice and tenons. A jig will be required to allow morticing to be carried out by machine.

Laminated hardwood cill

Laminated head

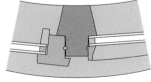
Solid mullion

Laminated construction with solid mullions

BUILT UP

The other method involves a series of flat frames (one for each facet). The ends of the cill are be mitred and handrail bolted together. The frames are then connected together with screws through the mullions, as shown.

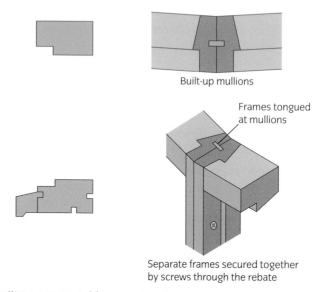

Built-up mullions

Frames tongued at mullions

Separate frames secured together by screws through the rebate

Built-up segmental bay construction

DOORS AND FRAMES SHAPED ON PLAN

Radiused corner

Any corner whose sharp point has been softened by a radius

Curved segmental doors and frames are rarely come across. In Victorian times, many public buildings built on the corner of two roads had a **radiused corner**. If the entrance to the building was on the corner, it required a segmental on-plan frame with a large single or pair of double curved doors.

Construction of the frames follows the same methods as identified in the previous section. The plan section of the jambs, however, can take two forms. These are known as 'radiating' or 'parallel' jambs. Whichever type is detailed by the architect, the rebates must be parallel; if they are not, the door will be trapped and will not clear the frame as it opens.

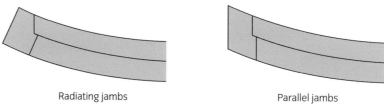

Radiating jambs

Parallel jambs

Jamb details for segmental on-plan door frames

The rails in the curved door construction should be built up in 50–75mm layers to make up the width required. The tenons can be produced by hand or by using a false bed on the tenoning machine. This will lift the rail to bring the tenon required parallel to the cutting circle of the tenoning heads.

The panel will be constructed by laminating thin short grained ply to make the thickness required. If the door is to be polished, before they are bent the faces of the ply will need to be veneered to match the species of timber.

ACTIVITY

Carry out research to identify a light-brown coloured durable hardwood that is suitable for external doors.

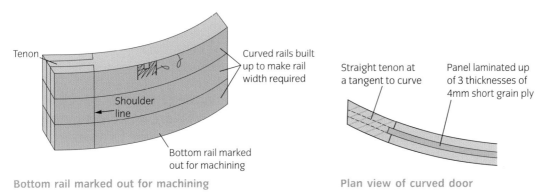

Tenon

Curved rails built up to make rail width required

Shoulder line

Bottom rail marked out for machining

Bottom rail marked out for machining

Straight tenon at a tangent to curve

Panel laminated up of 3 thicknesses of 4mm short grain ply

Plan view of curved door

SECTION PROFILES

Section profiles can vary considerably depending on whether what is being constructed is for new or restoration work. This applies particularly in window construction, where **high-performance** windows are required to conform to Part L of the Building Regulations (covering conservation of fuel and power). High-performance windows require much more complex sections which can sometimes be difficult to machine, particularly on curved sections.

High-performance

Modified to give superior performance

ACTIVITY

Go to www.trend-uk.com and look at their Modular Window System catalogue. Compare the high performace sections shown there with the traditional sections detailed in the illustrations in this section of the book.

External lining

Pulley style

Wagtail

Parting bead

Stash stile

Inside lining

Staff bead

Plan detail

Traditional sections for a box frame window

Sash stile

External lining

Frame jamb

Parting bead

Spring balances

Staff bead

Inside lining

High-performance box frame sections

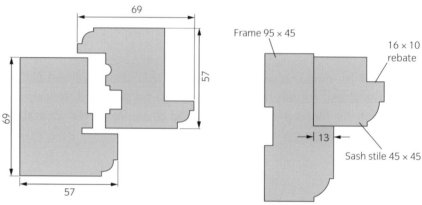

High-performance casement sections

Traditional casement sections

The sections required will have been detailed by the architect on their drawings. It is unlikely that they will be drawn from a tooling supplier's catalogue, therefore they will not be able to be exactly matched. Therefore tooling will have to be sourced to match as closely as possible the architect's details, and if necessary the architect will have to be consulted to agree that these will be okay. Generally, the architect will not be concerned with the fine detail, only that the finished product will look correct.

The sections for internal joinery components have not changed much over the years and should not pose any problems. (More information about them can be found in Chapter 4 of *The City & Guilds Textbook: Level 2 Diploma in Site Carpentry and Bench Joinery*.)

MACHINING JIGS FOR SHAPED COMPONENTS

As applied to woodworking machinery, PUWER outlines measures that manufacturers and employers must comply with. The wood machining ACoP, which is accompanied by guidance, gives safe practices to be used when machining timber that will ensure the PUWER requirements have been met. These should always be referred to before using any machinery.

The Health and Safety Executive (HSE) and the British Woodworking Federation (BWF) also produce guidance sheets on the safe use of machinery and show how the ACoP can be conformed to.

Using jigs and work holders is necessary for all stopped and curved work, unless the nature of the operation makes it impracticable. Even if workpieces are irregularly shaped and limited production runs are involved, jigs should be used.

The design of jigs and work holders is determined by the work to be done. They must be robust and they are typically made of hardwood and plywood. They should allow quick and accurate location of the workpiece, which should be held firmly in position. Jigs should have

INDUSTRY TIP

If there are any splits, they can be positioned to the back of the frame so that they are not obvious.

ACTIVITY

Visit www.hse.gov.uk and www.bwf.org.uk to look at their information sheets on safe use of woodworking machinery. Print them off: they are good to keep for reference.

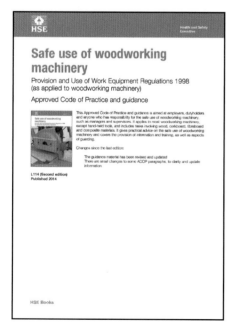

Wood machining ACoP

secure handles and wide bases so that machinists have a firm grasp at a safe distance from the cutters.

The workpiece should be clamped or secured within the jig. The most convenient method of holding the workpiece in the jig is to use manually operated quick-acting clamps which work with either a toggle or a cam action.

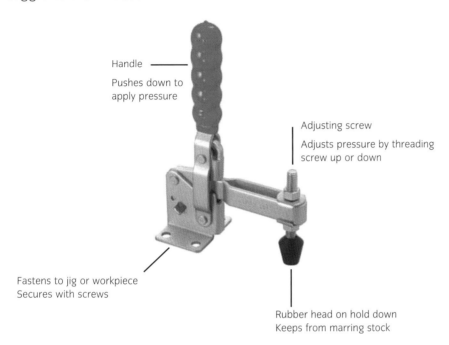

Handle ———
Pushes down to
apply pressure

Adjusting screw
Adjusts pressure by threading
screw up or down

Fastens to jig or workpiece
Secures with screws

Rubber head on hold down
Keeps from marring stock

A typical toggle cramp

With curved work, a combined templet and jig helps ensure that work is held firmly and correctly in order to produce the required shape and finish. Jigs can either be one or two sided, allowing both the internal and external curves to be produced using one jig. The templet should also be extended horizontally to be about 50–75mm longer at both ends of the curved component in order to provide better **lead-in** and **lead-out** control.

Lead-in and **lead-out**

Where a templet can come into positive contact with a ring fence before the cutter comes into contact with the timber. This prevents snatching and allows the work to be controlled as it is fed in and out

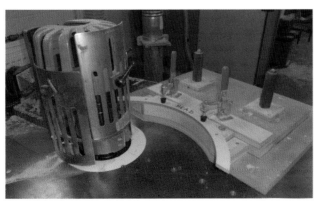

Typical single-sided jig

MARKING OUT AND MACHINING

MARKING OUT

Once all the materials have been machined to size, they can be marked out ready for jointing and profiling. The first process as ever will be to select the face side and edge for each of the components. You should look to remove any defects in the profile section where possible, such as having knots and shakes machined out in rebates or positioned to the back of the work. Components with large dead knots or that are twisted, bowed or sprung in shape should be discarded and replaced. In painted work, some **piecing in** is appropriate and avoids the need to replace components. This should be for minor damage or to replace loose knots, whether dead, face or edge knots.

Piecing in

Where diamond-shaped pieces are let in to the timber to repair a damaged area. Good piecing will be hardly visible

INDUSTRY TIP

When piecing hardwood, try to make the piece from the same timber component. This way the colour and characterisation of the grain will be as close a match as possible and the piece will be well disguised.

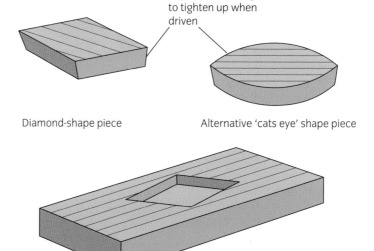

Slight taper to edges to tighten up when driven

Diamond-shape piece   Alternative 'cats eye' shape piece

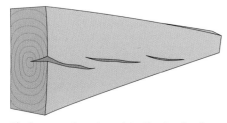

Shakes can be placed to the back of the component

Piecing up dead knots or damaged timber in hardwood

Snatching

In a machining context, this is where the rotating cutter hooks into the split, causing it to break or the work to be thrown back at the operative. The operative could lose control or have an accident

When inspecting the blanks for the curved components before applying the face marks, careful examination should be made for shakes. These can cause pick-up which, when being spindled to shape, could cause **snatching**, which is hazardous (one of the reasons a robust jig is required). Wherever possible try to lose knots on the part of the timber blank bandsawn away as waste. Knots are again a cause of snatching.

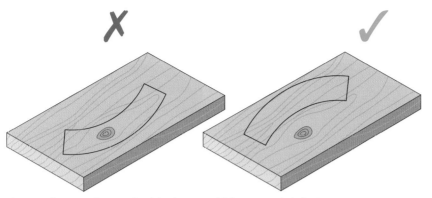

Placing the templet on the blank to avoid knots and defects

MARKING OUT AND MACHINING CURVED COMPONENTS FOR A REPLACEMENT SEMI-CIRCULAR HEADED BOX FRAME

In this example, marking out and machining curved components is being done for a replacement semi-circular headed box frame in a heritage building. The rod for this box frame is as shown on page 326.

The straight frame components can be marked out from the full-size height and width sections on the rod. The circular frame components will have to be machined to the curved shape (including the section shapes) before marking out from the full-size elevation.

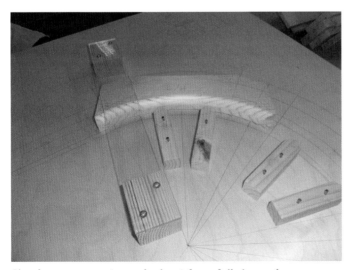

Circular components marked out from full-size rod

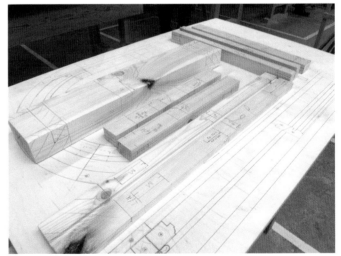

Straight components marked out from the height and width sections of the rod

PRODUCING CIRCULAR COMPONENTS BY MACHINE

STEP 1 Gather the templets that will be used to mark out the shape on the blank.

STEP 2 Bandsaw the blanks to shape, leaving about 2mm to be removed on the spindle to bring them to the correct width.

STEP 3 Remove the straight fences from the spindle and set up the ring fence and bonnet guard.

STEP 4 Set a planer or profile block flush to the ring fence.

STEP 5 Pin the templet onto the blank and cramp securely in the spindle jig.

STEP 6 Carry out a trial run to ensure the block is cutting flush to the templet.

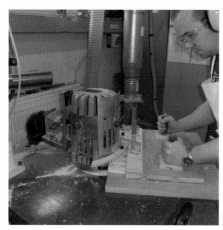

STEP 7 Machine all blanks for each templet.

STEP 8 Machine the profiles to the curved components.

The following are different types of profile block commonly used on the spindle moulder.

Moulding block

Grooving block

Rebating block

Router table

For small and lightweight sections, a router table could be used to profile and shape the section profiles. It must be remembered that the ACoP still applies when using this method.

A router table used for light work

Secondary machining
Jointing and profiling of planed components

FITTING UP COMPONENT PARTS

Once the components have had their **secondary machining** carried out, the various parts can be jointed using one of the methods shown and then fitted up. It is essential that the curved components are fitted up over the full-size rod in order to ensure the joints fit up tight and that it conforms to the shape and size required.

Curved head fitted up over rod

Any dowelled joints can now be draw bored ready for pinning in the assembly process. Draw pinning allows assembly to be carried out without the use of cramps. This method is very suitable for shaped work due to the difficulty of cramping positively against a curve's surface. See the following step-by-steps for this process.

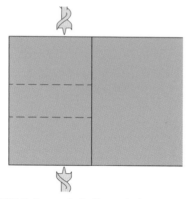

STEP 1 Bore a hole through the mortice.

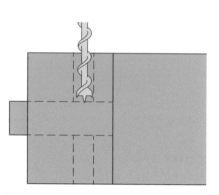

STEP 2 Push the tenon into the mortice and mark the centre of the hole onto the tenon using a boring bit.

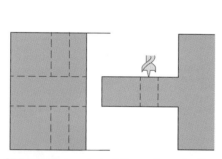

STEP 3 Bore a hole 1.5mm closer to the shoulder.

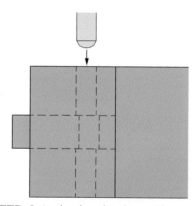

STEP 4 As the dowel is driven, the shoulders of the tenon are pulled up tight.

When you have made any adjustments required and are happy everything is correct, all the faces that are inaccessible after assembly can be papered up (see next section).

PAPERING UP

Sometimes called 'cleaning up', papering up is the process of using abrasive to remove any slight blemishes and machine or pencil marks, ready for the surface finish. It is essential in order to achieve a good finished appearance, otherwise any defects will appear magnified, particularly by a high gloss finish.

The grade of abrasive paper used will depend on the species of timber and the surface finish applied. Softwood requiring a painted finish would require a grit grade of between 80 and 100; any coarser and the scratches would show through the paint finish, and any finer would clog and not remove the machine marks.

Hardwood timber with a clear finish (polish or varnish) will require papering up by working through several grades of paper. Starting with coarse to remove defects, work through to finer papers to gradually lose the deeper scratches and provide a very fine finish suitable for polishing. A typical grade to start with is 80, then 100, 120, 150 and 180 grit grade.

The sanding can be carried out by hand on narrow surfaces where a sander would topple and ruin the surface, or on wider surfaces with a random orbital sander. This type of sander produces a far superior finish compared with that of a traditional orbital sander and does not leave discernible circular scratches.

ACTIVITY

Carry out research into open and closed coat abrasives and state which type is suitable for papering up softwood.

Random orbital sander

Orbital sander

All rebates and moulding profiles will also need papering up. The rebates can be papered up using a cork rubber with a sharp corner in order to make sure that all the machine marks are removed and the internal corner area is not missed. Moulded profiles need to be papered up using a purpose-produced cork or timber rubber block, made to the opposite profile of the moulding. A supple abrasive sheet is laid face down over the moulded profile and the shaped rubber is pushed into it. This ensures that the abrasive fits the profile and does not lose its sharpness while the machine marks are being removed.

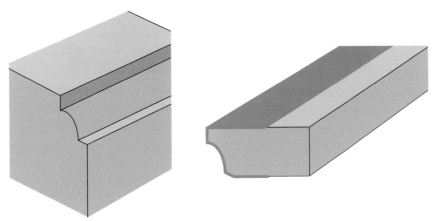

Abrasive paper wrapped around moulding rubber block shaped to the opposite profile of the moulding

ASSEMBLING SHAPED DOORS AND FRAMES

Successful assembly depends on good preparation. Everything required should be collected and set up in readiness for the assembly operation. Bench bearers will need to be levelled, the job dry fitted and laid on the bearers, and cramps with any necessary protection blocks set to the required size. The adhesive also needs to be to hand, along with a glue brush. A rule or squaring rod is required for carrying out the quality check, checking the diagonals for square, along with any stretchers required. Nails/screws and fixing tools will also be required for fixing the stretchers. The last thing we need is to be running around looking for these things when the adhesive is **going off** and we are fighting against time.

When all the equipment required has been collected, the assembly process can start. Adhesive is applied, the component parts assembled and the cramps applied. The frame should then be checked for square and wind before any wedges are driven.

INDUSTRY TIP

Have a clean, damp cloth available when assembling that you can use to clear any excess adhesive off the work.

Going off

A term used to describe the part drying of an adhesive

INDUSTRY TIP

It is good planning where possible to time assembly just before lunch or at the end of the day. This allows the adhesive to dry thoroughly before the cramps are removed and cleaning of the faces takes place.

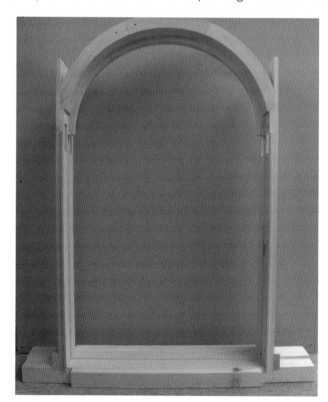

Semi-circular headed box frame assembled

ALTERNATE CRAMPING TECHNIQUES

Sometimes the shape of a job will require special cramping techniques. These could include the following.

Ratchet, band and web cramps

A useful addition to any joiner's toolbox, the web is available in various lengths and is made from a strong nylon. The cramp can be used to pull all sorts of odd shapes up when assembling. They can apply pressure in ways that traditional cramps cannot and can reduce the number of cramps required as they can completely surround the work.

Band cramp

Band cramps in use

Joiner's dogs

These are invaluable when assembling curved work. They can be driven into the back of a frame to pull the joints up. As they leave a hole, they are not suitable for face work unless it is to be painted when finished, as then the holes can be filled.

CLEANING OFF

Once the adhesive has dried, the cramps can be removed and any wedges trimmed off. It is good practice to have a pair of 'soft bearers' (carpet or felt covered) to replace the cramping-up bearers in order to minimise the possibility of surface scratching during the papering up of the surfaces. Traditionally we would clean up the back side of the frame first and the face last. The shoulders of joints and the heading joints should be flushed with a sharp and finely set smoothing plane before being sanded.

If **interlocking grain** is encountered when using the smoothing plane, a cabinet scraper can be used. The cutting action scrapes and will not tear up the grain. A course grit grade will also help clean out the tearing up of the grain. Always look to see which way the grain is running before planing. This will eliminate most grain tearing.

Joiner's dogs in use

INDUSTRY TIP

An orbital sander should not be used to level joints as it will not flatten the face and the undulations will show after the surface finish has been applied.

Interlocking grain

Also known as 'refractory grain', this is where grain spirals around the axis of a tree but reverses its direction regularly, causing a poor finish whichever way it is planed

A cabinet scraper in use

A cabinet scraper can generate a lot of friction-generated heat in use, so wear rigger gloves to protect your fingers from being burnt.

When the job is complete, the delivery address and any reference number should be placed on the back for identification purposes. The job should then be stored in a protected, dry and stable environment until it is ready to be delivered.

EXAMPLE ASSEMBLY METHOD

This section outlines the assembly of a semi-circular headed box frame.

1 Fit up and assemble the top sash over the full-size rod. The semi-circular headed sash and frame head then need to be built up directly over the rod to ensure conformity of shape and that it is constructed on a flat surface.

Top sash fitted up

2 Assemble the sashes, flush and leave to one side to fit to the box frame later.

3 Assemble the head to create one jointed curved component. It will need to have a stretcher tacked across the open end to maintain its size.

Circular head built up

4 Prepare the pulley stiles by fitting the pulley wheels and cutting the pockets. This consists of two cuts on the face and two on the back (as shown on the marked-out component). In addition, a rip cut will need to be made down the centre of the parting bead groove between the top and bottom cuts.

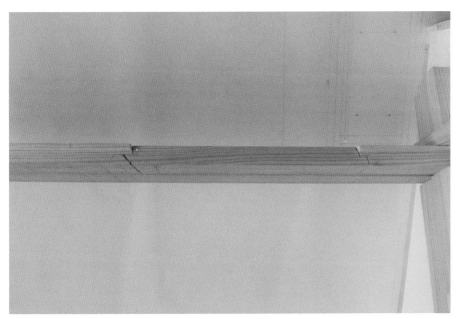

Pocket cuts and removal of tongue to the inside face of the pulley stile

5 Glue and wedge the pulley stiles into the cill and check for wind.

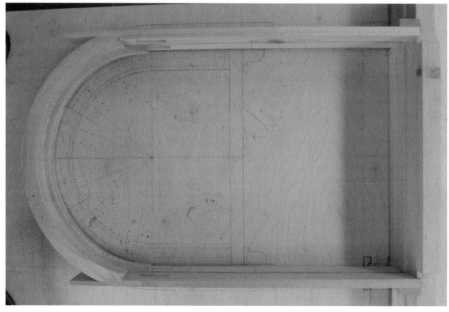

Pulley stiles glued and wedged into the cill

6 Assemble the frame (with the inside of the window facing downwards) across a pair of levelled bearers on the bench, directly over the rod.

7 Connect the shaped head by screwing through the back of the pulley stiles into the shaped head. The diagonals will need to be checked for square, then temporarily braced.

Frame assembled and ready for the fitting of the linings

8   Cut and fix the inside linings, flush the joints and sand the surface. Turn the frame over and repeat this for the outside linings.

9   **Shoot in** the top sash and fit the parting bead. The pockets are now knocked through from the back of the frame. The bottom sash is shot in and the staff heads fixed.

Shoot in

The process of fitting a door or sash to an opening with a parallel gap that allows for fitting and finish clearance. Allow 2–3mm for a painted varnish and 1–2mm for a polished finish

Completed box frame

Case Study: Eleanor

Eleanor's supervisor has asked her to make a replacement semi-circular headed box frame. Eleanor has never made one before but remembers learning about them in college. Her supervisor, Clare, has not made one either, but is keen to encourage Eleanor to develop. Clare has said that Eleanor can make it the way she feels is most appropriate so long as it is an accepted form of construction and that no special equipment needs to be bought to make it.

Put yourself in Eleanor's position. Remember that without planning and research, you'll encounter problems later, so thorough knowledge is required before any setting out takes place. The following are some of the key knowledge areas required before you start. See how you would get on.

- Research three different ways of constructing the head to the box frame.

- What special provision is made to prevent the crown of the sash from hitting the underside of the head to minimise the chance of the glass in the top sash breaking?

- Draw the construction details of the top sash showing the joint you would use. Explain why you have chosen this joint.

Work through the following questions to check your learning.

1   Detailed information collected from site is termed

   a   a check

   b   an audit

   c   a survey

   d   an appraisal.

2   The shape of a segmental opening is **best** obtained by

   a   drawing a sketch

   b   making a templet

   c   taking a photograph

   d   using a CAD program.

3   A large radius can be drawn using a

   a   flexi curve

   b   French curve

   c   trammel and beam

   d   springbow compass.

4   Which one of the following is the **most** pointed Gothic arch shape?

   a   Drop.

   b   Lancet.

   c   Segmental.

   d   Equilateral.

5   Which one of the following methods is **most** practical for drawing an ellipse in the workshop?

   a   Conic section.

   b   Pin and string.

   c   Auxiliary circle.

   d   Trammel frame.

6   Which one of the following is a line bounding a circle?

   a   Chord.

   b   Radius.

   c   Tangent.

   d   Circumference.

7   Which one of the following shapes is a three-centred arch closest to in shape?

   a   Elliptical.

   b   Segmental.

   c   Drop Gothic.

   d   Semi-circular.

8   False tenons are commonly used at

   a   heading joints

   b   crown joints

   c   meeting stiles

   d   meeting rails.

9   Components, timber lengths and the number of each required can be found on a

   a   specification

   b   cutting list

   c   invoice

   d   quote.

10 Any discrepancies discovered during a site survey should be reported to the

   a   estimator

   b   site agent

   c   supervisor

   d   marker out.

11 Which one of the following processes is carried out **first** when preparing planed timber?

a Ripping.

b Crosscutting.

c Planing faces.

d Planing edges.

12 Which one of the following is a regulation that covers the safe use of wood machines?

a PUWER.

b COSHH.

c WAHR.

d PASMA.

13 When machining curved components on a spindle moulding machine, they should be held in a

a jig

b carriage

c vacuum bag

d segment frame.

14 Which one of the following types of spindle tooling is used to trim curved components to the templet?

a Planer block.

b Scribing cutter.

c Moulding block.

d Expanding groover.

15 The process of checking that something is the correct curved shape is called

a fitting up

b inaccuracy

c error finding

d double checking.

16 Which one of the following cramping methods wraps the work?

a Bar.

b Web.

c Sash.

d Toggle.

17 What is the **last** grade of abrasive used when papering up hardwood?

a 40.

b 80.

c 120.

d 180.

18 Which of the following are both quality checks made after a joinery product has been assembled?

a Face side and face edge.

b Check for knots and splits.

c Check for square and wind.

d Job reference and address.

19 Which one of the following sanders produces the **best** finish on an assembled frame?

a Disc.

b Orbital.

c Random orbital.

d Oscillating bobbin.

20 A scraper is used to finish

a the end grain of mitres

b softwoods prior to painting

c wet timbers when laminated

d hardwood with interlocking grain.

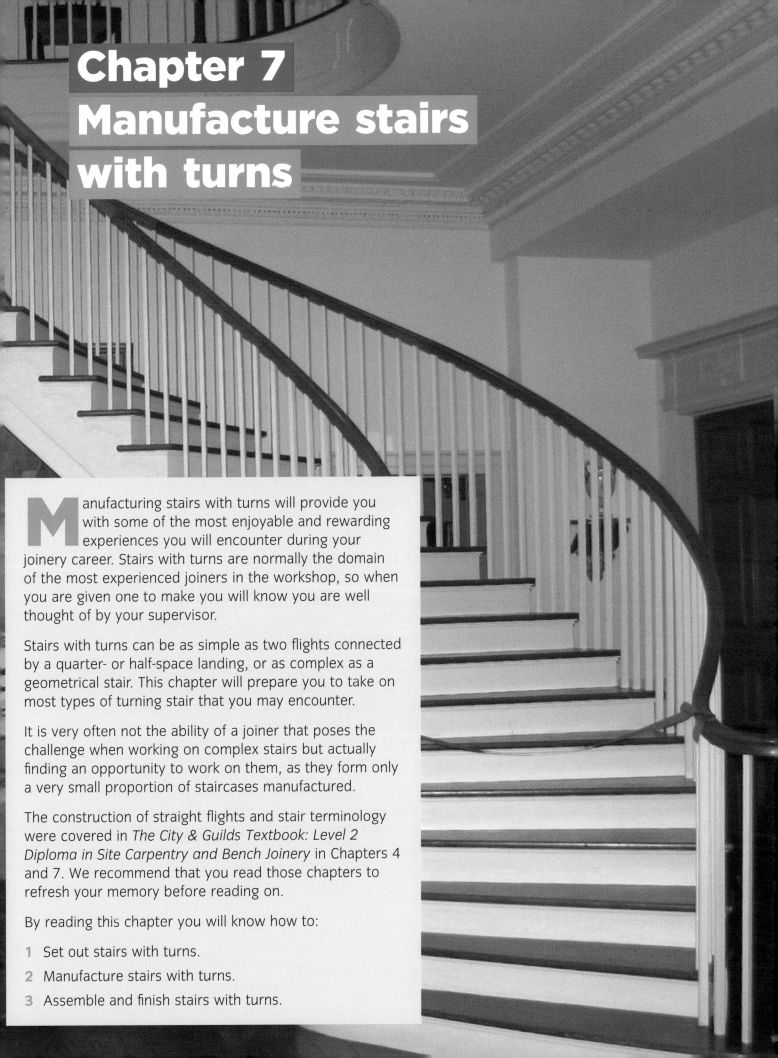

Chapter 7
Manufacture stairs with turns

Manufacturing stairs with turns will provide you with some of the most enjoyable and rewarding experiences you will encounter during your joinery career. Stairs with turns are normally the domain of the most experienced joiners in the workshop, so when you are given one to make you will know you are well thought of by your supervisor.

Stairs with turns can be as simple as two flights connected by a quarter- or half-space landing, or as complex as a geometrical stair. This chapter will prepare you to take on most types of turning stair that you may encounter.

It is very often not the ability of a joiner that poses the challenge when working on complex stairs but actually finding an opportunity to work on them, as they form only a very small proportion of staircases manufactured.

The construction of straight flights and stair terminology were covered in *The City & Guilds Textbook: Level 2 Diploma in Site Carpentry and Bench Joinery* in Chapters 4 and 7. We recommend that you read those chapters to refresh your memory before reading on.

By reading this chapter you will know how to:

1. Set out stairs with turns.
2. Manufacture stairs with turns.
3. Assemble and finish stairs with turns.

TYPES OF STAIRS WITH TURNS

The planned arrangement of a staircase will have been determined as a result of several considerations including:

- the need to divide the stair into more than one flight

- the need to change direction due to the shape of the building

- the demands of floor space.

Stairs are often named or described according to their plan shape or string construction type, and they fall into two classes:

- newel stairs

- non-newel stairs (commonly called geometrical).

Newels are generally used at both the bottom and the top of the stair. They allow both the string and the handrail to be jointed into them and provide a means of supporting the stair by fixing it to the landing trimmer. (See Chapter 7 of *The City & Guilds Textbook: Level 2 Diploma in Site Carpentry and Bench Joinery* and Chapter 3 of this book for further information on how stairs are fixed.)

STAIR ARRANGEMENTS

Stairs will be positioned as shown on the architect's drawings. In small residential properties, consideration is often given to minimising the space the stairs take up.

In the examples illustrated on the opposite page, each set of stairs contains 13 risers; this gives an idea of how much floor space each stair type will take. In examples A, B and C, note that by adding **winders** the **total going** required reduces. Example D shows winders at both ends of the stair and allows access to the stairs where the going is restricted by walls.

Stairs can turn a corner by using either winders or a landing. The turn is described by the method used; for example 'quarter-space landing' (as in example A) or 'quarter-space of three tapered steps' (as in example C).

Winders

These are tapered steps used to save space by allowing extra risers to be incorporated. They generally turn through 90° or 180°

Total going

The horizontal distance between the first and last riser in a straight flight

Below are examples of how much space a stair with 13 risers will occupy.

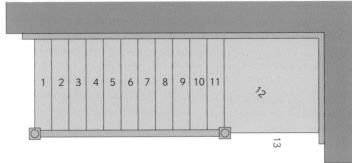

Example A: Quarter-space landing

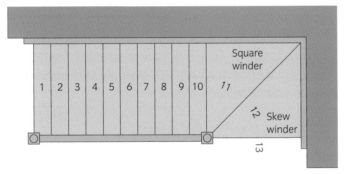

Example B: Half-space of tapered steps (made up of two quarters)

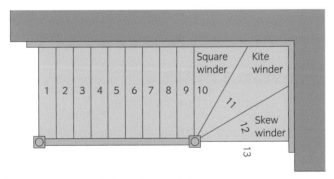

Example C: Quarter-space of three tapered steps

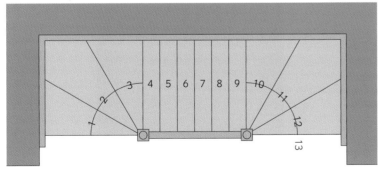

Example D: Quarter-space with two tapered steps

NEWEL STAIRS

Newel stairs can be one of the following types:

- straight flight
- quarter-space
- half-space
- winding.

STRAIGHT FLIGHT STAIRS

These have been dealt with in *The City & Guilds Textbook: Level 2 Diploma in Site Carpentry and Bench Joinery*. The illustration on the previous page (example A) shows a straight flight with a quarter-space landing. This is simply a straight flight with an additional riser turned through 90° and should present few difficulties to manufacture.

QUARTER-SPACE STAIRS

The addition of a quarter-space landing at the top of the flight saves one going space at the bottom of the stairs and gives more circulation space at the foot of the stairs. The building regulations state that the landing depth must be at least the width of the flight (Regulation 1.20).

HALF-SPACE STAIRS

A half-space landing can take two forms:

- *Dog-leg stair:* where width is restricted, a 'dog-leg' stair is used. This allows both the lower and upper flight to be as wide as possible as the string of the upper flight is directly over the lower, both being jointed centrally to the newel.

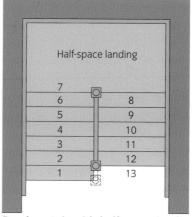

Dog-leg stair with half-space landing

ACTIVITY

Next time you come across a dog-leg stair, note how the handrail has been terminated at its upper end.

■ *Open-well stair:* if the opening is wider than the combined width of the two required flights, the gap between them is termed the 'well'. This type of stair is sometimes referred to as an 'open-newel stair' because two newels are used at landing levels.

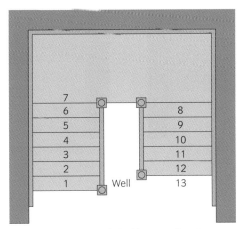

Open-well stair with half-space landing

Care must be taken when designing half-space stairs to ensure that there is sufficient headroom. Remember that the landing trimmer will run across the trimmed opening for the stair and therefore can cause a headroom obstruction. The building regulations require a minimum of 2,000mm headroom (Regulation 1.12). This will mean that there will need to be about 11 risers above the first step, depending on the step rise.

WINDING STAIRS

Where winders are used, it is most common to have two or three per quarter turn; in half a turn, four or six winders are common. (Any number is possible as long as the stair conforms to the building regulations.)

In straight flights the riser face is positioned at the centre of the newel; this is not possible with the winding risers as the regulations state that the minimum going must be 50mm (Regulations 1.25–29). This spreads the riser faces out around the face of the newel. To avoid complex construction problems where winders turn around half a turn, a double newel is used.

The joint used between wall strings will be a tongue and trenched joint. As stairs are always fixed from the top down, the tongue will need to be on the lower flight.

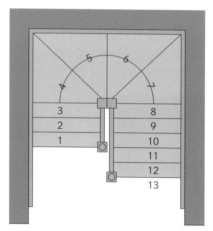

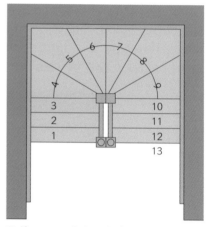

Half-space of four winders with a double newel

Half-space of six winders with a double newel

NON-NEWEL STAIRS

This type of stair is often termed a 'geometrical stair'. This term covers a multitude of varieties. The newels are replaced by a continuous string running from the bottom to the top of the flight, generally with some part of it being curved. The construction of this type of stair is far more complex, with the setting out requiring a degree of geometrical development to determine the 'stretchout' (true shape) of the string. This type of stair is only used for prestigious work as the time taken to construct it, due to its complexity, means that it is very expensive to produce.

Two examples are shown in the following illustrations. The first ('Geometrical stair with a continuous well string') has a continuous well string but straight wall strings. The complexity here over and above winding stairs is limited to the well string.

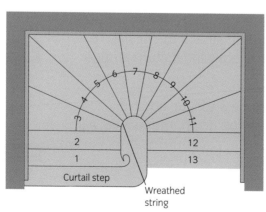

Geometrical stair with a continuous well string

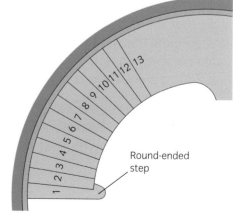

Circular or helical stair

The second ('Circular or helical stair') shows both the wall and the well stings radiating from a centre point. Each tapered step would be the same size and shape. Both strings here require building up.

STAIR TERMS

While this was covered in *The City & Guilds Textbook: Level 2 Diploma in Site Carpentry and Bench Joinery* in Chapter 4, the following illustrations and table serve as a reminder and add to those already mentioned.

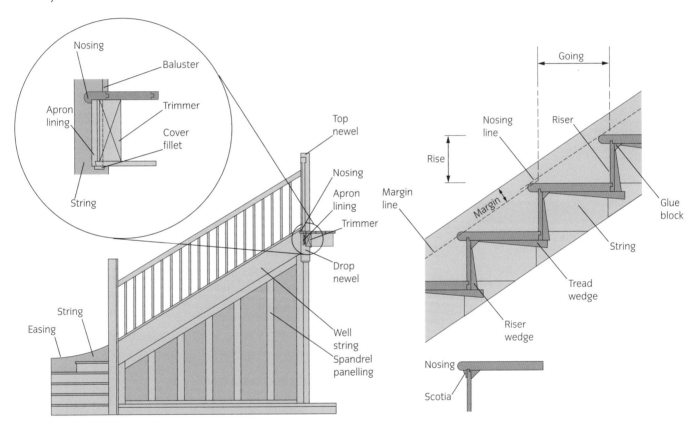

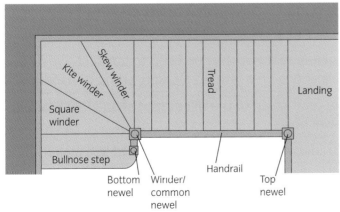

Stair terms

Term	Definition
Apron lining	A thin board (timber or MDF) that faces the sawn joists in the stairwell.
Balusters	Vertical components fixed between the string (or its capping) and the underside of the handrail on the stair, and the handrail and nosing on the landing. Provides guarding to the open side of the stairs and landing areas.
Balustrade	A collective term for the area between the handrail, string and newels.
Brackets	1 Timber used to provide interconnection between the carriage and the underside of the steps. 2 A decorative step end to cut string staircases.
Bulkhead	The intersection between the wall and the trimmer above a staircase. The headroom is measured to this.
Cap	The name given to the shaped top of the newel post. It can be worked on solid (a finial) or fixed on. The detail can also be applied to the newel drop.
Carriage piece	An inclined bearer that acts as a central support to wide stairs.
Cover fillet	A small section of timber used to cover the joint between two parts of the structure, eg between the spandrel and the string, or between the apron and the plaster finish.
Bullnose step	A common entry step with a radiused end found at the beginning of a flight.
Commode step	A step with a curved riser.
Curtail step	An entry step with a scrolled end, generally found with a handrail scroll and cage of balusters over.
Drop newel	The portion of newel that projects below the landing, often with a decorative end (finial).
Easing	A gentle radiused section used to connect two inclinations on a level to a raking string or handrail.
Finial	The term given to the decorative shaped end to newel posts.
Flier	A traditional term for a parallel tread in a staircase.
Flight	A set of steps running from floor to floor.
Glue blocks	Triangular blocks used to reinforce the joint between the tread and the riser.

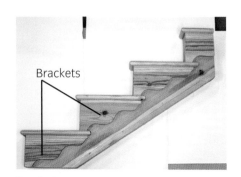

Brackets

Brackets

ACTIVITY

Sketch a detail of the next ornamental riser bracket you come across when visiting a building of interest. Ask your trainer/lecturer the best way of producing them.

Term	Definition
Going	1 *Step*: the horizontal distance between the face of one riser and the face of the next. 2 *Total*: the distance between the first and last riser face in the flight.
Handrail	A rail positioned at waist height to provide support and guarding to the stair. It may be required on one or two sides, depending on the width of the flight.
Headroom	The vertical distance between the step and the bulkhead above.
Landing	A platform at the top of the stairs or a resting place between one or more flights. May be quarter- or half-space.
Margin	The distance between the top of the string and the intersection of the tread and riser face.
Newel post	Top, bottom and common (to two interconnecting flights) newels are stout vertical components used at the top, bottom and at any change of direction. They provide a means of jointing the string and handrail and provide support to the flight.
Nosing	1 The moulded projecting front edge of the tread. 2 The top reduced-width tread that adjoins the landing floor boarding. It could also run around the stairwell, in which case it would be morticed to receive a tenon on the end of the landing balusters.
Nosing line	An imaginary line that touches each nosing in the flight. It is referred to in the building regulations but has no practical function.
Pitch	The angle of the staircase.
Pitch line	An imaginary line that touches all the nosings in the flight.
Rise	1 *Step*: the vertical distance between the top of one tread and the next). 2 *Total*: the vertical distance between finished floor line (FFL) at the bottom of the stairs and FFL at the top of the stairs.
Riser	The vertical component of a step.
Scotia	The shape of a small moulding used on the underside of a nosing in prestigious stair construction.

ACTIVITY

Draw two finial details commonly found on newels.

Term	Definition
Shaped bottom steps	These are intended to give the user more access to the bottom of the flight by providing a decoratively shaped bottom step that projects beyond the face of the bottom newel. Typical examples are bullnosed and curtail.
Spandrel	The triangular area below the string on the first flight of a staircase. It can be panelled in or made up of a series of doors, allowing the area below the stair to be used for storage.
Splayed step	An entry step with a splayed end.
Staircase	This is the complete stair structure including the flight, landings and the balustrade.
Stairwell	This is the opening formed in the floor layout to accommodate the flight (or flights) of stairs.
Step	The name given to an assembled tread and riser.
Storey rod	Traditionally, a lath of timber onto which the position of the landings was marked on site. The rod was taken back to the joinery shop and divided up to find the step rise for the stair. This is very seldom used in modern practice but the term is still used and shown on drawings.
String capping	This component is planted on the top of the string to increase its thickness, allowing balusters to be fixed where the balusters are thicker than the string.
Strings	1 *Wall*: the string fixed against the wall. 2 *Well*: the string on the open side of the staircase, sometimes called an 'open' or 'outer string'. 3 *Geometrical*: a string that runs continuously from the top to the bottom of a flight with part of it curved on plan. Used to change direction on stairs with no newels.
Tread	The horizontal surface of a step.
Well	The gap between two strings on a turning stair.

Term	Definition
Winders	1 *Square*: the first of the two or three winders encountered in a quarter turn. The nosing is square to the wall string. 2 *Kite*: the kite shaped second of the three winders encountered on a quarter turn. 3 *Skew*: the last of the three winders encountered in a quarter turn of two or three winders. The nosing is at a skewed angle to the wall string.
Wreathed handrail	Normally made in 90° sections, they are handrails that follow the plan shape of the string below. The handrail rises and turns at the same time.
Wreathed string	Found on geometrical stairs, these are curved strings that replace newels at a change of direction.

COMPONENT SIZES

The following are common component section sizes in a typical residential staircase; they may vary depending on the architect's details.

ACTIVITY

Produce a cutting list for a straight flight of stairs with five steps.

Component	Size	Material
Treads and winders	19–32mm thick depending on requirements. The larger the stair, the thicker the material required. Cut string stairs require thicker treads	▪ European whitewood/ redwood/MDF ▪ Hardwood as specified by the architect or client
Risers	9–18mm thick	▪ Ply/MDF ▪ Hardwood
Newels	70–120mm square	▪ European whitewood/ redwood ▪ Hardwood
Handrails	70mm × 45mm, 95mm × 45mm depending on section	
Balusters	22–45mm square, depending whether plain or turned	
Strings	26–45mm thick, depending on type	

Machining components to size

Refer to Chapter 5 for information about all machining operations.

TYPES OF STRING CONSTRUCTION

Chapter 4 of *The City & Guilds Textbook: Level 2 Diploma in Site Carpentry and Bench Joinery* concentrated on closed string stair construction, as shown in the photograph. But as the stairs become more decorative, 'cut string' stair construction is used more frequently for the well string. This is where the top edge of the string is cut to the shape of the step profile. The finished effect is designed to be aesthetically pleasing.

Closed string stair

Cut string stair

CUT STRING CONSTRUCTION

The work involved in this type of string construction is considerable. Much of it relies on having a high standard of hand tool skills and requires a great deal of fitting up on the bench. Badly fitting joints will be clearly visible and have an adverse effect on the appearance of the finished product, especially on a polished hardwood stair.

As much of the string is cut away, this will obviously reduce its strength. Thicker strings are used in order to compensate for this, typically between 38 and 45mm. The amount of parallel timber below the step profile (known as the 'waist') should be a minimum of 125mm. Treads are also generally thicker than on closed string stairs, typically 28–35mm.

There are two types of cut string stair:

- cut and mitred

- cut, mitred and bracketed.

CUT AND MITRED

In a cut and mitred string, the strings are cut to the profile of the step and the riser is jointed flush with the outside edge of the string. A square shoulder is produced on the riser line of the string in order to provide a positive location for the riser. This joint can be glued, screwed and pelleted. The back of the joint can be reinforced with triangular glue blocks or square blocks screwed in both directions.

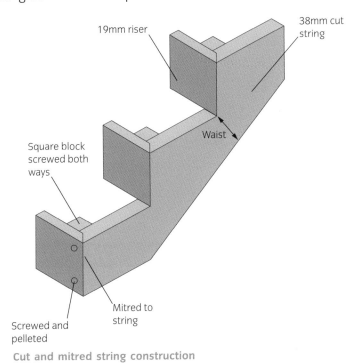

19mm riser

38mm cut string

Waist

Square block screwed both ways

Mitred to string

Screwed and pelleted

Cut and mitred string construction

CUT, MITRED AND BRACKETED

The cut, mitred and bracketed method is a more decorative version of the plain cut and mitred string. In this type of cut string construction, the strings are easier to cut as the riser cut on the step profile to the string is square. This leaves the riser lapping the face of the string and projecting by the thickness of the decorative bracket. This projecting part is mitred to the end of the bracket.

As the riser is reduced back to the thickness of the decorative bracket, fixing is limited to gluing and neatly pinning with three or four oval nails. These will be punched below the surface. The filling to the punched holes will depend on the finish. If it is painted, a standard filler can be used. If polished, a matching coloured filler or wax is used. Again, the back of this joint will need reinforcing.

Care must be taken after assembly of this type of stair not to 'rack' it during transportation as, until it is fixed, it is quite vulnerable and the joints on the cut string could fracture.

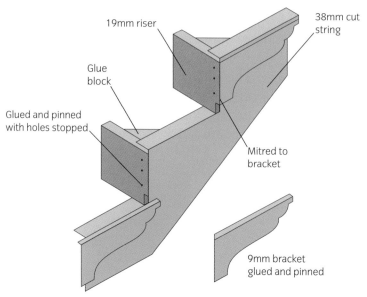

19mm riser

38mm cut string

Glue block

Glued and pinned with holes stopped

Mitred to bracket

9mm bracket glued and pinned

Cut, mitred and bracketed string construction

Adjustable pitch board

Steel square and fence

ACTIVITY

Research any proprietary jigs suitable for routing cut string stairs. Discuss in pairs the relative merits of each jig.

MARKING OUT CUT STRING STAIRS

Marking out cut string stairs is very similar to marking out a closed string stair – the main difference is that there is no margin.

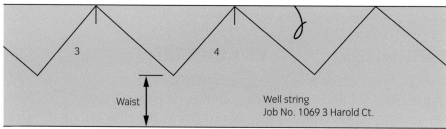

Waist

Well string
Job No. 1069 3 Harold Ct.

Tread and riser lines marked out with a pitch board or steel square

SHAPING THE STRING

Shaping the string can be done using a router jig made for the purpose. The bulk of the material can be cut away first using a jigsaw. The jig can also be used to produce the shoulder and the mitre, if the risers fit flush with the string.

INDUSTRY TIP

The term 'marking out' can be abbreviated to 'm/o'.

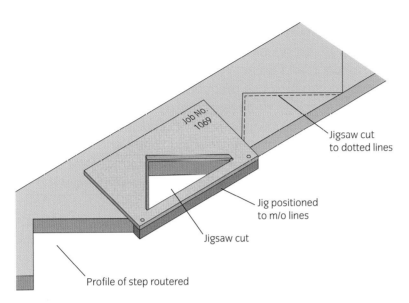

Jigsaw cut to dotted lines

Jig positioned to m/o lines

Jigsaw cut

Profile of step routered

Router jig for cut string stair

JOINTING BALUSTERS

Cut string stairs usually have the balusters jointed into the top of the tread face. There are two methods commonly used:

- *Dovetailed:* this method probably provides the best interconnecting joint to the tread. Commonly a barefaced dovetail is used and this is screwed into the end of the tread. The disadvantage of this method is that the return nosings have to be fitted to each step, numbered and sent to the site loose so that they can be fixed after the balusters have been installed. Typically the return nosing will be slot screwed to the end of the tread.

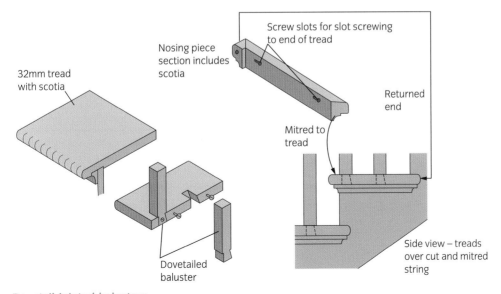

Screw slots for slot screwing to end of tread

Nosing piece section includes scotia

32mm tread with scotia

Returned end

Mitred to tread

Dovetailed baluster

Side view – treads over cut and mitred string

Dovetail jointed balusters

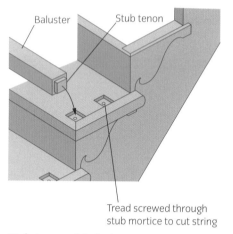

Baluster   Stub tenon

Tread screwed through
stub mortice to cut string

Stub tenoned balusters

- *Stub tenoned*: this method offers two main advantages. First, the tread can be screwed to the string during assembly. Second, the return nosings can be jointed, glued and flushed to the tread in the joiners' shop and do not have to be sent to the site loose. The return nosing can be loose-tongued or biscuit jointed to the tread. If the stair is specified to have a painted finish, MDF is ideal for the treads as the return nosing can be worked on the solid without the need for jointing the return nosing to the end of the tread. With this jointing method, the tenon on the end of the baluster can be secured on site by skewing a screw from below the tread into the end of the baluster.

BALUSTER ARRANGEMENTS

With closed string stairs it does not matter where the balusters are positioned in relation to the riser face, as long as there is not a gap exceeding 100mm between them (Regulation 1.39a). However, with cut string stairs there are two commonly used variants, shown in the illustration.

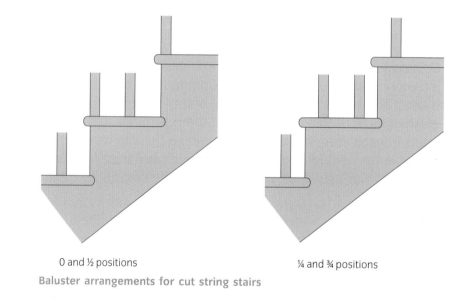

0 and ½ positions   ¼ and ¾ positions

Baluster arrangements for cut string stairs

STEP CONSTRUCTION

There are many ways of constructing steps. Four examples are shown in the illustration, including one bad example that should be avoided.

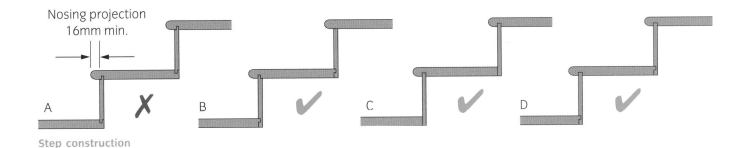

Nosing projection
16mm min.

A ✗   B ✔   C ✔   D ✔

Step construction

Example A is unsuitable because:

- the top shoulder of the tongued joint on the tread will open up on the face if shrinkage occurs

- a wider tread is required

- tread shrinkage will cause the riser to split

- the back of the tread will interrupt the riser wedge from driving the face home tight.

Example B is a traditional joint used between a solid timber tread and riser.

Example C is the most common method used in modern construction, using a solid timber tread and an MDF or ply riser.

Example D is an improved version of example C when jointing ply or MDF risers, but more expensive to produce.

Where treads are made from solid timber, they may require jointing in their width. If this is the case, it is good practice for the joint to be made at the back of the tread. In modern construction, where a painted finish is specified or the stair is to be covered in carpet, 25mm MDF may be used. This is much more economical and stable.

All fliers (parallel steps) will be 'boxed up' after the faces have been sanded. Boxing up is the process of assembling treads and risers, checking them for square and glue blocking them to reinforce the joint. A minimum of three glue blocks should be used (normally made from 50 × 50mm timber). They are then stacked on a flat surface. Three or four are placed, then on the second layer the steps are laid at 90° to the first and so on until they are all assembled. This allows the air to circulate around the faces of the steps and helps keep them stable and prevent them from **casting**.

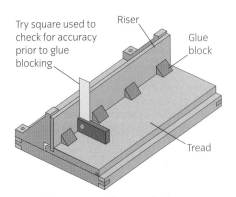

Try square used to check for accuracy prior to glue blocking

Riser

Glue block

Tread

Step being squared up and glue blocked

INDUSTRY TIP

Never nail glue blocks: the splinters produced as the nail pushes through the glue block creates a gap between it and the tread/riser, reducing the adhesion between the two.

Casting

A term used to describe the curling and movement of timber

Templet

Also known as a 'template', this is a thin piece of hardboard, MDF or ply that is cut to the shape required and is then used to help reproduce the shape

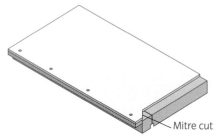

Templet used to profile mitred end of tread

Draw pin used to temporarily hold string to newels when fitting up

FITTING UP CUT STRING STAIRS

The mitred ends of cut string treads can be profiled using a router **templet**, as shown in the illustration.

The purpose of all fitting up is to ensure the joints fit, and the stair is the correct size and shape. Cut string stairs are difficult to fit up and assemble due to the return nosing of the tread and, where used, brackets that project beyond the face of the string. As with all assembly procedures, preparation is the key to success. The method is as follows.

1 If the stair has newels, these should be dry fitted to the strings and handrail first, before any other fitting up. Draw pins can be used as a temporary means of cramping the joints (these will be replaced by dowels on site).

2 A straight bearer is screwed to the far side of the bench to support the back edges of the steps.

3 The well string is cramped into the vice, level and at a height equal to the bearers (to ensure the stair is not assembled in wind).

4 Secure the well string at both ends to prevent it from slipping during the assembly operation.

5 Fit the steps to the cut string one at a time, and label by riser number. The bottom and top step will need to be fitted up with the newels on them to ensure they fit on site.

6 Once all the fitting up has been carried out, the parts can be disassembled and all the faces inaccessible after assembly cleaned up (with abrasive paper), ready for the surface finish.

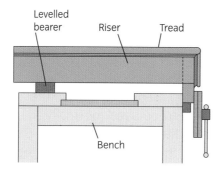

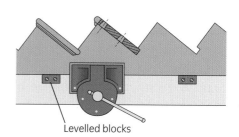

End view

Side view

Methods of fitting steps to a cut string

ASSEMBLING A CUT STRING STAIR

Assembling a cut string stair is a process of simply reassembling the stair as it was during the fitting up stage (but not the handrail or newel), but this time using adhesive and fixings.

The tread is screwed to the string through the stub mortices, and the riser face is pinned with ovals/lost heads or screwed and pelleted. The stair should be left to dry thoroughly overnight. The next day, the joints can be flushed and sanded as required.

The wall side of the stair can now be cramped and wedged. This is done in a similar fashion to a closed string stair, but additional care is needed as a packing bearer is required to be cut to the profile of the cut string, slightly thicker than the nosing projection. Lastly, the risers are screwed to the backs of the treads, and additional glue blocking can be added between the cut string and the steps for reinforcement.

QUALITY CHECKS

For most joinery items, this would include checking for square, wind, size and shape. Flights of stairs rely on the ends of the steps being cut square for the flight to be square. When both strings of the flight are assembled, the front edge of the nosings can be sighted for wind. The size and shape of any winders can be checked during the fitting up stage.

PREPARATION FOR DELIVERY

The stair should be adequately protected with bubble wrap and laid flat ready for delivery. The delivery address and contract number should be clearly marked on the wrapping. Remember that the top riser, the top nosing, the newels, the handrail, the balustrades, any winders and any shaped bottom steps are sent to site loose. These should also be well protected and labelled to ensure they are delivered to the correct address.

SHAPED ENTRY STEPS

Entry steps allow better access to the bottom of the flight and their use moves the newel at least one step further up the flight, giving a more open feel at the bottom of the stairs. There are a number of commonly used entry step shapes, including the following.

BULLNOSE

Probably the most common shape encountered, the end of a bullnose step has a 90° radiused end. When drawing this step, it is important to have at least 50mm of straight before coming into contact with the newel. A common mistake is to have the springing line on the face of the newel, which can look very awkward.

Traditionally the riser of this step would be reduced back to a veneer thickness around the shaped portion, and secured in place with a pair of folding wedges. In modern construction and in mass production, they are generally made of laminated ply. Once the former is made up they can be produced very economically, saving much hand work.

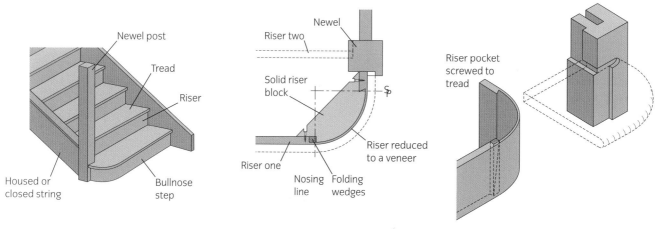

Construction of a traditional bullnose step including isometric detail

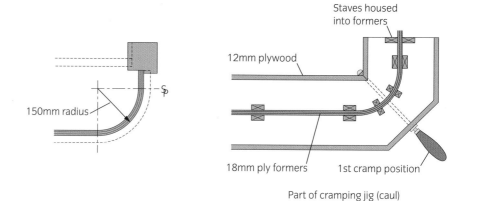

Laminated bullnose riser construction

SEMI-CIRCULAR ENDED STEP

Semi-circular ended steps are also known as 'round-ended' or 'D-ended' steps. This type of step turns through 180°. Again, there should be a short length of straight beyond the springing line before it meets the newel. Two methods of construction are shown in the following illustration.

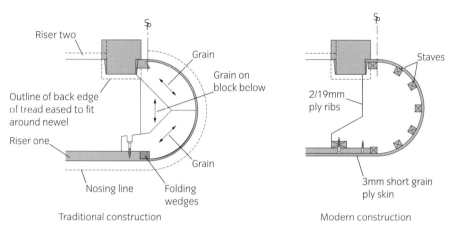

Semi-circular ended step: traditional and modern construction

SPLAY-ENDED STEP

The splay-ended step became popular in the 1960s and is still occasionally come across. The riser is faceted around a block or pair of formers. The end of the ply can either be loose-tongued or biscuit jointed, if thick enough.

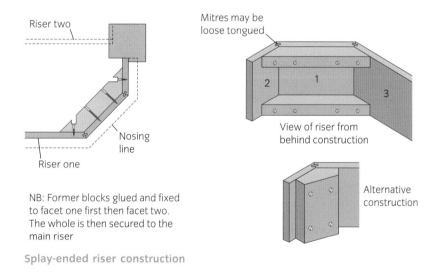

NB: Former blocks glued and fixed to facet one first then facet two. The whole is then secured to the main riser

Splay-ended riser construction

CURTAIL-ENDED STEP

Curtail-ended steps are also known as 'scroll-ended' steps. This type of step is only used with cut strings, generally when it forms the beginning of a geometrical stair. A handrail scroll will be formed over the scroll and a cage of balusters secured between.

ACTIVITY

What geometry is required to draw a curtail-ended step? Draw the plan shape of the riser with a scroll projection of 150mm.

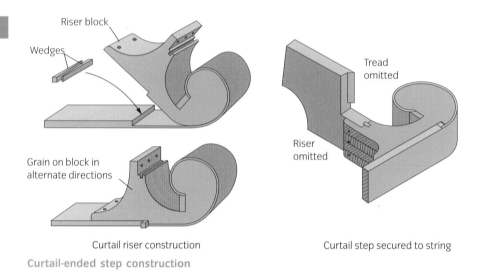

Curtail riser construction

Curtail step secured to string

Curtail-ended step construction

COMMODE STEP

This is the name given to any shaped step where the rise is continuously curved throughout the length of the riser. Any number of steps can have a commode shape. The shaped portion of the riser can be made in any number of ways. As with all entry steps, they are normally 'one-offs' and so a simple method of former construction is best.

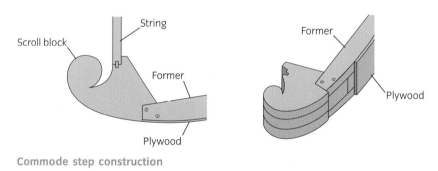

Commode step construction

WINDER STEP CONSTRUCTION

As we saw previously in this chapter, winders (now commonly known as 'tapered treads', as this is how they are referred to in the building regulations [Regulations 1.25–29]) can take any tapered shape. Traditionally in a 90° or 180° turn they are referred to as the square, kite and skew winder. Solid timber treads, whether hardwood or softwood, will require jointing in their width. Either a solid tongue-and-groove or a loose-tongued joint is used for strength, and biscuits should be avoided.

It is essential that the grain always runs parallel to the nosing, as this will minimise the effect of any shakes showing on the front of the nosing. Aesthetically it also looks better.

The shape of the winder can be obtained by making hardboard templets from the full-size rod or by making up skeleton frames using laths. These can be used to determine the most economical amount of timber required to joint up the winders. Once jointed, they are ploughed and nosed with the remainder of the treads and brought to shape from the plan. Care must be taken when profiling due to their shape, and the power feed on the spindle should be used for safety.

Templet used to minimise waste when jointing winders

The illustration shows how the winders and risers joint into the newel and the wall strings. Note how the nosing of the skew winder is squared off to joint into the wall string. This is to avoid an undercut scribe to the nosing. The narrow end of the kite winder is similarly treated. Note also how the newel is recessed for the risers.

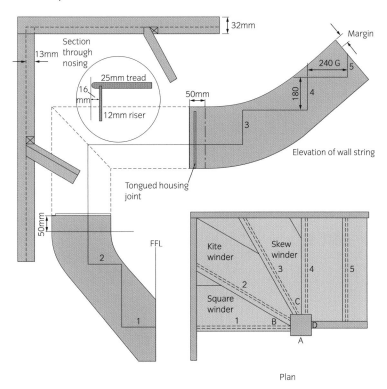

Quarter-space of three winders

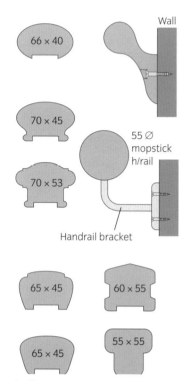

Handrail sections (measurements in mm)

FUNCTIONAL SKILLS

Research the cost of a 3.9m length of mopstick handrail required as a second handrail for a staircase. It is to be fixed using handrail brackets at approximate 1m centres. Handrails cost £7.50/m and handrail brackets cost £4.50 each. What would the total cost of materials be?

Work on this activity can support FM2 (C2.2).

Answer: £47.25

HANDRAILS

Handrails must be comfortable to grip and strong enough to guard the open side of the stair and any landing. Where the stair is 1,000mm or wider, the building regulations (Regulation 1.34c) state there must be a handrail on both sides of the stair. The wall-side handrail is generally of a smaller section and supported on handrail brackets. Sometimes a solid handrail section is specified for the wall side. The height that the handrail must be positioned is also determined by the regulations (Regulations 1.34–36). In domestic dwellings it should be 900mm over the pitch and 1,000 mm on a landing. In other buildings it is higher, and the regulations should be consulted for each stair to ensure conformity.

The illustration shows a range of common handrail sections.

CONTINUOUS HANDRAILS

Continuous handrailing is required with geometrical or concrete stairs. With concrete stairs, the handrail is supported by iron balusters bonded into the concrete steps. At the top, an iron 'core' rail connects the upper ends of the balusters. A wooden or plastic wrap handrail will finish this off. Where a wooden handrail is used, the handrail can be made to run continuously and to maintain guarding at the correct height, using a range of components shown in the illustration. These include ramps, knees and level bends.

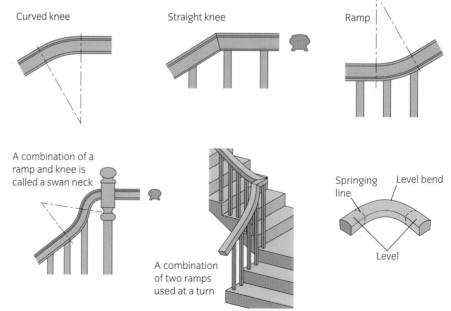

Common single curvature handrailing components

WREATHED HANDRAILING

As with wreathed strings, which rise and turn at the same time, handrails are required to do the same above them. A wreathed handrail will normally turn through 90° so two will be required to turn through 180°. The two most common types of wreathed handrailing are rake-to-level and rake-to-rake; the illustration shows examples of both of these types.

INDUSTRY TIP

The terms 'end elevation' and 'front elevation' are often abbreviated to 'E. Elev' and 'F. Elev' in industry.

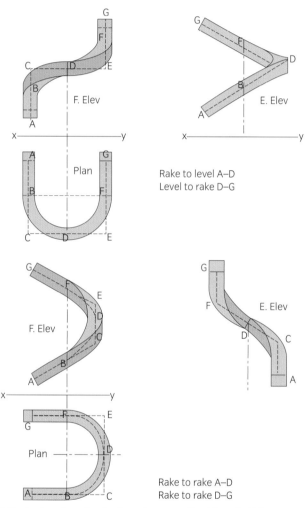

Rake-to-level and rake-to-rake wreathed handrail types

In addition to wreaths, other handrail shapes may be required to maintain guarding at the correct height.

Wreathed handrailing is a very complex subject and the full details are beyond the scope of this chapter.

TERMINATING SCROLLS

At the bottom of a staircase, the handrail will normally terminate with a level scroll. These are used at the lower terminating end of a handrail where a newel is not used.

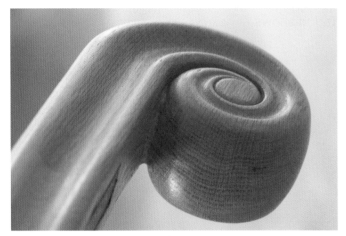

Vertical or monkey tail scroll

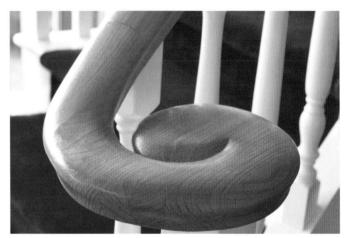

Horizontal scroll

JOINTING

INDUSTRY TIP

Tightening nuts are also known as 'castellated nuts', and square nuts are also known as 'captured nuts'.

Mortice and tenons are used to joint most handrails to newels. Where a handrail requires jointing in its length, such as at a scroll or handrail wreath, a handrail bolt is used.

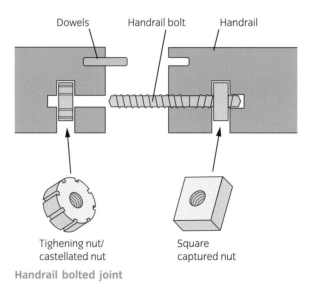

Dowels   Handrail bolt   Handrail

Tighening nut/ castellated nut

Square captured nut

Handrail bolted joint

Zipbolt™ used as a modern alternative to a handrail bolt

ACTIVITY

Research alternative methods of jointing handrails in their length. Discuss your findings in groups.

NEWELS

As mentioned earlier in this chapter, newels are used to form a turn between straight flights of stairs. Newels in domestic flights will generally be left square in shape for reasons of economy. The tops of square newels are finished with a **newel cap** or **turned finial**.

Turned newels

Turned newels are more expensive but will make the stair more visually appealing. Many companies now produce turned stair newels that can be inserted into a short newel.

In the best class of work, the full-length newels will be sent away for turning. This ensures consistency of grain colour and characteristics.

JOINTING STRINGS AND HANDRAILS TO NEWELS

When jointing strings or handrails to newels, mortice and tenon joints are used more or less always. On good-quality work, the tenon is centrally positioned and the shoulders housed into the face of the newel. This masks the effect of any subsequent shrinkage that may occur after fixing.

Newel cap

A moulded solid timber cap recessed and applied to the top of a newel

Turned finial

A turned decorative finish to the top of a post

ACTIVITY

Research in which direction timber shrinks the most: lengthwise, tangentially or radially.

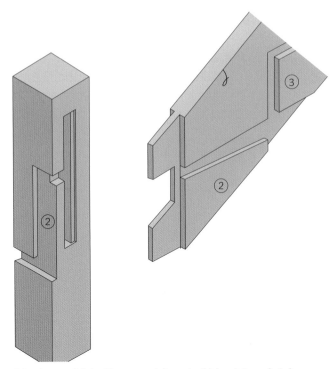

Tenon shoulder housed into the newel face to hide string shrinkage

Barefaced

Where there is a shoulder on one side of a joint only

In cheaper work, the tenons on strings are generally **barefaced** on the outside face and the practice of housing in the shoulder is omitted.

The following illustrations show common mortice and tenon arrangements for closed, cut and dog-leg stairs.

CLOSED STRING NEWEL JOINTING

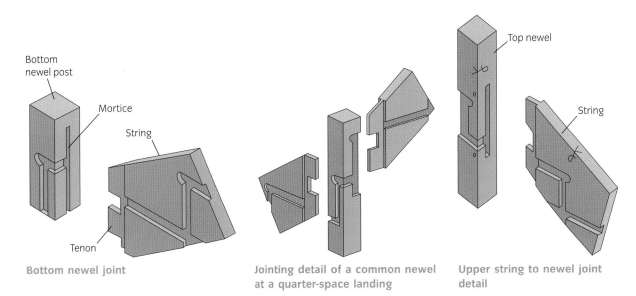

Bottom newel joint

Jointing detail of a common newel at a quarter-space landing

Upper string to newel joint detail

THE CITY & GUILDS TEXTBOOK

CUT STRING STAIR NEWEL JOINTING

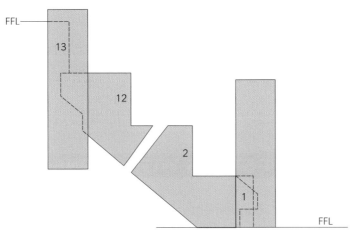

Top and bottom newel to cut string jointing details

DOG-LEG STAIR

Due to the fact that both strings intersect, the handrail is interrupted by the underside of the string in the flight above.

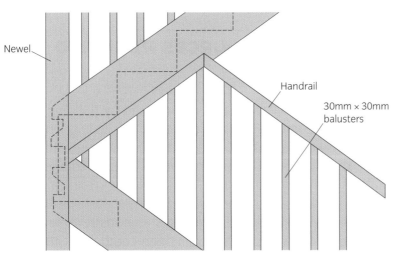

String to newel jointing detail in a dog-leg stair

BALUSTERS

As with the newels, depending on the quality of work these could be square or turned. Where they are turned they are often called 'spindles'.

There is a variety of patterns available to fit in with any architectural style. The turned decoration is limited to the centre portion, leaving both ends square for fitting into the groove on the underside of the handrail and the jointing to the string, its capping or the treads (depending on the construction of the stair).

Graduated balusters

Aesthetic

Relating to the sense of beauty

Building Regulations Part K

As has been mentioned earlier in this chapter, when fixing balusters the gap between them should not allow a 100mm sphere to pass through (Regulation 1.39a).

Generally, the balusters will be of the same length with the exception of where they are required for a dog-leg stair, where they are adjusted as shown in the illustration.

SETTING OUT STAIRS WITH TURNS

The main functional requirement of a staircase is to provide a safe means of passage from one floor to another. The **aesthetic** design of a staircase can also be very important and often is the dominant view when entering a prestigious building such as a public building or mansion. If this is the case, the stair will be constructed from a decorative hardwood rather than softwood, which would be used for an entirely functional residential stair.

INFORMATION SOURCES REQUIRED TO SET OUT STAIRS

Manufacturing of all staircases is regulated by Building Regulations Approved Document Part K (Protection from falling, collision and impact). The current version at the time of writing came into force in April 2013. Approved Document Part M (Access to and use of buildings) may also need to be consulted in certain circumstances. In addition to these regulations, we should also consult British Standard BS EN 15644, the standard covering stair construction. This provides additional information such as minimum component thicknesses, the minimum length of tenons on a string and details of fixings required to be used in stairs' construction.

All architect's drawings should conform to these information sources, but it is the stair manufacturer's responsibility to ensure that the finished product does. The drawings should be read in conjunction with the specification, which as usual will inform you about:

■ the precise description of the standard of materials and workmanship that the architect requires of the work

■ the species of timber required

■ the planned finish

■ any British Standards that need to be met (these will be those relating to the construction of the product and BS 1186, 'Timber for and workmanship in joinery').

Manufacturer's catalogues may need to be consulted for any off-the-shelf products that are required or specified, such as standard turned balusters.

Joinery workshops use a variety of administration documents. One that you may come across is a job sheet. Simply put, this outlines the main requirements and information for the job such as:

- contract number and client

- architect's details and specification

- any schedules that are relevant (baluster types, etc)

- finish required

- time allowed to complete and delivery date.

SITE SURVEY

As was covered in Chapter 6, before any setting out takes place a site survey is generally required to obtain accurate information. For stairs this could include:

- detailed measurements of the maximum available going and total rise from FFL to FFL

- positions of trimming/trimmer joists

- details and positions of any door and window openings

- templets of curved walls (for geometrical stairs)

- notes on any access problems to the premises for delivery and parking, etc.

Discrepancies in information

If the site survey highlights any discrepancies between the architect's drawings and what has been built, the details should immediately be passed on to your supervisor. For example, it is not uncommon for the trimmer joists to be put in the incorrect position, leaving insufficient room for a stair to be constructed that meets the regulations. Your supervisor will inform the architect, who will be required to make decisions as to how these issues are to be resolved. It is best that a written record is made and kept, email being ideal, noting these concerns and ensuring that the architect's instructions relating to these discrepancies are worked to as these will now form part of the contract.

Armed with the information collected from the site survey we can start to 'set out' the stair. To do this we need to have comprehensive knowledge of Building Regulations Approved Document Part K, particularly Section 1 of the current version. We recommend that

FUNCTIONAL SKILLS

Your supervisor has asked you to draft an email to the architect informing them that, following a site survey, it has been discovered that the stair planned for an opening will not fit while conforming to the regulations.

Work on this activity can support FE 2.3.1 and FICT2 2.

you read these pages now, as we will refer to these throughout the remainder of the chapter. You can find Part K at the Planning Portal website and download it for free: www.planningportal.gov.uk. You'll find a summary of this information in Chapter 3, pages 154–156.

SETTING OUT STAIRS WITH A QUARTER-SPACE OF WINDERS

The data collected from the site survey can be used to produce a scale drawing of the opening. The following sections will take you through a worked example of setting out a stair with a quarter-space of winders.

The illustration is an extract from an architect's drawing showing the requirements for a winding staircase in a residential property.

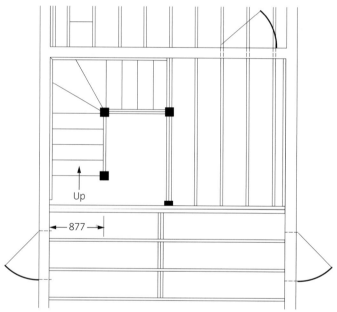

Extract from architect's drawing

Setter out

An experienced joiner who has proved their competency and is the person who produces the rods, cutting lists and orders for contracts

ACTIVITY

Make a list of what equipment will be required to carry out a site survey when measuring for stairs.

The **setter out** will be required to carry out a site survey to take accurate measurements. Both horizontal measurements will be the maximum going available in each direction. The vertical dimension is termed the total rise. It is important that this dimension is measured from the point where the flight will start to the point where it will finish at the trimming joist. To achieve this, a line is levelled around the wall from the trimming joist to above the start of the stair and a height taken at this point. If a vertical height was taken directly below the trimming, it would not be accurate as the ground floor may not be exactly level.

STAIR CALCULATIONS REQUIRED TO PROVE A STAIR WILL FIT

The following steps show the calculations required to fit the required stair to the opening in the survey (shown below).

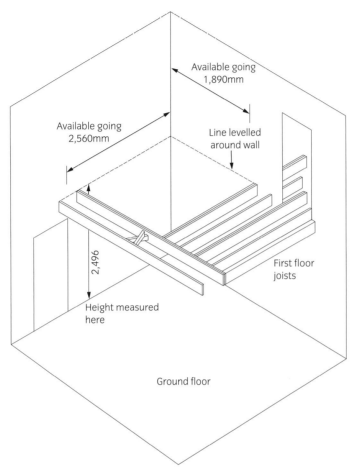

Available going
1,890mm

Available going
2,560mm

Line levelled
around wall

2,496

First floor
joists

Height measured
here

Ground floor

Survey details for stair measurements

Example

Step 1

Calculate the number of risers required. To do this we divide the total rise by the maximum step rise allowed of 220mm (Regulation 1.4):

total rise ÷ maximum step rise = actual step risers required

$$2{,}496 \div 220 = 11.35$$

Step 2

You cannot have 11.35 rises. If you reduce to 11 the rise will exceed that allowed (2,496 ÷ 11 = 226.9 step rise) so you round up to the next full number (12) and divide this into 2,496:

$$2{,}496 \div 12 = 208\text{mm}.$$

This will be the step rise.

INDUSTRY TIP

The term 'overall' can be abbreviated to 'O/A'.

INDUSTRY TIP

'Flier' is the traditional term for 'tread'.

Step 3

If there are 12 risers there will be one less going, therefore there will be 11 treads. We now need to work out how much space the winders will occupy. To do this we need to set out on sheet material such as ply or MDF the winders to full size, with the minimum going of 50mm at the newel (Regulation 1.18):

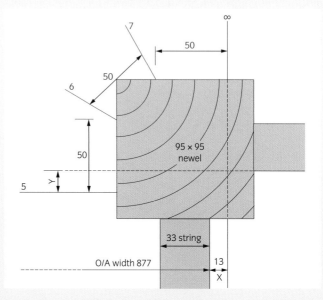

From this we can determine the total going of the upper flight. The measurement is needed from the well string to the radiating riser point ('X' distance).

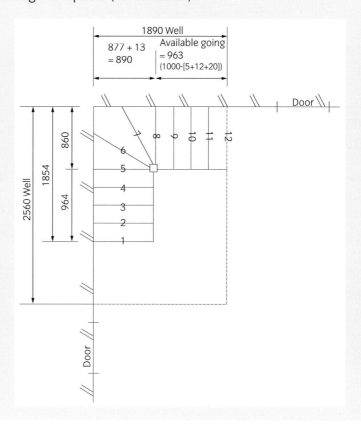

Step 4

The total going (horizontal distance) between the riser face of the first tread in the upper flight and the trimming face is then calculated as follows:

total going for upper flight − total available going − (width of flight + 'X' distance)

total going for upper flight = 1,890 − (877 + 13)

total going for upper flight = 1,890 − 890 = 1,000

total going for upper flight = 1,000.

Step 5

Deduct fitting allowances of 5mm at the wall, 12mm for the riser thickness and 20mm for clearance between the back of the riser and the trimming joist:

5 + 12 + 20 = 37mm.

We then deduct this from 1,000 to give the flight going:

1,000 − 37 = 963mm

Therefore the total going for the upper flight is **963mm**.

Step 6

Next we need to work out how many treads we have room for that will conform to the regulations for a private stair. The minimum going allowed is 220mm (Regulation 1.4). To calculate this, divide the total going by the minimum going allowed:

963 ÷ 220 = 4.37 goings

You cannot have 4.37 goings. If we were to round up to 5, the going would not be enough (963 ÷ 5 = 192.6) so we round down to 4 and divide 963 by this:

963 ÷ 4 = 240.75

This is rounded up to 241.

Step 7

When we have calculated a step rise and going for the upper flight, we need to confirm that it meets the **normal relationship** check (Regulation 1.5), which is that twice the riser plus the going should be between 550–700mm:

Normal relationship

This term, used in the Approved Document Part K, is considered the average person's step length

$$(2 \times \text{rise}) + \text{going} = (2 \times 208) + 241$$

$$= 416 + 241$$

$$= 657\text{mm}$$

As 657mm lies between 550–700mm, the check is positive.

Step 8

The normal relationship check being okay, we can complete the calculation. With four treads in the upper flight there will be five risers (so if there are 12 risers in total, in the upper flight they will be risers 8–12). There are three winder risers (5, 6 and 7), which leaves the lower flight containing the remaining four risers and four treads.

To calculate the total going of the lower flight we take the width of the flight, add the 'Y' distance (which will be the same as the 'X' distance, or 13mm in our example) and the four remaining goings.

$$\text{total going of the lower flight} = 877 + 13 + (4 \times 241)$$

$$= 877 + 13 + 964$$

$$= \mathbf{1,854mm}$$

This is well within our total going of 2,560mm so the stair will work for rise and going.

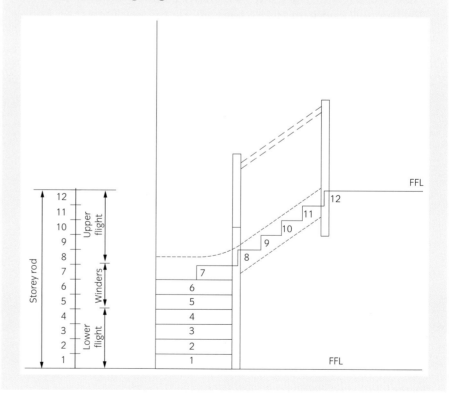

Headroom check

We need to check that there will be sufficient headroom (Regulation 1.10) of 2,000mm. We can do this by calculation but usually by scaling from the drawing. In our example, there are no clearance problems.

Pitch

The quickest way to check that the stair pitch will be less than 42°, the maximum allowed by the regulations for this type of stair (Regulation 1.4), is to draw a step full size and measure the angle with either a protractor or an adjustable set square.

INDUSTRY TIP

There should be at least 11 risers' clearance between the bulkhead and the step below.

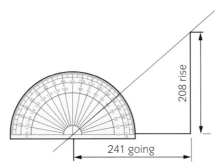

208 rise

241 going

Protractor being used to check the angle

If you prefer, trigonometry can be used to calculate the angle:

$$\tan(\text{pitch})° = \frac{\text{rise}}{\text{going}}$$

$$\tan(\text{pitch})° = \frac{208}{241}$$

$$\text{pitch} = \tan^{-1} = \frac{208}{241}$$

$$\text{pitch} = 40.68°$$

As this is below the maximum allowed of 42°, we have a proved solution that conforms to the building regulations.

DRAWING EQUIPMENT REQUIRED TO SET OUT AND DRAW STAIR DETAILS

The equipment needed to set out and draw stair details is very similar to that used and shown in Chapter 6 and includes:

- large set squares
- an adjustable set square
- trammel heads and beam
- dividers
- steel square
- large tee square
- line runner
- combination square
- 1m rule
- a parallel straight edge.

FUNCTIONAL SKILLS

Calculate the stair pitch where the step going is 250mm and the rise is 179mm.

Work on this activity can support FM2 (C2.2).

Answer: 35.6°.

DRAWINGS REQUIRED FOR A WINDING STAIR

With the information and equipment on the previous page, we can now produce a scaled line drawing (at 1:20 scale) showing the plan and the elevations of the wall strings. From these, the lengths of the strings and the additional width required for the housings for the winders can be obtained. The newels can also be added to this drawing to obtain their lengths.

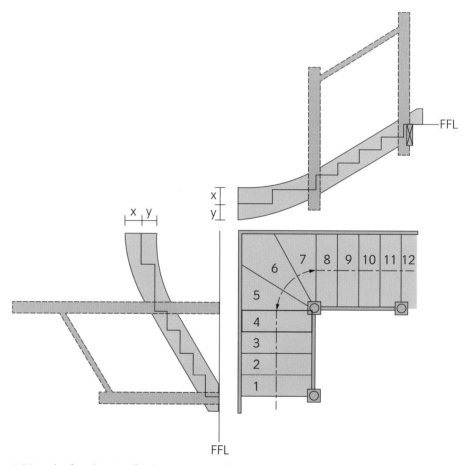

1:20 scale drawing to obtain component lengths

The only full-size details required are:

- a full-size section through one step: this will give the section sizes, profiles and the margin dimension

- full-size sections showing handrail, string capping and the finial detail (for moulding purposes)

- a full-size plan of the winders: from this the winder shapes and housings can be obtained. The corner joint between the two winder strings will be a tongued housing. Because stairs are always fixed top down, the tongue must be on the end of the lower flight.

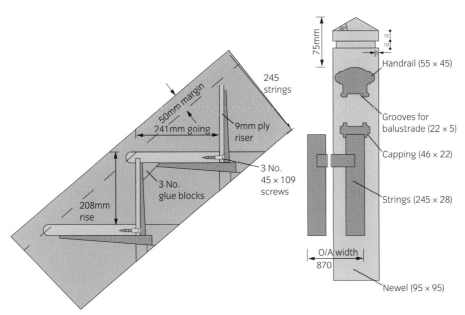

Full-size details required when setting out stairs

Labels in diagram:
- 245 strings
- 50mm margin
- 241mm going
- 9mm ply riser
- 208mm rise
- 3 No. glue blocks
- 3 No. 45 × 109 screws
- 75mm
- 30°
- 15
- 10
- Handrail (55 × 45)
- Grooves for balustrade (22 × 5)
- Capping (46 × 22)
- Strings (245 × 28)
- O/A width 870
- Newel (95 × 95)

OUR HOUSE

The owners of 'Our House' have some money to spend on a new hardwood cut string staircase. The total rise is 2,470mm and the available going is 3,200mm. Draw a full-size section through one step, showing:

- the joint you propose to use between the tread and riser
- the waist dimension suitable for the string
- an elevation of the decorative bracket
- the sizes of the components shown.

CUTTING SHEET

The required sizes can be obtained from both the scaled drawings and the full-sized details. Item numbers are usually put on the scaled drawings and duplicated where necessary on the full-size details for clarity.

JOINTING UP

Some components may require jointing up to increase their width or thickness. In winding stairs made from solid timber, often wall strings, winders and sometimes newels will require jointing in their width. A loose-tongued and grooved joint would be used in preference to biscuit jointing as the **glue line** is stronger. Where the end of the joint will be visible, a stopped loose tongue should be used.

Glue line

A term used to describe the amount of glue contact area of a joint. This is particularly important when designing widening joints. The greater the glue line, the stronger the joint

MARKING OUT COMPONENTS

MARKING OUT STRINGS

Camber

Slight curve in the depth of a joist

Once all the full-size details have been drawn and the timber has been prepared to size from the cutting sheet, marking out components can commence. The first operation as always is to select the face side and edge and mark them. When selecting these, the sides seen most should be as free from defects as possible. The strings should be sighted along their length for **camber**. If there is camber, the strings should both be marked so that the camber is uppermost.

TEMPLETS

Where strings are produced by hand we need to make the following templets:

- margin gauge
- pitch board
- riser and wedge templet
- tread and wedge templet.

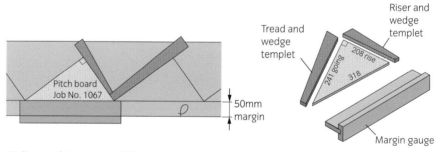

Stair templets required for marking out strings

The last two templets listed are not required if the strings are being housed out with a stair jig and router, as only the tread and riser face lines are required. Many joiners will use a steel square and a fence to mark out the strings and newels. This is quite heavy but very efficient. Lightweight plywood can be used as an alternative.

Steel square with a fence as an alternative to a pitch board and margin gauge

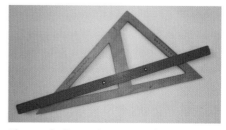

Plywood alternative to steel square

Errors can easily snowball when marking out the tread positions on the strings. This can occur when moving the pitch board along the string from one position to the next. To avoid this, the hypotenuse dimension of the pitch board is stepped along the margin line of the first string marked. These positions are then squared up to the top edge of the string and transferred to its pair.

In the case of stairs with winder strings, the straight flight treads are marked out first. The pair is then marked out from it as normal.

The wall strings of a winding flight require the winder housings marked out very accurately with a steel square. The measurements should be taken directly from the full-size rod. This informs you where additional material is required to increase the width of the string at this point.

Once these winder positions are marked, the tongued trench can be shown and finally the easings to the top and bottom edges of the string.

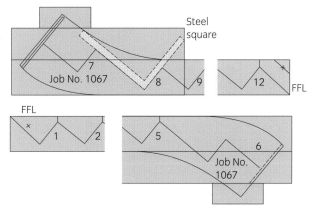

Winding strings marked out

MARKING OUT THE HANDRAILS

The handrails will be marked out directly from the well string shoulders, as shown in the illustration.

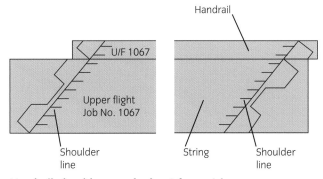

Handrail shoulders marked out from string

MARKING OUT THE NEWELS

The newels of winding stairs require quite a lot of careful marking out. If you are uncertain, the stretchout of the faces can be drawn on the full-size rod in order to familiarise yourself with what its final appearance will be. All the measurements are taken from the full-size plan of the newel.

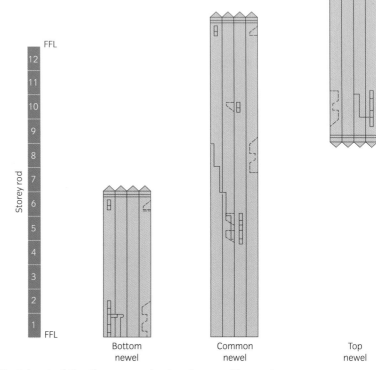

Stretchout of the three newels showing marking out

MARKING OUT THE TREADS AND RISERS

Marking out the treads and risers is a simple case of marking the face side and edge, indicating a length to be cut and the section details.

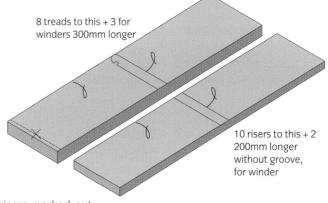

8 treads to this + 3 for winders 300mm longer

10 risers to this + 2 200mm longer without groove, for winder

Tread and risers marked out

MARKING OUT THE WINDERS

As mentioned earlier in this chapter, the winders should be cut to a templet. The final lines to be cut to are taken from the full-size plan of the quarter turn of the winders. Note on the drawing below that the direction of the grain is parallel to the nosing.

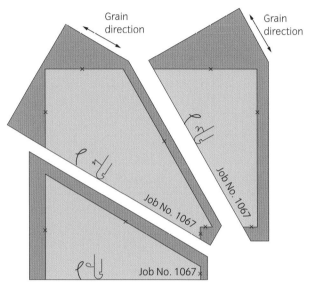

Templets placed on jointed boards (or MDF) marking shape of winders

SETTING OUT AND CONSTRUCTING GEOMETRICAL STAIRS

As covered earlier in this chapter, a geometrical string runs continuously from the bottom of the stair to the top, with generally a curved portion along its length. The curved portion is known as a wreathed string. This is a short section of string jointed to the straight flights and takes the place of a newel, allowing the stair to turn through generally 90° or 180°. The wreathed string rises and climbs at the same time. This type of string is normally cut and mitred or cut, mitred and bracketed.

To mark out the shape of the step profiles on the wreathed face of the string, we have to geometrically develop the actual shape. This is done by combining the step rises with their going around the curved face. The distance around the curved face is called the 'stretchout'.

The true length of a radius can be found geometrically as follows:

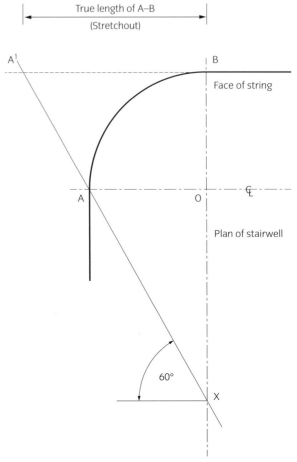

True length of radius A–B

1 The string radius 0–A is drawn through 90° to point B.

2 A tangent is drawn to the left of point B.

3 A 60° line is drawn through point A to touch the centre line at point X.

4 The 60° line is extended to meet the tangent at A¹.

5 The length from A¹–B is the true length of the radius A–B (the stretchout).

Any point along the radius line can be found as follows:

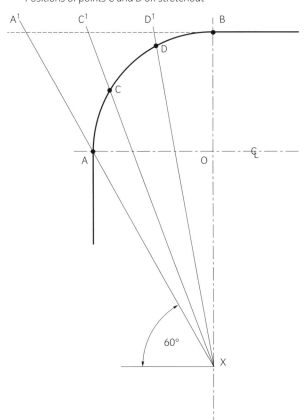

Positions of points C and D on stretchout

Finding the riser positions on the stretchout

1 Add any known points required, in this case points C and D (riser lines), onto the radius.

2 From point X, extend straight lines through points C and D to meet the tangent at C¹ and D¹. These are the true positions of C and D on the stretchout.

SETTING OUT A STRETCHOUT FOR WREATHED STRING

Using the information above and incorporating a storey rod for the wreathed string area, a stretchout can be produced as shown in the illustration on the following page. Note how one straight step is added at each end to allow for jointing to the straight string.

ACTIVITY

Draw a semi-circle with a diameter of 200mm. Measure the length of the stretchout.

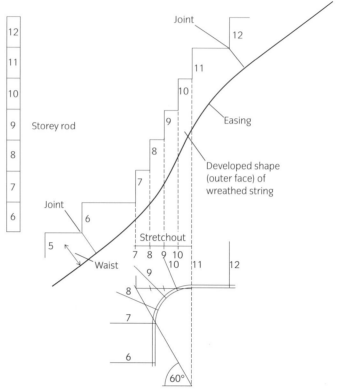

Setting out the geometrical development (stretchout) of wreathed string

WREATHED STRING CONSTRUCTION

The shaped portion can be constructed in three ways:

- built up (commonly termed 'staved')

- laminated

- solid.

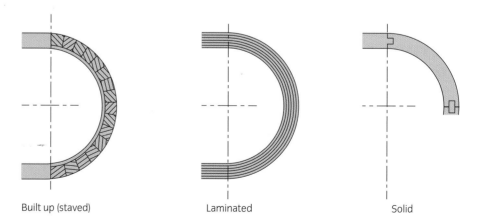

Built up (staved)   Laminated   Solid

Methods of constructing wreathed strings

BUILT-UP CONSTRUCTION

This is the most common method of constructing wreathed strings around a small radius turning through 90° or 180°.

Former

The former (traditionally known as a 'caul' and commonly called a 'drum') needs to be accurate and robust in construction. It is made so that the outside shape matches that of the well of the string. The former must be strong enough to take temporary screws through the staves.

If the geometrical development is drawn on 1.5mm birch ply, it may be used as the veneer for the wreathed portion. This will then already have the marking-out lines to enable you to cut the string to the step profiles (the ply can be pre-veneered with a matching hardwood, if required).

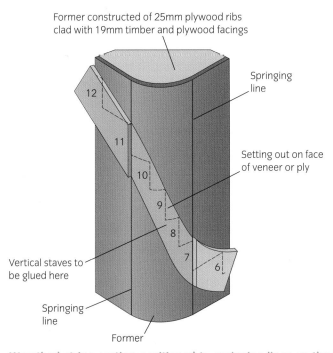

Former constructed of 25mm plywood ribs clad with 19mm timber and plywood facings

Springing line

Setting out on face of veneer or ply

Vertical staves to be glued here

Springing line

Former

Wreathed string section positioned to springing lines on the former

Staving the wreathed string

The staves have to be accurately profiled by hollowing their inside faces and tapering their edges. They must be produced from stable, knot- and shake-free material. The staves need to be about 200mm longer than required as the ends are screwed to the former to maintain an accurate shape and act as temporary cramps. The ready-prepared and marked-out veneer can be laid around the former and secured to it by the short straight ends of the wreathed string.

Accurate positioning is facilitated by aligning the springing lines marked on the string with those marked on the former. The prepared staves are then thoroughly glued and screwed through the waste parts into the former. The solid end portion adds strength where the wreathed portion is connected to the straight strings.

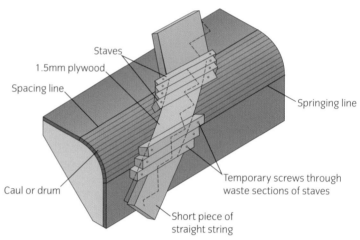

String in staving process

Shaping and fitting up

When thoroughly dry, the screws can be removed and the wreathed string cut to the marked-out lines. Much of this can be carried out on the bandsaw and finished by hand. The joints can then be shot, loose-tongued, counter cramped and fitted individually to the straight flights at either end. The steps can then be fitted to it, dismantled and sent to the site loose for fixing.

Counter cramp

The joint between wreathed and straight strings traditionally uses a counter cramp. This allows for assembling the string joint on site. The counter cramp consists of three morticed battens. The outside battens are secured to the straight string and the middle batten screwed to the wreathed portion. The mortice in the centre batten is staggered by 5mm. A pair of folding wedges is driven into the offset mortices, pulling the joint together. A loose tongue is fitted at the heading joint to increase the glue line of the joint. When the stair is fixed on site, the glued wedges are driven and the remaining screws inserted.

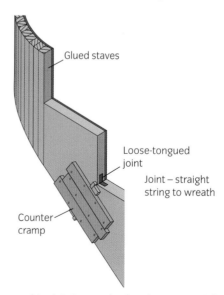

Counter cramp used to joint wreathed string to straight flights on site

LAMINATED CONSTRUCTION

The laminated construction method is generally used to manufacture geometrical strings with a large radius. Circular or helical stair strings are normally constructed in this way.

Setting out

When setting out this type of stair, make sure that the rise and particularly the going conform to the building regulations. In the drawing below you will see that two radii are drawn 270mm inside both strings: it is at both these points that the rise and going should conform (Regulations 1.25–28).

Helical stair

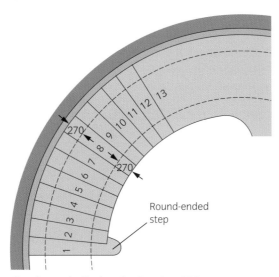

Round-ended step

Circular or helical stair showing 270mm walking line dimension

A full-size plan is required to obtain the tapered tread shapes. The development of the strings would normally only be drawn to scale to determine the length of the laminates required and the pitch of each string. The inside face of the laminate will be marked out prior to bending.

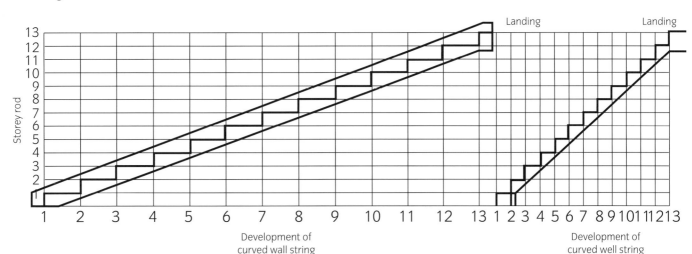

Development of curved wall string

Development of curved well string

Scale drawing required to determine string pitches and laminate lengths

A calculation can be made to double check the going for both the wall and the well strings. The following data is required:

- the radius of the inside face of the wall string
- the angle contained between riser 1 and rise 13.

Example

Let us say the radius is 4,000 mm and the angle contained between riser faces 1 and 13 is 60°.

Step 1

First we need to calculate the angle contained by each step. To do this we divide the total number of steps into the angle contained between steps 1 and 13. (Remember there is always one less step than riser.)

$$\text{angle contained by each step} = \frac{60}{12} = 5$$

Therefore each step contains 5°.

Step 2

Next, calculate the distance between steps 1 and 13. To do this we calculate the circumference length of the inside face of the wall string.

circumference = πD
circumference = 3.142 × 8,000m
circumference = 25,136mm

Step 3

Divide this circumference length by 360 (the number of degrees in a circle), then multiply the result by the number of degrees of one step as calculated previously (5 in our example).

$$\frac{25,136mm}{360} \times 5$$

$$= 69.82 \times 5$$

$$= \mathbf{349.1mm}$$

Therefore the going length on the development of the wall string should measure 349mm.

This method is more accurate than the geometrical development as it is difficult to draw accurately at this size.

Former

A former will need to be made for both well and wall strings. The pitch line should be put around the former to position the string while it is laminated.

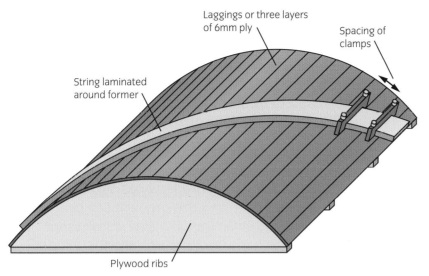

Laggings or three layers
of 6mm ply

Spacing of
clamps

String laminated
around former

Plywood ribs

Strings laminate around former

Manufacturing and assembly

The manufacturing method for this type of stair is similar to veneered and staved, in that a former is required and the laminates are held around it until the adhesive has set. A major consideration with stairs of this type is the weight of the string and being able to manoeuvre it while it is being worked on. The strings can be cut by hand or using jigs with a portable router. The tapered treads can be produced from a templet and each will be shot to fit the plan, being adjusted where necessary when fitting to the strings.

Stairs of this type are fitted up in a vertical position, generally directly over the full-size setting out. They require robust support and safe working platforms.

SOLID CONSTRUCTION

This method is only suitable when a quarter turn is required at a quarter-space landing or at the top of the stairs to the final landing. This method offers very little strength and is difficult to shape. As the grain will run vertically, it will not follow through from the strings.

Assembling a helical stair

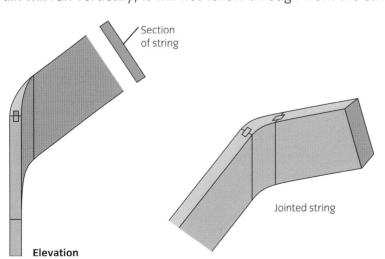

Section
of string

Jointed string

Elevation

Solid wreathed string construction

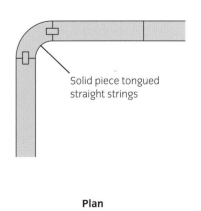

Solid piece tongued
straight strings

Plan

Case Study: Frank

A loft conversion is planned for a property. The architect has drawn up the plans and a builder has carried out the work. Now Frank has been asked to measure up and provide an estimate for a staircase to access the new loft conversion.

After carrying out a site survey it becomes clear that there is not enough available going to put in a straight flight of stairs as shown by the architect. The owner of the property has asked Frank to meet the architect to talk over how this problem can be overcome. The architect is unfamiliar with stair construction and is happy to be advised. She has asked Frank to provide a scale drawing of how he imagines the stair can be constructed to meet the regulations.

- List four considerations that Frank must bear in mind when working out a solution.

- State the areas of the building regulations that must be met for the stair to conform.

Work through the following questions to check your learning.

1   What part of the building regulations covers the construction of stairs?

   a   Part B.

   b   Part K.

   c   Part L.

   d   Part M.

2   The headroom and available going would be recorded during a site

   a   talk

   b   survey

   c   meeting

   d   induction.

3   Which one of the following is used to strike a 1,200mm radius?

   a   Dividers.

   b   Protractor.

   c   Trammel and beam.

   d   Springbow compass.

4   What is the **maximum** step rise allowed for a private stair?

   a   200mm.

   b   220mm.

   c   240mm.

   d   260mm.

5   Between what two dimensions **must** twice the rise plus the going (2R + G) equal?

   a   500–600mm.

   b   550–600mm.

   c   550–700mm.

   d   600–700mm.

6   What scale **must** winders be set out to?

   a   1:1.

   b   1:5.

   c   1:10.

   d   1:20.

7   Which one of the following states **two** methods of constructing a geometrical string?

   a   Built up and housed.

   b   Housed and built up.

   c   Laminated and housed.

   d   Built up and laminated.

8   Which one of the following is the joint used between both wall strings of a winding stair?

   a   Barefaced tenon.

   b   Tongue and trench.

   c   Tongue and groove.

   d   Barefaced dovetail.

9   What is the name of a string shaped to the profile of the steps?

   a   Cut.

   b   Well.

   c   Closed.

   d   Winding.

10   The length, width and thickness of a string is recorded on a

   a   schedule

   b   bar chart

   c   cutting list

   d   specification.

11 To who would discrepancies found in measurements shown on the drawing and those found during a site survey be reported?

a Site fixer.

b Supervisor.

c Marker out.

d Wood machinist.

12 The grain of winding steps should run parallel to the

a string

b newel

c back edge

d front edge.

13 What joint is used between the riser and the string of a cut string stair?

a Butt.

b Mitre.

c Dovetail.

d Housing.

14 What joint is used between the riser and the tread?

a Butt.

b Mitre.

c Grooved.

d Dovetailed.

15 The tread and riser joint is reinforced with a

a nail

b screw

c glue block

d counter cramp.

16 What is the **maximum** pitch allowed for a private stair?

a 37°.

b 38°.

c 42°.

d 43°.

17 What is the **minimum** height of a handrail on a private stair?

a 800mm.

b 850mm.

c 900mm.

d 950mm.

18 What is the name of the process of checking the winders for shape?

a Fitting up.

b Assembly.

c Setting out.

d Marking out.

19 What secures the steps into a closed string?

a Nails.

b Screws.

c Wedges.

d Glue blocks.

20 Which one of the following components should be sent to site loose?

a Step.

b Newel.

c Wall string.

d Well string.

TEST YOUR KNOWLEDGE ANSWERS

Chapter 1

1   c   Risk assessment.
2   d   Blue circle.
3   b   Oxygen.
4   a CO_2.
5   b   Control of Substances Hazardous to Health (COSHH) Regulations 2002.
6   c   75°.
7   c   Glasses, hearing protection and dust mask.
8   d   Respirator.
9   a   400V.
10   b   80 dB(A).

Chapter 2

1   b   Objects can be reproduced quickly.
2   c   1:1250.
3   a   A0.
4   b   Isometric.
5   b   Quotation.
6   b   Preliminaries.
7   d   Quantity surveyor.
8   a   Quote.
9   b   Pre-contract.
10   d   method statement.
11   c   lead time
12   b   Costs are increased in the long term.
13   c   Variation order.
14   c   method statement
15   d   during a toolbox talk.
16   c   Sender's address.
17   c   heat loss
18   c   Timber frame.
19   c   A–G
20   b   L.

Chapter 3

1   c   Part A.
2   b   Height of wall plate.
3   a   Hip.
4   a   hip rafter
5   d   at the gable end.
6   a   HRTL.
7   b   scarfing joint
8   b   along the valley of a dormer window
9   c   warm side of the wall behind the plasterboard
10   b   270mm
11   d   fitted around the floor joists.
12   a   joists forming the landing on which the strings sit
13   c   wreathed
14   c   550mm–700mm
15   a   42°
16   b   50mm
17   a   staircase turns and rises at the same time
18   a   quarter round
19   d   100mm.
20   d   550mm and 700mm.

Chapter 4

1   a   Additional resistance to distortion.
2   c   meeting stiles
3   d   Finish to a keyhole.
4   c   French doors.
5   a   Float.
6   c   To seal the door when the door is exposed to fire.
7   a   U-value.
8   c   both absorb and lose moisture
9   c   3mm.

10 b 60°

11 c one of the mouldings is curved

12 a Raking

13 c 1:1.

14 d Hawgood hinge.

15 c Parliament.

16 b Part N.

17 c They are fitted at an incline with the staircase.

18 a Roman

19 d 54mm.

20 d kerfing.

Chapter 5

1 b Provision and Use of Work Equipment Regulations

2 a isolate the power supply

3 d Crown guard

4 b 85 dB(A).

5 a Manufacturer's maintenance schedule, risk assessment, authorisation.

6 b 10 seconds.

7 b angled cuts

8 a strong, easily adjusted and always used

9 c 8mm.

10 c 3mm.

11 a Isolate machine, put a warning sign on the machine, inform your supervisor.

12 d 1.1mm.

13 c Fitted at the back of the band saw blade.

14 c Chisel, chisel bush, auger, auger bush.

15 a 1,200mm minimum.

16 a stop the out-feed operative from touching the saw blade and to stop the timber from closing up

17 b Maintenance log book.

18 b Negative.

19 d To restrict uncontrolled backward and sideways movement of the blade.

20 c increased risk of the blade coming off the pulley

Chapter 6

1 c a survey

2 b making a templet

3 c trammel and beam

4 b Lancet.

5 d Trammel frame.

6 d Circumference.

7 a Elliptical.

8 b crown joints

9 b cutting list

10 c supervisor

11 b Crosscutting.

12 a PUWER.

13 a jig

14 a Planer block.

15 a fitting up

16 b Web.

17 d 180.

18 c Check for square and wind.

19 c Random orbital.

20 d hardwood with interlocking grain.

Chapter 7

1 b Part K.

2 b survey

3 c Trammel and beam.

4 b 220mm.

5 c 550–700mm.

6 a 1:1.

7 d Built up and laminated.

8 b Tongue and trench.

9 a Cut.

10 c cutting list

11 b Supervisor.

12 d front edge.

13 b Mitre.

14 c Grooved.

15 c glue block

16 c 42°.

17 c 900mm.

18 a Fitting up.

19 c Wedges.

20 b Newel.

INDEX

PICTURE CREDITS

Every effort has been made to acknowledge all copyright holders as below and the publishers will, if notified, correct any errors in future editions.

123RF: © amornme p178; **Ace Fixings:** pxxvi; **acornwoodturning.co.uk:** p388; **Ainscough:** p78; **APL:** p25; **Art-Drawing Office:** p324; **ashbrookroofing.co.uk:** p150; **Axminster Tool Centre Ltd:** pp xiv, xv, xvi, xviii, xix, xx, xxi, xxiii, xxiv, xxv, xxviii, xxix, xxx, xxxi, xxxii, xxxiii, xxxiv, xxxv, xxxvi, xxxvii, xxxviii, xxxix, xl, xli, xliii, xliv, xlv, 20, 22, 23, 34, 36, 267, 349, 384; **Bambuh:** pxxvi; **BM TRADA:** p191; **bulmashini.com:** pxliii; **Burton and South Derbyshire College:** pp xxiv, xxxv, 169, 207, 232, 235, 256, 259, 260, 266, 270, 271, 272, 273, 284, 289, 290, 298, 299, 366; **carllswoodproducts.com:** p343; **Carpentry-Tips-and-Tricks.com:** p181; **Central Sussex College:** pp xvi, xxx, xxxvii, xxxviii, xli, xlv, 119, 130, 131, 132, 133, 134, 135, 129, 242, 259, 260, 270, 297, 298, 303, 347; **ChiefArchitect.com:** p407; **Construction Photography:** © Adrian Greeman pp xxxiv, 21; © BuildPix pp xlv, 11, 33; © Chris Henderson p34; © Grant Smith p13; © Image Source pp 15, 16; © MakeStock p43; © Xavier de Canto p1; **Craftsman Studio:** pxli; **Croft Architectural Hardware Ltd:** p204; **Crown copyright:** pp xiv, 93, 103, 342; **D&E Architectural Hardware Ltd:** p196; **Distinctive Doors Ltd:** p185; **Everglade™ trade mark of Pilkington plc:** pp xxxii, 308; **Forest Stewardship Council®:** pp xxv, 103; **Fotolia:** © Alan Stockdale p39; **Gary Katz, THISisCarpentry.com:** p409; **Gedore UK Ltd:** p351; **Gowling Stairs:** p384; **H+H UK Limited:** p100; **Hackney Community College:** p34; **Häfele UK Ltd:** p202; **Hawes Plant Hire:** p18; **Hellopro.co.uk:** pp xlii, 276; **HSE.gov.uk:** © p9; **idealsealants.com:** pxxxix; **iDigHardware.com:** pp xxv, 192; **If Images:** pxxxiii; **Ironmongery Direct:** pp xxviii, 194, 195, 197, 199, 200, 203, 204, 210; **Ironmongery2u Limited:** p195; **iStock:** © 4x6 p91; © arturasker p109; © BanksPhotos pp 98, 100; © drbimages p143; © stocknroll p106; **Jacob Butler:** p376; **Jake Fitzjones Photography for Bisca:** p384; **Kitchen Surplus:** pxlii; **Laxton's Publishing Limited** p61; **Leaderflush Shapland Laidlaw:** p178; **lockmonster.co.uk:** p198; **Lutes Custom Woodworking:** p319; **Martin Burdfield:** pp 372, 398; **Martin Burdfield/Building Crafts College:** pp xlii, 310, 315, 316, 322, 324, 326, 327, 328, 239, 335, 343, 345, 346, 347, 350, 351, 352, 353, 354, 355; **Matt Tibbles:** p158; **Mediscan:** p23; **Meredith Corporation:** p288; **Meteor Electrical:** p35; **Mike Jones** p53; **PAT Training Services Ltd:** pp xxxv, 37; **Power Tools Direct:** pp xxiii, xxxix; **RICS:** p58; **Science Photo Library:** © Dr P. Marazzi/Science Photo Library p23; **shawstairs.com:** p385; **Shutterstock:** © 3DDock pxviii; © Alena Brozova pp xxvii, 98, 150; © Alexander Erdbeer p13; © alterfalter p100; © Andrew Bassett pxlv; © aragami12345s p18; © Arina P Habich p183; © arturasker pxx; © auremar pp 18, 88; © Barry Barnes p41; © Bokic Bojan pp xxxviii, 48; © chaoss p172; © Christina Richards p359; © Claudio Divizia p54; © Coprid pp xx, 100; © Corepics VOF p51; © Cynthia Farmer p18; © dabldy pxxiii; © Dario Sabljak p94; © Darkkong p15; © daseaford p41; © DeiMosz p41; © demarcomedia p15; © DenisNata p22; © Dmitry Kalinovsky p31; © Elena Elisseeva p186; © Epsicons p101; © Fireco Ltd p40; © Ford Photography p41; © foto infot p410; © Goodluz p45; © HABRDA p177; © Igor Sokolov (breeze) p3; © Israel Hervas Bengochea p25; © ivvv1975 p72; © Joe Gough p70; © John Kasawa p98; © Kaspri p41; © kavalenkau p37; © KIM NGUYEN pxxxiv; © Kraska p23; © Krzysztof Slusarczyk pxxxiv; © kzww p83; © LAI CHING YUEN p307; © Leah-Anne Thompson p77; Marbury p31; © Mark Atkins p73; © matthew Siddons p105; mike.irwin p16; © Mr.Zach p39; © Naiyyer p23; © Naphat Rojanarangsiman p213; © Natsmith1 pxxxvi; © Nikita G. Sidorov pxxxvi; © objectsforall p18; © Offscreen pxlii; © Phiseksit p15; © Phuriphat p15; © ppart pxxxii; © Pressmaster pxiv; © Rafael Fernandez Torres p16; © RAGMA IMAGES p85; © ravl pxxxix; © richsouthwales pxlvi; © Rob Kints p24; © RTimages p177; © SergeBertasiusPhotography p30; © Sergej Razvodovskij p84; © sbarabu p36; © Shipov Oleg p102; © smuay p200; © StockLite p166; © SueC p99; © Tribalium p41;